Communications
in Computer and Information Science 715

Commenced Publication in 2007
Founding and Former Series Editors:
Alfredo Cuzzocrea, Xiaoyong Du, Orhun Kara, Ting Liu, Dominik Ślęzak,
and Xiaokang Yang

More information about this series at http://www.springer.com/series/7899

Jerzy Mikulski (Ed.)

Smart Solutions
in Today's Transport

17th International Conference
on Transport Systems Telematics, TST 2017
Katowice – Ustroń, Poland, April 5–8, 2017
Selected Papers

Springer

Editor
Jerzy Mikulski
Polish Association of Transport Telematics
Katowice
Poland

ISSN 1865-0929 ISSN 1865-0937 (electronic)
Communications in Computer and Information Science
ISBN 978-3-319-66250-3 ISBN 978-3-319-66251-0 (eBook)
DOI 10.1007/978-3-319-66251-0

Library of Congress Control Number: 2017952384

Printed on acid-free paper

This Springer imprint is published by Springer Nature
The registered company is Springer International Publishing AG
The registered company address is: Gewerbestrasse 11, 6330 Cham, Switzerland

Preface

This book contains extended and revised versions of a set of selected papers from the 17th International Conference on Transport Systems Telematics, TST 2017, held in Katowice and Ustroń, Poland, April 5–8, 2017. The TST conference aims to bring together researchers, engineers, and practitioners interested in intelligent transport and other aspects of information and telecomunication systems and technologies involving advanced applications.

The book presents knowledge from the field of intelligent transport systems, specific solutions applied in this field, and their influence on the efficiency of transport systems improvement.

This year we met for the 17th time and the Polish Association of Transport Telematics was the main conference organizer. The conference was related to the intelligent transport systems development program in the area of all modes of transport infrastructure and enabled the exchange of experience and knowledge of the latest technologies and trends in the development of ITS systems.

The publication constitutes a collection of the scientific achievements presented by their authors during the conference sessions. Nearly 130 papers were submitted to the conference, focussing on all of the different modes of transport infrastructure, which were presented - apart from the plenary session - in 6 scientific sessions, 2 poster sessions, and 2 sessions of scientific announcements.

This book comprises approximately 30% of the submitted papers. The papers were selected by the Program Committee members and their selection was based on a number of criteria that included the classifications and comments provided by the session chairs and also the program chairs' global view of all papers included in the conference program. The authors of selected papers were then invited to submit a revised version of their papers.

I would like to thank all the authors for their contributions and also the reviewers, who helped to ensure the quality of this publication.

February 2017

Jerzy Mikulski

Organization

Organizers

Polish Association of Transport Telematics, Poland

Co-organizers

Transport Committee of the Polish Academy of Sciences, Poland
University of Economics in Katowice, Poland
Katowice School of Technology, Poland

Co-operating Universities

Gdynia Maritime University, Poland
Maritime University of Szczecin, Poland
University of Bielsko-Biała, Poland
Warsaw University of Technology, Poland
Wrocław University of Science and Technology, Poland
University of Technology and Humanities in Radom, Poland
WSB Schools of Banking in Wrocław, Poland
University of Lodz, Poland

Scientific Program Committee

J. Mikulski (Chair)	Polish Association of Transport Telematics, Poland
M. Bregulla	Ingolstadt University of Applied Sciences, Germany
A. Bujak	WSB Schools of Banking in Wrocław, Poland
M. Bukljaš-Skocibušic	University of Zagreb, Croatia
A. da Silva Carvalho	University of Porto, Portugal
M. Chrzan	University of Technology and Humanities in Radom, Poland
R. van Duin	Delft University of Technology, The Netherlands
M. Franeková	University of Žilina, Republic of Slovakia
G. Gentile	Università di Roma "La Sapienza", Italy
P. Groumpos	University of Patras, Greece
S. Iwan	Maritime University of Szczecin, Poland
M. Jacyna	Warsaw University of Technology, Poland
A. Janota	University of Žilina, Republic of Slovakia
J. Januszewski	Gdynia Maritime University, Poland
U. Jumar	Instit. f. Automation & Kommunikation, Germany
A. Kalašová	University of Žilina, Republic of Slovakia

R. Kozłowski	University of Lodz, Poland
J. Krimmling	Technische Universität Dresden, Germany
O. Krettek	RWTH Aachen, Germany
D. Laskowski	Military University of Technology, Poland
A. Lewiński	University of Technology and Humanities in Radom, Poland
M. Luft	University of Technology and Humanities in Radom, Poland
A. Maczyński	University of Bielsko-Biala, Poland
R. Madleňák	University of Žilina, Republic of Slovakia
M. Michałowska	University of Economics in Katowice, Poland
D. Peraković	University of Zagreb, Croatia
Z. Pietrzykowski	Maritime University of Szczecin, Poland
A. Prokopowicz	Instit. of Global Innovation, Economics & Logistics, USA
K. Rástočný	University of Žilina, Republic of Slovakia
M. Siergiejczyk	Warsaw University of Technology, Poland
J. Skorupski	Warsaw University of Technology, Poland
L. Smolarek	Gdynia Maritime University, Poland
J. Szpytko	AGH University of Science and Technology, Poland
R. Thompson	University of Melbourne, Australia
R. Toledo-Moreo	Universidad Politécnica de Cartagena, Spain
Y.A. Vershurin	Coventry University, UK
R. Wawruch	Gdynia Maritime University, Poland
A. Weintrit	Gdynia Maritime University, Poland
K. Wydro	University College of Technology and Business in Warsaw, Poland
E. Załoga	University of Szczecin, Poland
J. Ždánsky	University of Žilina, Republic of Slovakia

Honorary Committee

Tatiana Corejová	University of Žilina, Slovak Republic
Milan Dado	University of Žilina, Slovak Republic
Andrzej Grzybowski	Katowice School of Technology, Poland
Tomasz Kamiński	ITS Polska, Poland
Barbara Kos	University of Economics in Katowice, Poland
Tatiana Kováčiková	ERA Chair of ITS, University of Žilina, Slovak Republic
Anna Križanova	University of Žilina, Slovak Republic
Zbigniew Łukasik	University of Technology and Humanities in Radom, Poland
Laimuté Sladkeviciene	Vilnius College of Technologies and Design, Lithuania
Juraj Spalek	University of Žilina, Slovak Republic
Roman Srp	Network of National ITS Associations, Czech Republic
Miroslav Svitek	Czech Technical University in Prague, Czech Republic
Wojciech Ślączka	Maritime University of Szczecin, Poland
Robert Tomanek	University of Economics in Katowice, Poland
Wojciech Wawrzyński	Warsaw University of Technology, Poland

Contents

Selected Problems of ITS Project Development – Concept Exploration and Feasibility Study

Grzegorz Karoń[1(✉)] and Jerzy Mikulski[2]

[1] Silesian University of Technology, Akademicka 2A, 44-100 Gliwice, Poland
grzegorz.karon@polsl.pl
[2] Katowice School of Technology, Rolna 43, 40-001 Katowice, Poland
mikulski.jurek@gmail.com

Abstract. This article presents some problems that can occur while defining configuration and architecture of intelligent transport systems (ITS) at the stage of concept development and design of these systems for urban areas. The authors propose a system procedure (systems engineering) during the development of the documentation of ITS systems and their implementation in urban areas. Rounding out the description of the proposed policies and procedures are examples of methodology of concept exploration and feasibility study of ITS for the city of metropolitan area.

Keywords: Systems engineering · ITS intelligent transport systems architecture · Deployment programme · Feasibility study · Concept exploration

1 Introduction

One of the more frequently used by municipalities and authority's agglomeration solutions, which aim is to improve traffic conditions in cities are intelligent transport systems ITS. The implementation of these systems is associated with a number of interdisciplinary problems, because ITS systems primarily use telematics systems and software:

- to collect data from many different sources,
- to process these data into control signals and useful information,
- to transmit these signals to the various devices and to distribute the information to end users.

One of the basic problems is the development of a suitable variant of ITS configuration, adapted to the needs of the city and its users [8, 23]. In addition, an important task is to prepare a high-level design of ITS [3, 17] – ITS architecture containing the description of the ITS functions and data flows between the ITS sub-systems and ITS users to ensure implementation of relevant ITS services [1, 20, 21]. The article presents the functional assumptions of ITS systems for urban areas and selected elements of the design of ITS systems with the use of systems engineering. Examples are taken from the case study for the city of Katowice in Poland, in which the authors of the article were involved using the presented methods of systems engineering [6, 7, 9, 10].

© Springer International Publishing AG 2017
J. Mikulski (Ed.): TST 2017, CCIS 715, pp. 1–15, 2017.
DOI: 10.1007/978-3-319-66251-0_1

2   Functional Assumptions of ITS for the Urban Area

ITS framework architecture is an architecture (inter alia with a uniform terminology), from which a subset of elements may be selected to form the ITS architecture dedicated to a specific urban area [1, 3, 17, 20, 21]. The benefits of ITS:

- efficient use of capacity transport network elements, especially the road network through the use of management systems and road traffic control [23],
- an increase in traffic flow, among others by reducing the number of unnecessary or forced stops (e.g. traffic lights),
- reduction of negative impact on the environment by increasing the flow of traffic,
- providing information:
  - reliable, useful,
  - with a clear content and adapted to the location,
  - with the content about complementary and substitute modes of transport (transport connections and interchange),
  - different channels of information distribution (internet, radio, mobile devices and applications, VMS signs),
  - in different languages,
- increasing the efficiency, usefulness and attractiveness of public transport and multi-modal journeys:
  - priority and privilege to public transport vehicles,
  - increase punctuality and thus the reliability of a journey public transport,
  - increase the scope and distribution channels of information concerning: timetables, transfers, travel planning before (pre-trip) and during its duration (en-route),
  - integrated solutions for the tariff and the variety of forms of distribution of tickets, from traditional paper tickets in the stationary and mobile (inside vehicle) ticket machines to electronic tickets using smartphones and wireless readers.
- increasing efficiency in the use of car parks and reducing "idle running" - in search of free parking space.

3   Systems Engineering in the Design of ITS

Assuming V model in the methodology of the design of ITS (Fig. 1) [12, 13, 17] process to be followed during the planning and design begins with ITS concept exploration on the basis of a feasibility study. The result is, among others:

- analysis of the current state of urban area in terms of:
  - existing and anticipated transport problems in the near future,
  - factors affecting the transport processes in transport systems of the analysed area,
- defining strategic objectives for measures to address or reduce the transport problems,

- identification of weaknesses and threats as well as strengths and chances of solving transport problems,
- a proposal strategies, including inter alia:
  - type of action,
  - the order of their implementation in the form of an action schedule,
  - recipients: all stakeholders, including in particular the beneficiaries and end-users.
- a proposal options for action, implementing the proposed strategy and the defined goals.

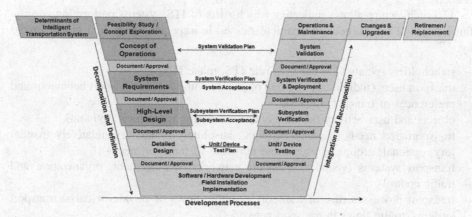

Fig. 1. Model V – the use of systems engineering for ITS systems design. (Own study based on [12, 13, 17, 20, 21])

The detailed structure of this stage is shown in Fig. 2. This is high-level design stage, followed by the second stage - the detailed design of the particular real system.

It should be emphasized the important role of ITS Architecture indicating the basic benefits of its development:

- involvement of all stakeholders,
- support for the project by the local authorities,
- a clear view and precise objectives,
- realistic aspirations of stakeholders,
- a complete project team, both at the contractor and the customer side,
- transparent and understandable ITS services for ITS users and other ITS stakeholders,
- sharing of information and ITS services by various ITS stakeholders,
- the agreed common assumptions concerning the interoperability of future telematics systems, IT systems and ICT systems,
- developed communication between future ITS sub-systems [5, 18], which will take advantage of the usefulness of these systems in an optimal way, in order to ensure the implementation of the adopted ITS services for ITS stakeholders,

- the development of ITS architecture is the stage of strategic planning, taking into account the budget planning and planning of project implementation of ITS subsystems and other complementary projects,
- proper ITS system planning, consistent with planning the area development,
- transparent and understandable requirements for contractors in the next stages of the project, i.e. detailed design stage and the subsequent stages of the development, integration and recomposition processes (see Fig. 1).

The main document, enabling the start of this detailed design stage is a deployment programme, and the ITS architecture developed for the chosen variant of the ITS concept.

ITS architecture allows indicating which parts of ITS systems performing specific functions of ITS are needed in a particular case - in a specific urban area, characterized by:

- stakeholders (wants ITS, use ITS, rule ITS, make ITS),
- transport users (individuals or group of people with specific transport behaviour and preferences of transport modes),
- objects and user activity (absolutely and relatively mandatory, optional),
- transportation needs (passenger, freight; absolutely mandatory, relatively mandatory, optional; groups of people and purposes of transport needs),
- transport systems (vehicles, transport networks, management, organization and traffic control),
- transport processes (use of transport systems, flows of people or cargo transport systems, traffic flows in transport networks).

Taking as an example the European ITS Frame Architecture [3], The Functional Areas of ITS Functions, which include about 350 ITS functions are:

- Provide Electronic Payment Facilities,
- Provide Safety and Emergency Facilities,
- Manage Traffic,
- Manage Public Transport Operations,
- Provide Advanced Driver Assistance Systems,
- Provide Traveller Journey Assistance,
- Provide Support for Law Enforcement,
- Manage Freight and Fleet Operations,
- Provide Support for Cooperative Systems.

Based on the analysis of the functional needs for a particular variant of the ITS concept the designed ITS architecture includes, among others:

- ITS specific functions,
- data storages,
- certain ITS services,
- defined end-ITS-users,
- data flows diagrams between ITS functions, ITS sub-systems, ITS users and ITS data stores.

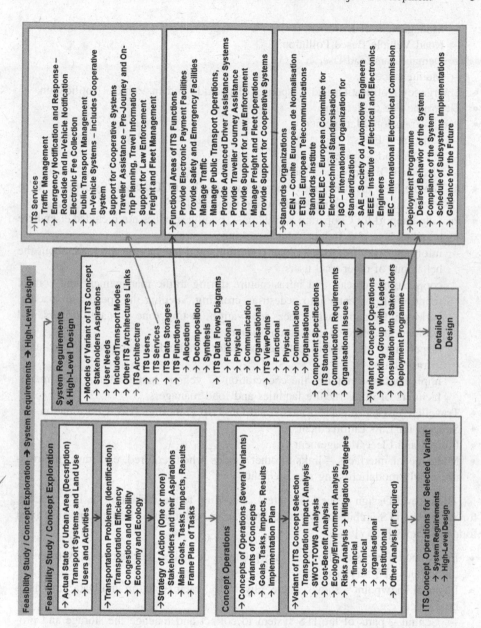

Fig. 2. Feasibility Study/Concept Exploration/System Requiremens/High-Level design of ITS – based on model V of systems engineering. (Own study based on [7, 20])

Scope of ITS services may include - for example ITS FRAME - among others, the following functionality [3, 17]:

- Traffic Management:
  - Urban, Inter-Urban,
  - Parking, Tunnels and Bridges,

- Management of Incidents,
- Road Vehicle Based Pollution,
- Demand for Road Use,
- Maintenance,
- Emergency Notification and Response – Roadside and In-Vehicle Notification,
- Electronic Fee Collection,
- Public Transport Management
  - Schedules, Fares,
  - On-Demand Services,
  - Fleet and Driver Management;
- In-Vehicle Systems – includes Cooperative Intelligent Transport Systems (C-ITS) [18] with Internet of Things (IoT) in the future [5],
- Provide Support for Cooperative Systems – specific services not included elsewhere such as bus lane use, freight vehicle parking:
  - informing/warning drivers about conditions and dangers on the road, about the behaviour of other road users,
  - cooperation vehicles and infrastructure during traffic management and control (controlling the entry into pedestrian crossing, keeping the speed and spacing, parking information with places reservation, resting zone for cargo drivers, and booking the place of loading/unloading),
  - dispersal of the system between vehicles, traffic lanes, objects of technical infrastructure,
  - implementation requires the cooperation of vehicle manufacturers with manufacturers of infrastructure facilities and road managers,
- Traveller Assistance – Pre-Journey and On-Trip Planning, Travel Information,
- Support for Law Enforcement,
- Freight and Fleet Management,
- Multi-modal interfaces – links to other modes when required, e.g. travel information, multi-modal crossing management.

High-Level Design of the ITS system, according to the variant of the ITS concept, for which the ITS architecture was designed, can be presented in the form of at least four ITS views [3, 17]:

- Functional Viewpoint (Logical Viewpoint) – describes the functionality to create the various ITS services,
- Physical Viewpoint – shows where, physically, each Function and Data Store is to be located, e.g.:
  - Central – parts of the ITS system to collect and manage the storage and processing of traffic data, toll payments, freight shipping orders, and/or the generation of traffic management measures, or fleet management instructions, with or without human intervention, e.g. a traffic control/information centre, or a freight and fleet management centre,

- Roadside – parts of the ITS system for the detection of traffic, vehicles and pedestrians, or the collection of tolls, and/or the generation of traffic management measures, and/or the provision of information and commands to drivers and/or pedestrians,
- Vehicle – parts of the ITS system can be installed during manufacture or can be added on later into passengers vehicles (bicycles, motorcycles, cars, public transport vehicles) and into cargo vehicles (vans, trusks etc.),
- Personal Device – part of the ITS system installed into mobile devices (e.g. smartphones, tablets etc.) and can be used by travellers,
- Freight Device – part of the ITS system installed into device as integral part of a freight carrying unit, e.g. freight container, trailer, or vehicle body,
- Kiosk – part of the ITS system installed into device, usually located in a public place to enable travellers to have limited and controlled access to some of its facilities,
- Communications Viewpoint – is the result of an analysis of the Physical Data Flows and describes the kinds of communications links that will be required, taking into account:
  - minimum data transfer rate – time between the creation of the data and quantity of data in each item,
  - data sharing – time required before a new value,
  - wireless links – from mobile devices,
- Organisational Viewpoint – considers the ownership and business issues, e.g. who owns what, who manages what, and the business/contractual relationships between the various parties involved.

The viewpoints in the ITS architecture describe the architecture, but a single view describes only selected aspects of architecture [1, 3].

4 Feasibility Study for ITS – Example for the Capital City of the Metropolitan Area

When planning ITS system at the stage of conception problems can occur due to circumstances and relationships [12]:

- interactions occurring in the city treated as separated urban system from the environment,
- interactions occurring at the interface between urban system with its environment, resulting from the interdependencies, which in terms of the system are mapped as a set of inputs and outputs and impacts/information flowing between them in both directions; wherein there are at least three cases of mutual relations:
  - city ← → dispersed environment,
  - city ← → environment integrated on the same hierarchical level – city neighbouring conurbation (polycentric agglomeration), adjacent directly and indirectly,

- city ← → environment integrated on various hierarchical levels, city neighbouring agglomerations (monocentric agglomeration):
  - overriding city ← → subordinate cities,
  - subordinate city ← → overriding cities,
  - subordinate city ← → subordinate cities,
- interactions occurring in the vicinity:
  - due to internal conditions of surrounding towns,
  - resulting from the impacts of surrounding cities on the analysed city,
  - resulting from the impact of the analysed city on the surrounding city.

Type of interaction results from the type and nature of the interrelationship – and dependences conditions – which are present in planes:

- functional connections of transport systems (network structure of transport subsystems),
- functional connections of transport needs (relations in the activity ← → users),
- functional relationships of socio-organizational (relationships in a social system management),
- functional relationships of socio-economic (relations in a socio-economic management system),
- functional relations of ecological (relations in the system of ecological impact).

These factors and relationships are taken into account when drawing up the Feasibility Study for the city of Katowice [16], at levels: Regional Level of City Location, Sub-regional Level of City Location, Metropolitan Level of City Location. In the Feasibility Study for ITS Katowice [10, 16], used transportation model for the city of Katowice. Scheme of feasibility study for ITS Katowice is shown in Fig. 3. Is currently being developed on updating the feasibility study for the ITS Katowice 2017, which will be drawn up on the basis of the methodology for the design of ITS for the city of Katowice [7], which includes:

- use of the transportation model for the city of Katowice,
- the use of ITS conception for metropolitan level [9, 10, 19],
- the use of ITS conception for national roads [4] with architecture compatible with FRAME [3],
- the use of FRAME tools [3] to develop an ITS architecture for the city of Katowice in accordance with the FRAME methodology,
- a two-stage project of the ITS according to the V model for ITS [6, 7]:
  - first step is the Feasibility Study/Concept Exploration, High-Level Design for variant of ITS concept with ITS architecture for Katowice city and Deployment Programme,
  - the second step is the emergence of contractors of the ITS system project, who will develop ITS detailed design and will carry out the development, integration and recomposition processes.

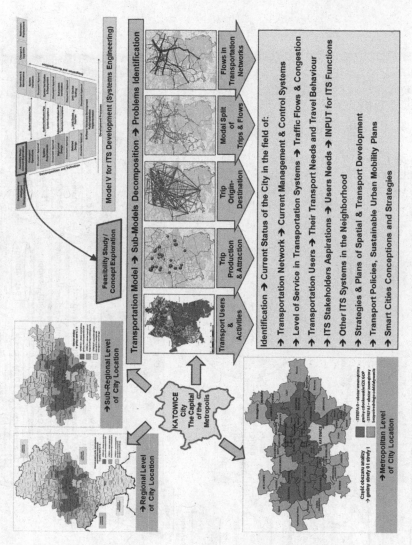

Fig. 3. First Stage of ITS development based on V-Model: Feasibility Study/Concept Exploration – Example for Katowice city – the Capital of Upper Silesian Metropolis. (Own study based on [6, 7, 9, 10])

Expected results, planned in the methodology for the design of ITS for Katowice city is schematically shown in Fig. 4. These are the Feasibility Study/Concept Exploration, High-Level Design for Variant Concept of ITS with ITS Architecture for Katowice City and Deployment Programme.

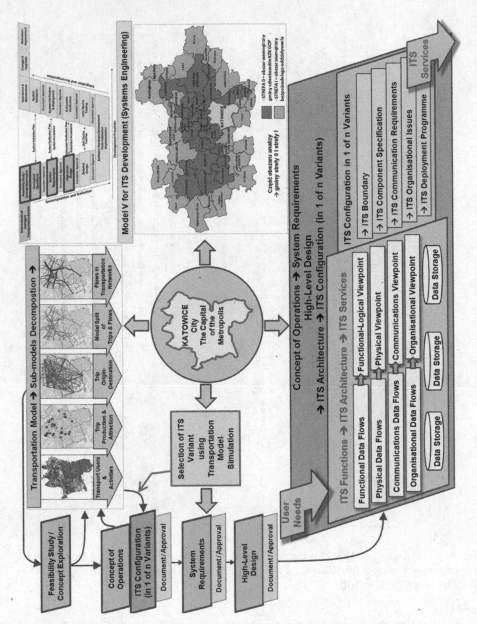

Fig. 4. Selection proces of ITS architecture in 1 of n variants in concept of operations. (Own study based on [6, 7, 20, 21])

The rest of this article presents the key assumptions and results of Feasibility Study for the ITS Katowice [10, 16], which will also be included in the new Feasibility Study for the ITS Katowice 2017 [6, 7].

4.1 Targets in the ITS Concept

The design of ITS Concept for the city of Katowice [16] the main objective of the project of ITS is to increase the investment attractiveness: city, agglomeration [9, 16], region [14, 15] and country through the development of technical infrastructure while protecting and improving the environment, health, preserving cultural identity and developing territorial cohesion. This objective will be achieved through improvements in traffic management in the form of modern ITS – intelligent traffic management and traffic of public transport with a view to enhancing the attractiveness of public transport in the area of the capital city of the largest metropolitan area in the country, which is the agglomeration of Upper Silesia [2, 22] and its capital – the Katowice city.

Have been adopted following further specific objectives, which are consistent with key lines of action in the framework of sustainability of urban mobility [16]:

– to control car traffic on the road network and decrease its congestion,
– more efficient use of existing road and transport infrastructure,
– to streamline and thus to increase the attractiveness of transport offer in public transport,
– to improve the safety and comfort of traffic through efforts to increase the competitiveness of public transport to alternative forms of transport, which is the car traffic and its reduction especially in the area of the project,
– reduce travel time by public transport, by introducing priorities of travel and passenger information system,
– inhibition of passengers the outflow of public transport observed not only in Katowice and in the whole area of the Upper Silesian agglomeration, but also in many other metropolitan areas in Poland,
– to improve traffic safety by reducing transport work in car transport resulting from the improvement of the offer in public transport.
– to decrease in car traffic by running the information systems about congestion and parking facilities to eliminate redundant (idle) car traffic,
– to improve the environment by reducing transport works in car transport resulting from the improvement of the offer in public transport.

4.2 Variants of ITS Concept

An example of ITS concept variants example for the city of Katowice, located in the urban area with a conurbation nature is [10, 16]:

• variant **V1** – assuming the realization of a system facilitating traffic in the public transport sector (tramway transport) – increase communication speed of trams in the main corridors of the city:
  – system of a priority in traffic control signalling at intersections (variable-time coordination, with acyclic control with priority for trams),
  – at stops boards of dynamic passenger information system,
  – web application and mobile devices application for travel planning by public transport,
  – information kiosks and electronic tickets,

- variant **V2** – assuming additionally (in relation to variant V1) implementation of a system facilitating traffic, partly in public transport (bus service) as well as car transport in the city centre:
  - area-traffic control system in selected transport corridors in the downtown area (variable-time coordination, with acyclic control the area, in cooperation with the tram corridors),
  - guidance system for free parking places,
  - the introduction of a downtown bus-lanes and locks for buses at intersections,
  - changes in the organization of car traffic involving the lack of continuity in passing through downtown,
  - extension of the existing paid parking zones,
- variant **V3** – assuming additionally (in relation to variant V2) implementation of the information system for drivers of traffic congestion and implementing thus balancing the network loads at rush hours:
  - determining the average travel time on the sections by using: cameras with automatic license plate recognition with their encryption (ANPR - automatic number plate recognition) and identification the movement of taxis (by using GPS),
  - system of variable message signs (VMS),
  - transmitting drivers information about traffic conditions,
  - extension of the web/mobile application for travel planning of individual transport,
- variant **V4** - assuming (in relation to variant V3) further implementation of the system to streamline car traffic through the inclusion of a control system of the main access routes to the city centre:
  - area traffic control system will adjust the intensity of traffic flows flowing laterally to privileged transport corridors in order to maximize the flow of traffic on these corridors, and above all, the flow of traffic in the downtown area (which is the primary goal),
  - the introduction of the variant V4 improve traffic flow on those car users, for whose implementation of the previous variants V1 ÷ V3 was not a sufficient incentive to change the way of travel,
  - the early implementation of V4 variant can be rather undesirable competition for the promoted public transport, so it seems appropriate to be temporary located the implementation of this stage in the final phase of the ITS project.
- variant **V5** – variant analysed only in the functional aspect - requiring further study and analysis of traffic mainly in terms of freight and rail, especially in the broader aspects: agglomeration [9, 14, 19] and regional [15]; it is a and development option of the target size of the project assuming the implementation of additional subsystems and functionality relating to:
- freight management:
  - automatic safety inspection (installation of weights for freight vehicles)
  - control the transport of hazardous materials (ANPR camera with signs detection of hazardous materials),

- management of regional rail transport in the area of Katowice city:
  - automatic vehicle location of Rapid Urban Railway (RUR),
  - travel information on nodes and railway stations with the RUR,
  - travel information in vehicles of RUR,
  - dispatch function for the operator of RUR (traveller information concerning changes in the circulation of RUR, vehicle breakdowns, information about possible connections, etc.)
- travel information about the transport links of the airport (International Airport Pyrzowice) with the transport system of the Katowice city:
  - information kiosks located at the airport terminals with the current information on multi-modal connections with the possibility of travel planning.

4.3 Result Indicators for the Expected Variants

Conducted cost-benefit analysis for ITS variants V1 ÷ V4 in relation to the option of abandoning the project (V0 variant) indicates the extremely high cost-effectiveness of all ITS variants especially two of them, that is V3 and V4 (Table 1). Implementation of the ITS project in all variants of investment leads to a return on investment (at the basic rate of return = 5%) as early as 1 year after the start of the investment. Definitely the largest shares in the economic benefits of all variants have the following factors [10, 16]:

- savings resulting from the reduction in the cost of users travel time (remaining with the current transport mode and abandoning the car in favour of public transport),
- benefits resulting from travel cost savings of users (running costs of cars reduced by the cost of purchasing tickets due to increased use of public transport); mainly due to the resignation of the car in favour of public transport,
- the benefits of reducing pollution and reducing the number of accidents in the metropolitan network - relatively small, a total range up to a few percentage.

Table 1. Economic efficiency of ITS project for Katowice city (in Poland) [10, 16]

Socio-economic index	ITS functional variants			
	V1	V2	V3	V4
ENPV – Economic net present value	270 200 000 [EUR]	212 700 000 [EUR]	2 517 700 000 [EUR]	3 079 700 000 [EUR]
ERR – Economic rate of return	125.76%	64.56%	244.96%	252.42%
BCR – Benefits to costs ratio	19.73%	7.26%	64.10%	72.47%
Total investment cost of project	10.0 mln [EUR]	24.7 mln [EUR]	29.8 mln [EUR]	32.4 mln [EUR]
Annual operating costs	1.2 mln [EUR/year]	2.7 mln [EUR/year]	3.3 mln [EUR/year]	3.6 mln [EUR/year]

5 Conclusion

The experience of the authors [6–11] as well as many projects [1, 3, 17, 24–26] and reports containing the so-called good practices, indicate that presented in the article approach to the design of ITS using systems engineering and ITS architecture is appropriate but not fully utilized in the urban ITS projects in Poland. Good example of using this methodology is the project ITS for national roads in Poland in which, as shown in the article approach of systems engineering has resulted in useful documentation [4], on the basis of which is being currently under design and implementation of individual ITS subsystems. Prepared methodology of ITS project development for the city of Katowice [6, 7, 10] also assumes such a procedure, as outlined in the article, which should also result in a useful pre-project documentation: Feasibility Study for the ITS Katowice 2017, Deployment Programme and ITS Architecture for optimal variant of the ITS Concept.

References

1. Bělinová, Z., Bureš, P., Jesty, P.: Intelligent transport system architecture different approaches and future trends. In: Düh, J., et al. (eds.) Data and Mobility, AISC 81, pp. 115–125. Springer, Berlin (2010). doi:10.1007/978-3-642-15503-1_11
2. Dydkowski, G., Urbanek, A.: The implementation, maintenance and development of IT systems: selected problems of designing contracts. Arch. Transp. Syst. Telemat. 9(3), 9–15 (2016)
3. European Intelligent Transport System (ITS) Framework Architecture (E-FRAME). http://frame-online.eu/. Accessed 12 Feb 2017
4. Feasibility study of National Traffic Management System, Kraków, General Directorate for National Roads and Motorways (GDDKiA in Poland) (2012). http://www.kszr.gddkia.gov.pl/. Accessed 12 Feb 2017
5. Gubbia, J., et al.: Internet of things (IoT): a vision, architectural elements, and future directions. Future Gener. Comput. Syst. 29(7), 1645–1660 (2013)
6. Karoń, G.: Analysis of the Current Status in the Development of Pre-design Documentation for the Project Entitled Intelligent Transport Management System for Katowice City. Report of Research Work for City Hall of Katowice, Poland (2016). (in Poland)
7. Karoń, G.: Guidelines for the selection of the contractor of project and documentation of Intelligent Transport Management System for Katowice City. Report of Research Work for City Hall of Katowice, Poland (2016). (in Poland)
8. Karoń, G., Janecki, R.: Development of various scenarios of ITS systems for urban area. In: Sierpiński, G. (ed.) Intelligent Transport Systems and Travel Behaviour. AISC, vol. 505, pp. 3–12. Springer, Cham (2017). doi:10.1007/978-3-319-43991-4_1
9. Karoń, G., Mikulski, J.: Expertise of The Concept and Architecture of the Intelligent Traffic Management System (ITS) for the Territory of the Public Transport Operator in Upper Silesian Industrial Region. Report of Research Work for Municipal Transport Union of the Upper Silesian Industrial District (KZK GOP). Katowice, Poland (2015)
10. Karoń, G., Mikulski, J.: Expertise of The Functional Analysis for Intelligent Transportation Management System (ITS) for Katowice City. Report of Research Work for City Hall of Katowice. Katowice, Poland (2011)

11. Karoń, G., Mikulski, J.: Forecasts for technical variants of ITS projects—example of upper-silesian conurbation. In: Mikulski, J. (ed.) Activities of Transport Telematics, TST 2013. CCIS, vol. 395, pp. 67–74. Springer, Heidelberg (2013). doi:10.1007/978-3-642-41647-7_9

12. Karoń, G., Mikulski, J.: Problems of ITS architecture development and its implementation in upper-silesian conurbation in Poland. In: Mikulski, J. (ed.) TST 2012. CCIS, vol. 329, pp. 183–198. Springer, Heidelberg (2012). doi:10.1007/978-3-642-34050-5_22

13. Karoń, G., Mikulski, J.: Problems of systems Engineering for ITS in large agglomeration—upper-silesian agglomeration in Poland. In: Mikulski, J. (ed.) TST 2014. CCIS, vol. 471, pp. 242–251. Springer, Heidelberg (2014). doi:10.1007/978-3-662-45317-9_26

14. Karoń, G., et al.: The Methodology and Detailed Concept of Traffic Surveys and Measurements and Transportation Model Development for the Territory of the Public Transport Operator in Upper Silesian Industrial Region. Report of Research Work for Municipal Transport Union of the Upper Silesian Industrial District (KZK GOP). Katowice, Poland (2016). (in Poland)

15. Karoń, G., Żochowska, R., Sobota, A.: The Methodology and Detailed Concept of Traffic Surveys and Measurements and Transportation Model Development for Transportation Study of Central Subregion Territory. Report of Research Work for Union of Municipalities and Counties of Central Subregion of Silesian Region. Katowice, Poland (2016). (in Poland)

16. Mikołajczyk, M., et al.: The Functional Analysis for ITS in Katowice. Report of Research Work, BiT for City Hall Katowice in Poland (2011). (in Poland)

17. National ITS Architecture 7.1. http://local.iteris.com/itsarch/. Accessed 12 Feb 2017

18. Sun, L., Li, Y., Gao, J.: Architecture and application research of cooperative intelligent transport systems. Procedia Eng. 137, 747–753 (2016)

19. Szarata, A., et al.: The Concept and Architecture of the Intelligent Traffic Management System (ITS) for the Territory of the Public Transport Operator in Upper Silesian Industrial Region. Report of Research Work for Municipal Transport Union of the Upper Silesian Industrial District (KZK GOP). Katowice, Poland (2015)

20. Systems Engineering Guidebook for Intelligent Transportation Systems. U.S. Department of Transportation (2009)

21. Systems Engineering for Intelligent Transportation Systems. U.S. Department of Transportation. FHWA-HOP-07-069 (2007)

22. Urbanek, A.: Pricing policy after the implementation of electronic ticketing technology in public urban transport: an exploratory study in Poland. In: Mikulski, J. (ed.) TST 2015. CCIS, vol. 531, pp. 322–332. Springer, Cham (2015). doi:10.1007/978-3-319-24577-5_32

23. Żochowska, R., Karoń, G.: ITS services packages as a tool for managing traffic congestion in cities. In: Sładkowski, A., Pamuła, W. (eds.) Intelligent Transportation Systems – Problems and Perspectives. SSDC, vol. 32, pp. 81–103. Springer, Cham (2016). doi:10.1007/978-3-319-19150-8_3

24. Kalašová, A., Černický, L., Hamar, M.: A new approach to road safety in Slovakia. In: Mikulski, J. (ed.) TST 2012. CCIS, vol. 329, pp. 388–395. Springer, Heidelberg (2012). doi:10.1007/978-3-642-34050-5_44

25. Černický, L., Kalašová, A.: Microscopic simulation of the coordinated signal controlled intersections. In: Mikulski, J. (ed.) TST 2013. CCIS, vol. 395, pp. 15–22. Springer, Heidelberg (2013). doi:10.1007/978-3-642-41647-7_3

26. Kubíková, S., Kalašová, A., Černický, L.: Microscopic simulation of optimal use of communication network. In: Mikulski, J. (ed.) TST 2014. CCIS, vol. 471, pp. 414–423. Springer, Heidelberg (2014). doi:10.1007/978-3-662-45317-9_44

Equilibrium Method for Origination Destination Matrix Estimation Exploited to Urban Traffic Simulation Calibration

Marek Bazan[✉], Tomasz Janiczek, and Łukasz Madej

Faculty of Electronics, Wrocław University of Science and Technology,
ul. Janiszewskiego 11/17, 50-372 Wrocław, Poland
{marek.bazan, tomasz.janiczek}@pwr.edu.pl,
lukasz-madej@live.com

Abstract. One of methods used to find better light signalization plans for urban regions is a traffic flow numerical simulation. In this paper we present solution of two main issues the user of traffic flow simulation tool encounters while simulating traffic continuously registered in ITS by counting detectors, namely: (a) origination-destination routes setting, (b) calibration of the driver model. We consider two methods for origination-destination matrix estimation. The first is classical Spiess gradient method and the second one is the equilibrium method of Casseta which is suitable to congested scenarios. Resulting origination destination matrices from both methods are input to the traffic simulator to generate traffic between nodes in the city. The traffic generated in the simulator in this manner is used to calibrate the average driver model, to achieve as good recovery of vehicle counts on detectors as possible. As a simulation tool we use an open source ArsNumerica Execution Environment running on top of SUMO simulator. The driver model that is used is stochastic Krauss model. Calibration is done for one of the most congested areas in the city of Wrocław in the morning peak.

Keywords: Origination-destination matrix estimation · Congested traffic simulation · Traffic simulation calibration

1 Introduction

A crucial component of the road traffic simulation process that is based on real data is calculation of the origination destination matrix for the city/region. The input for the methods to estimate origination destination matrix are vehicle counts on locations of the detectors in a road network within the prescribed time.

The output of the estimation process i.e. an estimated origination destination matrix together with the assignment matrix contains the information how to distribute routes of the vehicles between any two nodes of the network to obtain counts of vehicles on the counting detectors that recover real data within the prescribed time.

In this work we consider two methods to calculate origination-destination matrix. The first one is the classical Spiess method [1] and the second one is Cascetta equilibrium method [2]. A review of classical methods of the origination destination matrix estimation can be found in [9] and more modern ones in [8].

© Springer International Publishing AG 2017
J. Mikulski (Ed.): TST 2017, CCIS 715, pp. 16–27, 2017.
DOI: 10.1007/978-3-319-66251-0_2

The third issue for best possible recovery of the counts on detectors for the simulation is calibration of the driver model. In this work we consider the stochastic Krauss model with such parameters as acceleration, deceleration, delay, maximum speed and an average driver reaction time.

In this paper we consider three different methodological areas of the simulation [8] that recovers traffic counts:

- traffic assignment – i.e. supply model,
- origination destination matrix calculation – i.e. demand model,
- simulation with a use of both assignment matrix and the origination destination matrix – i.e. an average diver model. The employment of the numerical calculations in the above three aspects of the road traffic simulation enables us to propose a framework to compare results of the origination destination matrix estimation methods which is nowadays being sought by many researchers (c.f. [9]). A novelty of our approach is usage of the road traffic simulation and the analysis of the recovery of the ground truth traffic. This approach is simpler than the general framework proposed in [9] – giving nevertheless possibility of quantitative comparison of the methods.

We consider two methods for the origination destination matrix estimation – the classical Spiess [1] and more adequate to congested networks equilibrium method by Cascetta et al. [2]. In the notation of [8] both of these methods require link counts and initial OD matrix i.e. they are both updating methods. For the full static load of traffic from the morning peek hour (from 7:00 till 8:00) the Cascetta method as expected works better for saturated networks – it however gives a recovery error about 23% - which is in, as expected, accordance with the literature [7]. The difference between the quality of the obtained matrices by the two methods is shown by checking the hypothesis on the difference between a mean error of the recovery of the ground truth counts.

The remainder of the paper is organized as follows: in the next section we present a notation together with a demonstration of dimensionalities of the vectors and matrices involved, the third section recall the Spiess and Cascetta methods respectively. The fourth describes a procedure of creation of the simulation process to calibrate the driver model which in turn is described in the sixth section. In the fifth section describes a measure to calculate link counts recovery quality which is crucial for comparing results of both methods. In the seventh section we describe a genetic algorithm which is used for a drivel model calibration. The eight section describes numerical results. And finally we conclude the paper.

2 Origination Destination Matrix Problem Formulation

An O-D matrix is a set of values describing traffic from a source point i to a target point j. In the equation below, it becomes vector g, with $c = n \cdot (n - 1)$ elements. The main equation binding the vector od measurements with the origin-destination matrix presented below will become

$$v = Ug \tag{1}$$

where:

m – a number of nodes (intersections) in the given topology, actually a number of locations counting detectors,

c – a number of all O–D pairs equal to $m \cdot (m - 1)$.

v – a vector containing values of traffic acquired based on measurements of vehicles traveling between intersections: $v = (v_1, v_2, \ldots, v_m)' \in R^m$,

g – origin destination matrix in the vector form. Each matrix element is equal to the number of vehicles moving from intersection (area) i to intersection (area) j. Index alongside the element means a number of a pair $(i \rightarrow j)$: $g = (g_1, g_2, \ldots, g_c)' \in R^c$.

$U \in R^{c \times m}$ – assignment matrix, where element u_{ia} describes the fraction of total traffic for O–D pair of i number on a edge where $i = 1, \ldots, c$ and $a = 1, \ldots, m$.

Despite the low complicity of Eq. (1), calculation of g is usually difficult even in spite of the exact knowledge of the measurement vector v and U matrix. The main problem while solving Eq. (1) is dimensionality of its components. That is, the v vector contains m elements, whereas the g vector contains c elements. As the number of all possible combinations of intersection-intersection pairs is usually much bigger than the number of all intersections in the net $(c \gg m)$, it is clear, that Eq. (1) is strongly underdetermined. This fact is a reason why there is no unique solution of the equation. The number of possible solutions is infinite.

3   Two Methods of the Origination Destination Matrix

In this section we recall two methods of the origination destination matrix estimation: the Spiess gradient method [1] and Cascetta equilibrium method [2]. For the Spiess method we used assignment method based on Poisson distribution using always a prescribed number of the shortest paths between nodes. As in [6] it was up to 5 shortest paths for each OD pair. For the Cascetta method an assignment process is done by the method itself.

3.1   Spiess Gradient Method

One of the methods used in this paper to determine the matrix OD is gradient method described by Heinz Spiess [1]. This is an iterative method for determining the minimum or maximum of the test function defined by the formula:

$$f : R^n \rightarrow R, \quad x \in R^n$$

The vector x is assigned to several individual values. To start operation of the algorithm is necessary to define the starting point x^0, which allows you to calculate the value of a point in the next iteration, as shown in the following formula:

$$x^{i+1} = x^i - \alpha \nabla f(x^i),$$

where:

l – number of iteration,

x^l – argument value for l-th iteration,

α – parameter specifies the length of the iteration step, usually values are selected from the interval $(0,1)$.

∇ – gradient operator.

This figure is a general formula describing this method. For the study related to the matrix OD, you must modify the model functions as follows:

$$Z(g) = \frac{1}{2} \sum_{a \in A} \left(v_a - v'_a\right)^2$$

where:

v_a – the measured traffic on the edge of a,

v'_a – the estimated value of the intensity at the edge of a,

a – the edge of the graph,

A – set of the edges,

g – OD matrix in a vector form.

Then, in the estimation of matrix flows we use the steepest descent algorithm:

$$g_i^{l+1} = \begin{cases} \hat{g}_i, & l = 0 \\ g_i^l \left[1 - \lambda^l \left[\frac{\partial Z(g)}{\partial g_i}\right]\right], & l = 1, 2, \ldots \end{cases}$$

where:

$\hat{g}_0$ – chosen start point,

g_i – i-th component of the vector,

l – number of the iteration,

λ – length of iteration.

You should also designate pattern derivative of the cost function $Z(g)$ in dependence on the length of iteration step:

$$\frac{dZ(\lambda)}{d\lambda} = \sum_{a \in \hat{A}} \frac{dv_a}{d\lambda} \frac{\partial Z}{\partial v_a} = \sum_{a \in \hat{A}} v'_a \left(v_a - \hat{v}_a + \lambda v'_a\right)$$

Equating to 0 the previous equation we are looking for the value of λ it meets, which will be determination of the optimal length iteration:

$$\lambda^* = \frac{\sum_{a \in A} v'_a (\hat{v}_a - v_a)}{\sum_{a \in A} v'^2_a}.$$

3.2   Starting Point for Spiess Method

In order to circumvent the problem of lack of the starting origination destination matrix $\hat{g}_0$ the Tikhonov regularization method is employed. Despite high under determination of the system (1) it is possible to calculate such so that is satisfy it approximately with a certain error. For this purpose the More-Penrose inverse solution to (1) may be derived by multiplying (1) both sides by U'

$$U'v = U'Ug \tag{2}$$

and then by adding nonzero component to a diagonal of $U'U$ on the right side of the above equation one gets

$$U'v = (U'U + \lambda I)g, \tag{3}$$

which allows already to approximate g since the matrix on the right hand side is nonsingular even for very small positive λ's. Therefore one gets

$$g_0 = U'(U'U + \lambda I)^{-1}v \tag{4}$$

In this form however g_0 cannot be used in the process since it may contain negative components. This is not interpretable in the context of the origination destination matrix estimation. To resolve this problem $\hat{g}_0$ is calculate from g_0 by substituting 0 value where negative component is encountered. This approach gives a starting point which provides a stability of the Spiess iteration process.

3.3   Equilibrium Method

Another method is the one described by Cascetta [2]. We consider situation in which an O–D connections and their location is known and constant. Load of connections and the cost of traffic are dependent of each connection flow and function of travel costs. On the other hand, the flow rate on a particular combination depends on the probability of selection of the specific movement path connecting OD pair in the model. Generally speaking, this model has the task to regulate the interaction between the "supply" of vehicles, and the "demand" of roads, or the number of vehicles that will be able to drive the route most fluent way. The model is presented by following equations (Fig. 1):

$$g_{od}^* = \Delta_{od}^T c \left(\sum_{od} \Delta_{od} h_{od}^* \right) + g_{od}^{NA} \quad \forall od$$

$$V_{od}^* = -g_{od}^* \qquad\qquad\qquad \forall od$$

$$h_{od}^* = d_{od} p_{od}(V_{od}^*) \qquad\qquad \forall od$$

Δ_{od} in the above equations means a row from this matrix corresponding to the pair with an index od.

$$\Delta = \begin{array}{c|cccccc} & 1 & 2 & 3 & 4 & 5 & 6 \\ \hline 1,2 & 1 & 1 & 0 & 0 & 0 & 0 \\ 1,3 & 0 & 0 & 1 & 0 & 0 & 0 \\ 2,3 & 1 & 0 & 0 & 1 & 0 & 0 \\ 2,4 & 0 & 1 & 0 & 0 & 1 & 0 \\ 3,4 & 1 & 0 & 1 & 1 & 0 & 1 \end{array}$$

Fig. 1. Example of connections matrix, columns – number of junction, rows – OD pair [2]

After substituting:

$$g^*_{od} = \Delta^T_{od} c \left(\sum_{od} \Delta_{od} h^*_{od} \right) + g^{NA*}_{od} \quad \forall od$$

$$h^*_{od} = d_{od} p_{od} \left(V^*_{od} \right) \qquad \forall od$$

We have the following variables:

g^*_{od} – vector movement cost for a given path OD, interpreted as travel time, index NA means non additive costs,

h^*_{od} – flow on the path for an OD pair, the number of vehicles on the track,

d_{od} – the expected flow on the path for a given pair of OD.

p_{od} – the probability of choosing a path for a given pair of OD

c – the cost of connections,

Δ^T_{od} – connections matrix.

The values of variables marked with an asterisk are determined at the time of calculation, while the others are known.

3.4   Starting Point for the Equilibrium Method

As a starting point of the equilibrium method we used the solution of the Spiess iteration process. As it may be seen from the analysis of the results of the worked out examples the equilibrium method can substantially improve this solution.

4   Procedure of Simulation Creation

At the beginning it is necessary to prepare the map of the area. As nodes was the intersection where the cameras are located in Wroclaw Intelligent Transport System, and thus it is possible to read data on the movement of vehicles in these places. The number of vehicles was read during the morning rush, on the work day beyond the summer break. Intersections main, were considered transit centers, that means those where new vehicles cannot appear – they can only enter and leave. Powering the model performed with the nodes of up to major intersections, is declared to them the flow of vehicles. Amount of vehicles on specified connections can be calculated by one of described models. After that it is possible to perform simulation. In our work we used traffic simulator SUMO [3] and the ArsNumerica wrapper [4, 5] for it enabling to perform calibration. We prepared simulation map based on one used to calculations. There was used constant traffic lights settings.

5   Measuring Link Counts Recovery Quality

Basic way to establish quality of model was calculating difference between number of vehicles which should arrive to specific output and number of vehicles which actually arrived to this point. However, this method isn't very accurate. That's why we used also Krauss model to more specific calculation of quality function for both methods:

$$F = \sum_j \left(\frac{DDD_j - DSS_j}{DDD_j} \right) \Big/ N \tag{5}$$

where:
DDD_j – read data in simulation for detector,
DSS_j – known data for detector,
K – tested detector,
N – number of detectors.

6   Krauss Calibration Model

In order to compare the quality of action set out methods of determining the matrix flows were carried out a survey to calibrate Krauss model [3]. This will determine the optimal parameters of the model, which are interpreted as the most appropriate way driver behavior in traffic. However, the main benefit of this study is to answer the question which come to mind: which method provides more capacity in traffic?

To carry out this study has been prepared runtime environment, consisting of several sub-programs:

- M++atlab script - is responsible for calculating successive input values for the calibrated model, using a genetic algorithm global optimization
- Traffic Program - a program implemented in C++, whose task is to type in the startup files simulator SUMO model parameters, as well as the calculation of the value of the objective function for each iteration,
- SUMOConnector program - a program implemented in C++, corresponding to the communication with the simulator. Its function is to read in a dynamic way during the simulation of the traffic measurement detectors and write to a file,
- SUMO - simulator processing received data on the traffic, the model parameters Krauss and road network. It allows you to obtain the necessary output.

The present model is implemented in the simulator used, which allows for testing using the developed runtime without additional implementation. The survey will proceed by iteratively calculating successive values of model parameters Krauss on the basis of which will be carried out simulation.

7 Genetic Algorithm to Solve Calibration Task

Like it was mentioned in previous section during calculations with usage of the Krauss model the genetic algorithm was used. It was implemented in the same way as in [4, 5]. The Krauss model simulation was performed every iteration for new values of this model parameters i.e. acceleration, delay, maximum speed, reaction time.

8 Tested Area

Performed tests was executed on part of city of Wroclaw. It is known as *area3* and covers Grunwaldzki Square and surrounding area. Schematic picture of this part of city is presented below (Figs. 2, 3).

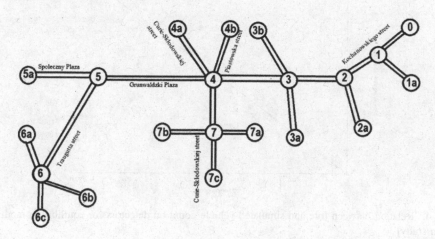

Fig. 2. Schematic view of the simulated network – it corresponds to the one of the most congested region in the center of Wrocław city (Poland) – surroundings of the Reagan roundabout [own study]

8.1 Tests of OD Matrix Calculating Methods

At first we performed calculations of OD matrices by both methods. Next, we prepared simulations with obtained results like it was described in previous sections. In this test we measured number of vehicles, which achieved their targets in one simulation. Each simulation was depicting 1 h of traffic. Results are presented in Fig. 4.

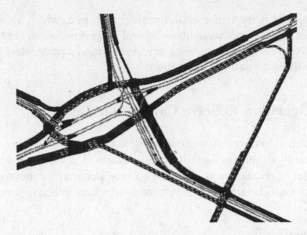

Fig. 3. Example view of the simulation - a zoom of the Reagan roundabout and its surroundings from SUMO [3]

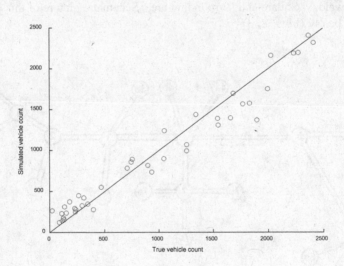

Fig. 4. Relation between true and simulated vehicle count on detectors for equilibrium model. [own study]

Second test was performed with usage of Krauss calibration. In this test we performed 30000 continuous simulations for each method. Krauss model parameters was set by genetic algorithm for every iteration. Results are presented by plot in Fig. 6.

8.2 Test of Hypothesis

In last test we checked hypothesis that calibration error is significantly smaller for equilibrium method than Spiess method. We take 30 consecutive results of objective function for each method to check if result achieved for equilibrium calibration is significantly smaller. For trust level 0.95 and 28 degrees of freedom we get critical value of 1.701 and critical area $\langle 1, 701, +\infty \rangle$. Using Eqs. (3) and (4) as result we obtained value 122.50, which is higher than critical value and belong to critical area. On this basis we can tell that calibration error is significantly smaller for equilibrium method:

$$S_{X1-X2} = \sqrt{\frac{(N_1 - 1) * s_1^2 + (N_2 - 1) * s_2^2}{N_1 + N_2 - 2} * \left(\frac{1}{N_1} + \frac{1}{N_2} \right)} \qquad (6)$$

$$T = \frac{X_1 - X_2}{S_{X1-X2}} \qquad (7)$$

where:

X_1 – mean value for Spiess method,
X_2 – mean value for equilibrium method,
S_1 – variance for Spiess method,
S_2 – variance for equilibrium method,
N_1 – size of first group of values,
N_2 – size of second group of values,
$T_1, S_{X_1-X_2}$ – estimators.

8.3   Plots

See Figs. 5, and 7.

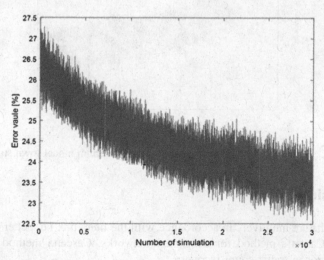

Fig. 5. Error value in Krauss calibration for equilibrium model. [own study]

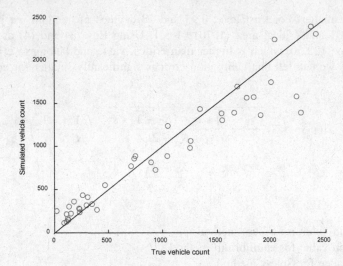

Fig. 6. Relation between true and simulated vehicle count on detectors for Spiess *model* [own study]

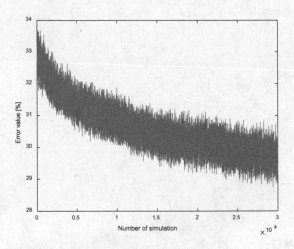

Fig. 7. Error value in Krauss calibration for equilibrium model [own study]

9 Conclusion

Obtained results are intuitively in accordance with the literature, i.e. better results were obtained for Cascetta method for congested networks. Cascetta method approaches 20% global error in traffic count recovery.

The outlook for future research to beat 20% barrier is to implement two things

– fully dynamic operation of intersections signalization to be in accordance with ITS microprograms that work on most of the intersections,

- dynamic assignment model more tightly coupled with origination-destination matrix estimation method, this is may be achieved with bi-level programming methods (c.f. [10]).

In this paper we introduced a methodology to compare methods for estimation of the origination destination matrix. The approach consists of calculation of the origination destination matrix by the methods being compared and then to start the calibration process of the driver model using a genetic algorithm. The comparison is performed by calculation of the relative mean error of the link counts recover on the calibrated models. This approach seems to be simpler than proposed in [9] and gives as a spin-off the behavior of the origination destination matrices calculated by the methods in the calibrated simulation.

Acknowledgement. The work presented in this paper was partially financed from Grant 0401/0230/16.

References

1. Spiess, H.: A Gradient Approach for the O-D Matrix Adjustment Problem, EMME/2 Support Center, Switzerland (1990)
2. Cascetta, E.: Transportation Systems Analysis. Model and Applications. Springer Optimisations and ITS Applications, 2nd edn. Springer, Heidelberg (2009)
3. Krauss, S.: Microscopic modeling of traffic flow: investigation of collision free vehicle dynamics. DLR, Ph.D. thesis (1998)
4. Bazan, M., et al.: Intelligent transportation system auditing using micro-simulation. Arch. Transp. Syst. Telemat. **8**(4), 3–8 (2015)
5. Bazan, M., et al.: Green wave optimisation. Arch. Transp. Syst. Telemat. **9**(3), 3–8 (2016)
6. Bazan, M., et al.: Two methods of calculation of the origination destination matrix of an urban area, Raport W04/P-007/15, Wrocław University of Technology (2015)
7. Treiber, M., Kersting, A.: Trafic Flow Dynamics, Data, Models and Simulation. Springer, Heidelberg (2013)
8. Toledo, T., et al.: Network model calibration studiem. In: Traffic Simulation and Data: Validation methods and Applications, pp. 141–162. CRC Press, Boca Raton (2015)
9. Antoniou, C., et al.: Towards a generic benchmarking platform for origin–destination flows estimation/updating algorithms: design, demonstration and validation. Transp. Res. Part C **66**, 79–98 (2016)
10. Tavana, H.: Internally consistent estimation of dynamic network origin–destination flows from intelligent transportation systems data using bi-level optimization. Ph.D. thesis, University of Texas, Austin (2001)

Availability Factors in Delivery of Information and Communication Resources to Traffic System Users

Ivan Cvitić[✉], Dragan Peraković, and Tibor Mijo Kuljanić

Faculty of Transport and Traffic Sciences, University of Zagreb,
Vukelićeva 4, 10000 Zagreb, Croatia
{ivan.cvitic,dragan.perakovic}@fpz.hr,
tibor.kuljanic@gmail.com

Abstract. Availability, as one of the basic elements of IC (information and communication) system security, plays an important role in its work and in operation of services that are provided through it. IC system and related services (IC resources) must be available to end users when needed. Availability of IC resources is especially evident in IC systems implemented in a traffic environment. The emergence of new concepts such as Internet of Things or Cloud Computing and technologies such as 5G resulting in an increasing processes automatization in the traffic system. In doing so, the unavailability of individual IC resource can result in negative user experience in usage of certain classes of services, but can also represent a potentially large security risk and compromised user safety. In such environment required level of IC resources availability are often extremely high so it is essential to know which factors affect it. This research aims to identify these factors, and the results of this research will provide guidelines by which factors needs to be managed to achieve the required level of IC service availability.

Keywords: Availability assurance · Security threats · Information and communication security · Cloud computing · Internet of things

1 Introduction

Availability, along with the confidentiality and integrity is a basic requirement when considering the security of information and communication (IC) system. The development of IC technology in recent years has resulted in a high degree of their integration within the traffic system. In such traffic system, enhanced with IC technologies, a high level of availability of IC resources is required. The term resources imply on IC system, all its elements, data, information and services that are delivered to end users.

Due to the specificity of the traffic system, unavailability of IC resources can cause many problems. An example can be negative user experience in the use of certain services up to compromising safety of the traffic system users. It is the reason that the IC resources in the traffic environment need to ensure the required level of availability. The level of availability depends on the class and service that is provided to the user through the IC system. The motivation for this research arises from the importance of

© Springer International Publishing AG 2017
J. Mikulski (Ed.): TST 2017, CCIS 715, pp. 28–41, 2017.
DOI: 10.1007/978-3-319-66251-0_3

the availability requirements in environments such as traffic system as well as the lack of research focused at factors that can negatively affect the availability of the IC resources in mentioned environment. The aim of the research shown in this paper is to identify the factors that can affect availability of IC resources. The research results will provide a clearer insight into the factors which need to be managed to ensure the required availability level at delivery of IC resources to the traffic system users. The purpose of the research is to increase the security of IC resources in a complex environment that enables the delivery of various ITS (Intelligent Transport System) services to end users.

1.1 Previous Research

Security of IC technologies in the traffic field is the subject of a growing number of research which is caused by the rapid development and application of ITS and related services.

Comprehensive research carried out in [1] identify current research challenges in the development of secure cooperative ITS services relating to V2X (Vehicle-to-Vehicle/Infrastructure/User) communication. Security threats assumed to V2X communication environment were analyzed and classified. The research has, among other, covered only some of the existing factors that may affect the availability of IC resources in the V2X communication. Factors affecting the availability analyzed by research are related primarily to the illegitimate human influence on availability (DDoS, jamming, malware, and others). Lack of research the insufficient analysis of other factors affecting the level of availability.

In the research [2], the impact of DDoS (Distributed Denial of Service) attacks on the availability of ITS services based on IoT (Internet of Things) concept were analyzed. Case study on the electronic toll collection in Croatia were conducted and the point of possible implementation of DDoS attacks has been identified. In addition, the number of users that would not be able to use the electronic toll collection service in case of DDoS attacks were determined. Based on the research methods for protection against DDoS attacks per IoT architecture layers have been proposed. This research is restricted exclusively to the DDoS attack as one of the possible factors disrupting the IC resources availability which have been identified as a lack of research.

Research of cooperative ITS architecture security is shown in [3]. The research has analyzed the basic security requirements (confidentiality, integrity and availability) and have been expanded with additional requirements specific to the cooperative ITS architecture (plausibility, freshness, scalability, and others). The research focus on data privacy, the risk assessment of the threats and at proposed asymmetric cryptographic algorithms protection through different PKI (Public Key Infrastructure) which, according to [1], has no effect at the availability protection. Lack of the research is seen in a very limited review and analysis of the IC resources availability as well as factors that affect the availability of the cooperative ITS architecture.

Threats to security of connected vehicles are covered by the research presented in [4, 5] and includes threats to communication link, data validity, devices, identity of vehicles, access control, and privacy of vehicles and drivers. The research [5] includes only the availability of the communication element (VANET) while research [4] does not include the availability of IC resources.

The analysis of the available previous research noted the lack of identification of factors affecting the IC resources availability implemented in the traffic system. This research aims to complement detected lack of previous research due to importance of availability requirements in complex environment where timely delivery of data, information or services have key importance for the satisfaction and safety of the traffic system users.

1.2 Research Methodology and Limitations

Intelligent transport systems in this research were assessed from the aspect of application in road traffic system. The research analyzes the available scientific-technical literature. Through the results synthesis of conducted researches new knowledge was presented in the field of security and protection of IC system in a traffic environment. Researches directed to security and protection of various IC technologies and new concepts were analyzed, as an addition to research related to security of IC resources in traffic environment. The reason for this is their utilization or potential utilization in the traffic environment which implies on possibilities of mirroring availability factors between different IC environments.

2 Information and Communication Resources Availability

Availability represents one of the key security requirements. Every IC system have the purpose to deliver the data, information or service to end user. User can be human, but it can also be a process, machine or any other entity that can benefit from received data, information or service. If IC system cannot perform delivery of them, then it loses its purpose [5]. If IC resource is not available at requested time, then security requirements such as integrity or confidentiality losses their importance as well. Availability is usually described in "number of nines notation" where one nine denotes 90% availability level or 36.5 days/year of downtime and six nines denotes 99.9999% of availability level or 31.5 s/year of downtime [6].

According to [7], availability is the probability that a system or system element is performing its defined function at a given time. That is why it is important to observe factors that can affect availability of IC resources in ITS system, which implies every element of IC system and not just communication segment. Elements of IC system whose availability needs to be considered are shown in Fig. 1.

Motivation for conducting an attack can vary, but according to [8] and Fig. 2, 24.43% of organization stated that motivation for generating security incident is disruption of service. That information is supporting importance of research oriented towards identification of factors that can have negative affect on IC resources availability.

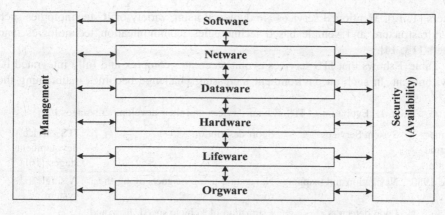

Fig. 1. Elements of IC system [own study]

Fig. 2. Motivation for attacks on IC resources [8]

3 ITS and Connected Services Evolution

Today, road traffic system has the growing trend of vehicle number and according to [1], that number in 2050 will reach two billion. The mentioned increase in the number of vehicles implies an increase in the number of challenges in the field of optimization of transport processes, congestion management in a traffic network, traffic safety, informing the participants of the transport system and ecology [9].

Therefore, in recent decades ITS is increasingly developing and implementing. The concept of ITS indicates continuous upgrading and improvement of the classic traffic system using IC technologies. The integration of these technologies provides opportunities for the effective solving the identified problems as well as optimization of transport processes. In addition, more frequent is the application of IC technologies in informing people (pedestrians) with disabilities about the traffic environment as well as their guidance [10, 11].

The problem of the observed domains is trying to be solved with one or more services such as vehicle information and navigation system, air pollution control, traffic control and management services, road safety services, parking services, and many others [1, 12]. Mentioned services are deployed using variety of IC technologies such as infrastructure and vehicle based technologies, communication technologies, and others [13, 14].

Table 1 shows that ITS services evolution creates complex and fully integrated IC environment. In such an environment increasing challenge becomes maintaining the

Table 1. Evolution of ITS system services and their development stages [15]

Time period	System/Services	Short description	ITS service development stages [16]
Pre-1980	Navigation and mapping	Navigation using magnets along the road	A single system
	Loop detectors	Estimation of vehicle speed, flow and occupancy	
	Dynamic message signs	Providing information and warnings to users	
	Ramp management	Freeway safety problem solution	
	Traffic management systems	Weather, speed, congestion, incidents information processing and taking actions on given information	
	Global positioning system	Speed, position and time determination	
1980s	Automated traffic surveillance and Control system	Integration of vehicle detectors, closed-circuit television, and coordinated signal timing	Partially integrated system
	Traffic congestion management	Improving traffic operations and reducing fuel consumption and emission	
	DARPA Autonomous land vehicle	Use of artificial intelligence for autonomous vehicle	
	Electronic License Plate	Automated vehicle identification	
1990s	FAST-TRAC	Small-scale traffic control system	
	TravTek	In-vehicle navigation system	
	Pathfinder	In-vehicle information system	
	Electronic toll collection	Toll collection without need to stop the vehicle	
2000s	Driver Assistance system	Collision warning, lane departure warning, lane change warning	
	Clarus	Relevant information on road condition to transportation managers and users	
	Integrated Corridor management	Traffic management based on current and predicted traffic condition	
Post-2010	Vehicle-to-vehicle	Cooperative ITS (C-ITS)	Fully integrated system
	Vehicle-to-infrastructure		
	Vehicle-to-everything		

required level of access to IC resources to traffic system user. The importance of the IC resources availability excels in a new paradigm of cooperative ITS communication, which seeks to connect IC resources on transport infrastructure, vehicles and other traffic entities. Such systems require continuous communication between all entities that is achieved using wireless communication technologies [16].

3.1 ITS Services Classification

Today there is a number of ITS services that are delivered to the traffic system user. Those services can be categorized in various ways. According to [17, 18], it can be divided in six main classes:

- Demand and access management (infrastructure use pricing, carbon credit scheme, restricted traffic zones, pay-as-you -drive strategy)
- Traffic management and control (isolated controlled intersection, ramp metering, dynamic speed limits)
- Travel and traffic information (on-board navigation system, intelligent parking, pre-trip information service)
- Driver assistance and cooperative systems (intersection organization between vehicles, electronic stability control, blind spot detection, lane departure warning)
- Logistics and fleet management (commercial fleet management, parking/loading/ delivery management, slot management, automated vehicle management system for public transport)
- Safety and emergency systems (on-board accident prevention system, infrastructure-based incident prevention system, incident management system).

Availability requirement for mentioned services and their categories can vary depending on potential consequences of their untimely delivery. There are only few researches and technical documents that define required availability for services such as autonomous vehicle control (99.999%), collision avoidance (99.999%), and traffic jam control (95%) [19, 20]. Towards mentioned, services that can have direct impact on safety of traffic system user should maintain highest possible availability level (five or six nines).

3.2 Information and Communication Infrastructure for ITS Services Delivery

Delivery of ITS service to traffic system users requires application of various IC technologies. Progress in their development emphasize development and delivery of new service or upgrading current ones. According to [14, 21], there are three main classes of enabling IC technologies for ITS service:

- Data collection and sensing technologies – includes infrastructure based technology (inductive loops, sensors, CCTVs and others), and vehicle-based technology (GPS, OBD, and others).

– Communication technologies - includes mainly wireless technology for short and long range communication such as mobile data, VANET, ZigBee, WiFi, Bluetooth, WiMAX, RFID.
– Database management and computational technology – managing collected traffic data, artificial intelligence, etc.

Such complex IC environment is shown in Fig. 3 that visualize various objects and communication technologies implemented for their communication. Besides that, there are illustrated elements for data processing and storage such as central information system and Cloud Computing which plays key role in service delivery to the users of traffic system.

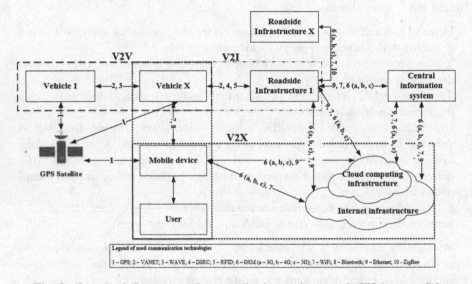

Fig. 3. Complex information and communication environment in ITS [own study]

With further technology development, new concepts are used to deliver ITS services to traffic system user. Machine-to-Machine communication, Internet of things and Cloud Computing concepts are implementing for distribution of variety of services such as parking services, electronic toll collection service [2, 12, 22]. Every new technology implemented in IC environment for service delivery to traffic system users is increasing complexity and heterogeneity of that environment and resulting in growing number of security issues. Geographic dislocation of IC resources, especially sensors and roadside communication equipment and their exposure to physical and natural threats causes additional security issues.

4 Identification of Factors Affecting Availability of IC Resources

There are many factors that can cause security incidents, and disrupt one of three basic security requirements (confidentiality, integrity and availability). IC resources implemented in traffic system represent a critical infrastructure and needs to be assured with high availability [23]. Inadequate level of availability in observed environment can negatively affect users in terms of their dissatisfaction with quality of service (QoS) or even violation of their safety in traffic network. Based on before mentioned, it is of great importance to identify factors affecting IC resources availability so it can be managed in future with goal to obtain its required level.

Factors that affect IC resources availability are classified according to multidimensional classification model [24]. Motive for choosing mentioned model is its applicability in complex and dynamic environment. The factors classification criteria are:

- Source (internal, external),
- Agent (human, environmental, technological), and
- Impact (in this research impact is always availability oriented).

Factors that have potential to negative affect IC resources can be internal or external. Further internal or external factors can be segmented by human, environmental or technological agent. Mentioned classes of factors will be discuss further in this paper.

4.1   External Factors Affecting IC Resources Availability

External factors, shown in Table 2, that can have negative affect on IC resource availability in this research are segmented on human, environmental and technological. Mentioned factors are located outside of the system or organization boundaries [25].

Human factor includes users who do not have authorized access to information and communication resources including facilities, system components, network or software elements [24]. They can use various types of intentional or unintentional activities that can affect availability of IC resources. Additional problem is the lack of exact perimeter of today ITS environment which makes more difficult to protect IC resources.

Attack caused by human factor that is often mentioned in numbers of research is DDoS attack, and its subclasses. It is classified as number one out of top 15 critical factors that can impact security in intelligent public transport [23]. It can be referred as attack that can occur in every IC resource that has connection to public communication network with intention to disable its availability [2, 26, 27]. Those IC resources can be Cloud Computing, IoT, on-premises servers (web, application, database, etc.), routers, switches, and others. Besides DDoS, other factors that can affect availability of IC resources are jamming, implementation of malicious code, theft, vandalism, terrorism, hacktivism, and others.

Jamming is factor that can cause interference in frequencies used for ad-hoc communication between traffic entities (V2V, V2I, V2X) in M2M or IoT environment and cause unavailability of data communication [2]. Another key factor that have potential impact on IC resource availability is unauthorized physical access to one of the IC system elements. Examples can be seen in M2M or IoT environment [28] where data collection and identification devices are often limited dimension and deployed without any physical protection. They can be stolen or intentionally destroyed with purpose to cause their communication and data collection unavailability [29]. Unauthorized physical access can also occur in other environment such as Cloud Computing (as a vital

Table 2. Identified external factors that can negatively affect IC resources [own study]

External		
Human	DDoS attacks	
	Jamming	
	Theft	
	Vandalism	
	Malicious code	
	Terrorism	
	Hacktivism	
Environmental	Natural disasters	Natural disasters
		Floods
		Earthquakes
		Hurricanes
		Lightning strike
		Tsunami
		Electromagnetic storm
	Human triggered	Fire
		Dangerous radiation leaks
		Pollution
		Dust
		Corrosions
		Major events in the environment
Technological	Equipment failures	
	Power failures	
	Telecommunication service provider failure	
	Cloud computing service provider failure	

IC resource for service delivery) where can cause disruption of cloud provider, customer or third party services [30].

Terrorists and cyber terrorists are another rising external factors whose actions are aimed at attaining political, economic, religious or other goal. Their intention is to intimidate or attack people by physical or cyber-attack on infrastructure (physical or IC). In year 2015 from total 12204 terrorist attacks, 381 (3.12%) was conducted in transportation sector, and 46 (0.36%) in telecommunication sector [31]. Those factors can cause the degradation or unavailability of services delivery vital for user safety in traffic system through physical destruction or cyber-attacks.

Implemented in-vehicle IC systems enabling V2V, V2I and V2X communication for delivery of services such as autonomous driving, car steering, car usage restriction, vehicle health monitoring and management require connection to public communication network. Till year 2018, it is expected 35 million vehicles on the market that have embedded internet connectivity, which is shown on Fig. 4 [32]. Every device connected to the public communication network has the great potential from malicious code infection. Such malicious code can cause degradation of service quality or even unavailability of before mentioned service that can have great influence on driver safety.

Along external human, important factor that can affect availability level of IC resources is external environmental factor. This factor can be divided into natural disasters and those triggered by human, such as earthquakes, floods, wildfires, pollution, dust, and corrosion. It can have negative effect on all IC resources at once or individually [23]. According to [33], mentioned factor needs to be priority in every disaster recovery plan because of potential of damaging a large geographic area which can include IC facilities, human resources, IC elements for data storage, processing and transmission.

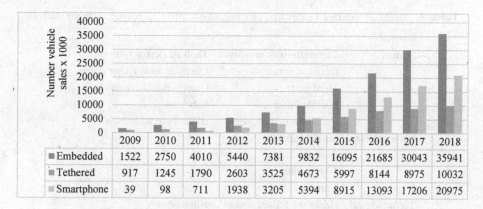

	2009	2010	2011	2012	2013	2014	2015	2016	2017	2018
Embedded	1522	2750	4010	5440	7381	9832	16095	21685	30043	35941
Tethered	917	1245	1790	2603	3525	4673	5997	8144	8975	10032
Smartphone	39	98	711	1938	3205	5394	8915	13093	17206	20975

Fig. 4. Growth of connected vehicle number [32]

External technological factors imply on power sources, telecommunication services and infrastructure, and other equipment that is not under the domain of the observed IC system. Outages of such equipment can lead to unavailability of IC resources [34]. Also, growing number of services are delivered to traffic system user using Cloud Computing based platforms [35, 36]. They can be a subject of numbers of attacks that can cause their unavailability which also means unavailability of ITS services that are delivered via Cloud Computing [37].

4.2   Internal Factors Affecting IC Resources Availability

Internal factors that can affect availability level of IC resources are segmented on human, environmental and technological. Internal human factor can be intentional or accidental. According to [38], it can be a current or former employee, business partner or third party who has or had authorized access to IC resources and can cause harm with or without malicious intent to confidentiality, integrity or availability of mentioned IC resources. Common characteristic of internal human factor is that they have authorized physical or remote access to IC resources [39]. Examples of internal human factor are shown in Table 3 [40]. This factor can affect availability in various ways such as sabotage, theft or destruction of equipment, misconfiguration, black hole formation, and operator or user errors. Unintentional human errors are factor that cause approximately 80% of internal security incidents [38]. They can affect availability by hardware or software misconfiguration. Intentional human factor has potential to affect availability trough theft or destruction of hardware equipment or destruction of data.

Example is unsatisfied current or former employee that can use insider knowledge of system for causing IC resources unavailability [41].

Internal environmental factor needs to be considered to ensure required level of availability. Hardware equipment placed inside system facility (servers, network equipment, etc.) is sensitive to environmental factors such as humidity, temperature, liquid, dust, etc. Specificity of ITS environment additionally rise influence of this factor on availability level for equipment placed externally [42].

Table 3. Identified external factors that can negatively affect IC resources [own study]

Internal		
Human	Current/former employee	Theft or destruction
	Business partners	Operator and/or user errors
	Third party	Configuration errors
	Reckless user	Sabotage
	Unsatisfied employee	Black hole formation
Environmental	Liquid leakage	
	Dust	
	Temperature	
	Humidity	
Technological	Hardware failure and/or malfunctions	
	Software failure and/or malfunctions	
	Interruption and/or disruption of facility electrical supply	
	Interruption and/or disruption of frequency	

Internal technological factors that can cause unavailability of IC resources are often related to accidental faults. According to [43] there are three main classes of accidental faults in critical infrastructure:

– cascade failures (failures in one part of IC resources causing the failure of other part of IC resources),
– escalating failures (existing failure in on part of IC resource deteriorates failure of other IC resources), and
– common cause failures (two or more IC resources are affected simultaneously).

Mentioned classes of failures can occur under the influence of common technological factors shown in Table 3, hardware or software failure, interruption or disruption of electrical supply in facility [44].

5   Conclusion

Presented paper shows research that is aimed on identification of factors that can negatively affect availability of IC resources in traffic environment. Identified factors are classified per multidimensional classification model in two main classes (internal and external) and three subclasses (human, environmental and technological). Factors were identified through analysis of available scientific and technical literature and can represent valuable information to future research directed toward risk analysis, identification and management in IC systems implemented in specific transport and traffic environment.

Depending on specific environment, individual factors that are identified in this research will or will not represent the risk for IC resources availability. For example, if IoT concept is not implemented then it is not necessary to observe factors that are affecting availability of IoT specific IC resources. Factors that have influence on observed environment need to be managed through mitigation methods or implementation of available protection methods that can reduce their influence.

Future research will be oriented on methods that can reduce the influence of identified factors with focus on IoT and Cloud Computing environment. It will be considered possibilities of fog computing and edge intelligence implementation for achieving and maintaining required level of IC resources availability.

References

1. Hamida, E., Noura, H., Znaidi, W.: Security of cooperative intelligent transport systems: standards. Threats Anal. Cryptogr. Countermeas. Electron. **4**(3), 380–423 (2015)
2. Cvitić, I., Peraković, D., Periša, M., Jerneić, B.: Availability protection of IoT concept based telematics system in transport. In: Mikulski, J. (ed.) TST 2016. CCIS, vol. 640, pp. 109–121. Springer, Cham (2016). doi:10.1007/978-3-319-49646-7_10
3. Boudguiga, A., Kaiser, A., Cincilla, P.: Cooperative-ITS architecture and security challenges: a survey. In: 22nd ITS World Congress, pp. 5–9 (2015)
4. Othmane, L.B., Weffers, H., Mohamad, M.M., Wolf, M.: A survey of security and privacy in connected vehicles. In: Benhaddou, D., Al-Fuqaha, A. (eds.) Wireless sensor and mobile ad-hoc networks, pp. 217–247. Springer, New York (2015)
5. Sumra, I.A., Hasbullah, H.B., AbManan, J.B.: Attacks on security goals (confidentiality, integrity, availability) in VANET: a survey. In: Laouiti, A., Qayyum, A., Saad, M.N.M. (eds.) Advances in Intelligent Systems and Computing, vol. 306, pp. 51–61. Springer, Singapore (2015)
6. Hoda, R., Roosta, A.K.: Calculating total system availability. Universiteit van Amsterdam (2014). https://staff.science.uva.nl/c.t.a.m.delaat/rp/2013-2014/p17/report.pdf. Accessed 13 Dec 2016
7. Barabady, J.: Improvement of System Availability Using Reliability and Maintainability Analysis, Luleå University of Technology (2005)
8. ISACA: State of Cybersecurity: Implications for 2015, Rolling Meadows (2015)
9. Cardoso, R.M., Mastelari, N., Bassora, M.F.: Internet of things architecture in the context of intelligent transportation system: a case study towards a web-based application deployment. In: 22nd International Congress of Mechanical Engineering (COBEM 2013), pp. 7751–7760 (2013)
10. Periša, M., Peraković, D., Šarić, S.: Conceptual model of providing traffic navigation services to visually impaired persons. Promet Traffic Transp. **26**(3), 209–218 (2014)
11. Peraković, D., Periša, M., Remenar, V.: Model of guidance for visually impaired persons in the traffic network. Transp. Res. Part F: Traffic Psychol. Behav. **31**, 1–11 (2015)

12. Ji, Z., Ganchev, I., O'Droma, M., Zhang, X.: A cloud-based intelligent car parking services for smart cities. In: 2014 XXXIth URSI General Assembly and Scientific Symposium (URSI GASS), pp. 1–4 (2014)
13. Qi, L.: Research on intelligent transportation system technologies and applications. In: 2008 Workshop on Power Electronics and Intelligent Transportation System, pp. 529–531 (2008)
14. Zear, A., Kumar Singh, P., Singh, Y.: Intelligent transport system: a progressive review. Indian J. Sci. Technol. 9(32), 1–8 (2016)
15. Lockwood, S., Auer, A., Feese, S.: History of Intelligent Transportation Systems. Washington, USA (2016)
16. United Nations ESCAP: Intelligent Transportation Systems for Sustainable Development in Asia and the Pacific, Bangkok, Thailand (2015)
17. Jonkers, E., Gorris, T.: Intelligent Transport Systems and Traffic Management for Urban Areas (2015)
18. Benz, T., et al.: Inception report and state-of-the-art review (2011)
19. 5G systems (2017). https://www.ericsson.com/assets/local/publications/white-papers/wp-5g-systems.pdf. Accessed 26 Jan 2017
20. The Road to 5G: Drivers, Applications, Requirements and Technical Development (2015). http://www.huawei.com/minisite/5g/img/GSA_the_Road_to_5G.pdf. Accessed 18 Jan 2017
21. European Commission, Intelligent Transport Systems – EU-funded research for efficient, clean and safe road transport, Brussels (2010)
22. Ashokkumar, K., Sam, B., Arshadprabhu, R.: Cloud based intelligent transport system. Procedia Comput. Sci. 50, 58–63 (2015)
23. Levy-Bencheton, C., Darra, E.: Cyber Security and Resilience of Intelligent Public Transport: Good Practices and Recommendations. Heraklion, Greece (2015)
24. Jouini, M., Rabai, L.B.A., Aissa, A.B.: Classification of security threats in information systems. Procedia Comput. Sci. 32, 489–496 (2014)
25. Gibson, D.: Managing Risk In Information Systems (2010)
26. Somal, L.K., Virk, K.S.: Classification of distributed denial of service attacks – architecture, taxonomy and tools. Int. J. Adv. Res. Comput. Sci. Technol. 2(2), 118–122 (2014)
27. Mittal, A., Shrivastava, A.K., Manoria, M.: A review of DDOS attack and its countermeasures in TCP based networks. Int. J. Comput. Sci. Eng. Surv. 2(4), 177–187 (2011)
28. Cvitić, I., Vujić, M., Husnjak, S.: Classification of security risks in the IoT environment. In: 26th Daaam International Symposium on Intelligent Manufacturing and Automation, pp. 0731–0740 (2016)
29. Rose, K., Eldridge, S., Chapin, L.: The Internet of Things: An Overview. Geneva, Switzerland (2015)
30. Jansen, W.A.: Cloud hooks: security and privacy issues in cloud computing. In: 2011 44th Hawaii International Conference on System Sciences, no. 4, pp. 1–10 (2011)
31. National Consortium for the Study of Terrorism and Responses to Terrorism, Country Reports on Terrorism 2015, Maryland, USA (2016)
32. GSMA: Connected Car Forecast: Global Connected Car Market to Grow Threefold Within Five Years, London, UK (2013)
33. Cisco Systems Inc.: High Availability Disaster Recovery: Best Practices. http://www.cisco.com/en/US/technologies/collateral/tk869/tk769/white_paper_c11-453495.html. Accessed 12 Jan 2017
34. Reuter, C.: Power outage communications: survey of needs, infrastructures and concepts. In: Proceedings of the 10th International ISCRAM Conference, pp. 884–889 (2013)
35. Gupta, E.: Process mining a comparative study. Int. J. Adv. Res. Comput. Commun. Eng. 3 (11), 17–23 (2014)

36. Meneguette, R.I.: A vehicular cloud-based framework for the intelligent transport management of big cities. Int. J. Distrib. Sens. Netw. **2016**, 1–9 (2016)
37. Ertaul, L., Singhal, S., Saldamli, G.: Security challenges in cloud computing. IEEE Trans. Intell. Transp. Syst. **14**(1), 36–42 (2013)
38. CERT® Division: Unintentional Insider Threats: A Foundational Study (2013). http://www.sei.cmu.edu/reports/13tn022.pdfedu. Accessed 18 Dec 2016
39. Marinos, L., Belmonte, A., Rekleitis, E.: Threat Landscape 2015. Heraklion, Greece (2016)
40. Casey, T.: A Field Guide to Insider Threat (2015). http://www.intel.com/content/www/us/en/it-management/intel-it-best-practices/a-field-guide-to-insider-threat-paper.html. Accessed 11 Oct 2016
41. Ophoff, J., et al.: A descriptive literature review and classification of insider threat research. In: Proceedings of Informing Science and IT Education Conference (InSITE) 2014, pp. 211–223 (2014)
42. Australia Government: Physical security of ICT equipment systems and facilities. https://www.protectivesecurity.gov.au/physicalsecurity/Pages/PhysicalSecurityOfICTEquipmentSystemsAndFacilitiesGuidelines.aspx. Accessed 07 Jan 2017
43. Montanari, L., Querzoni, L.: Critical Infrastructure Protection: Threats, Attacks and Countermeasures (2014). http://www.dis.uniroma1.it/~tenace/download/deliverable/WP4-Deliverable4b-TENACE.pdf. Accessed 21 Jan 2017
44. ENISA: Threat Landscape and Good Practices for the Internet Infrastructure. Heraklion, Greece (2014)

Modern Methods of Image Processing in Safety-Critical Applications within Intelligent Transportation System

Emília Bubeníková[✉], Mária Franeková, and Alžbeta Kanáliková

Faculty of Electrical Engineering, University of Žilina, Univerzitná 8215/1,
010 26 Žilina, Slovak Republic
{emilia.bubenikova,maria.franekova,
alzbeta.kanalikova}@fel.uniza.sk

Abstract. Modern collision avoidance systems implemented into the Intelligent Transportation Systems work on the base of the image information analysis. For the control of unintentional departure from the lane, the vehicles in Cooperative-intelligent Transportation Systems use surveillance systems—Lane Departure Warning. A very important parameter of these systems is the reliability of the method of digital image information processing, which depends on the choice of the optimum algorithm for the detection of road markings in different light conditions. The parameters of image processing algorithms should be set on the base of software simulation with real collected traffic data. The contribution is focused on finding and testing effective methods of digital image processing algorithms to monitoring the crossing of lane with a focus on reliability evaluation.

Keywords: Cooperative-intelligent transportation system · Advanced driver assistance system · Lane departure warning system · Image processing · Detection algorithms · Reliability

1 Introduction

Intelligent Transportation Systems (ITS) are advanced applications which gratefulness of development more powerful processors, communications technology, sensors enable in the means of transport take to more control, monitoring, safety functions and functions that increase comfort in vehicles. In the first place, Intelligent Transportation Systems exert to minimize consequences vehicles accidents and primarily to prevent them [1].

Information and Communications Technologies (ICT) are very employed in the means of transport where are embedded elements of intelligence. Applying of ICT into the area of road transport and its interfaces by others type of transports [2] significantly contributed to restriction to influence of road transport to the environment as well as to improvements in effectiveness including energy effectiveness, safety, and prevention of road traffic.

While e.g. in railway transport from the point of safety is required the quantitative assessment of all element of the transmission system, which can occur hazard (dangerous event) [3, 4], look at the security within road transport is a bit different.

© Springer International Publishing AG 2017
J. Mikulski (Ed.): TST 2017, CCIS 715, pp. 42–54, 2017.
DOI: 10.1007/978-3-319-66251-0_4

Cooperative-intelligent transportation system (C-ITS) belong to the new generation of ITS, which offer in the local wireless network valuable information to the driver directly at vehicles namely through transmissions of various types of warning messages [5]. But have to be considered an assurance of credibility of received messages i.e. authorization of source messages on what techniques on the base of cryptography (for generation and verification of digital signature) are being exploited [6, 7].

Exists a number of C-ITS applications. Many of them are founded beside sensors - based techniques also on image data processing with using methods of computer vision, which allows controlling of transport processes on the basis of image information [8].

Applying of computer vision algorithms and digital signal processing methods is possible to achieve not only raising of comfort by the drive a vehicle increasing continuity of transport, reducing of the ballast of environment but mainly reducing the traffic accidents.

On the present on the sphere of automobile industry constantly runs through improvement of algorithms which are able from image data to determine if an unin- tentionally vehicle is leaving own lane just yet before as if may come out an accident.

Applications of Advanced Driver Assistance System (ADAS) used different types of sensors for data acquisition by over vehicles and its neighbourhood. After the acquisition of data object detection, methods of identification and technique of pro- cessing and evaluate of threats is applied.

Identification process, monitoring, and driving evaluation are demanding. Car drivers have control over their vehicles in the various meteorological conditions (sunshine, rain, snow, fog).

Lane Departure Warning System (LDWS) using detector CMOS onto night vision, infra-red sensors, adaptive cruise control, and park-line assistant. LDW systems warn the drivers in the event if the vehicle moves outside alert zone (safe area) i.e. the vehicle retreats from center line.

A quality of data scanning by sensors is affecting by fare styles and luminous conditions whereby can to hide important information needs for identifications and monitoring of objects. Even through data processing has to be realized within real time and delay may not be bigger like the 30 ms. Each step from data collection to action performance requires considerable accomplishment of signal processing exactly and in time. Digital signal processors specially designed and optimized for automobile safety applications provide sufficient efficiency.

LDW systems are designed so in order to minimize accident damage thereby they solve generally reasons of collisions (e.g. mistake of the driver, diversion or sleepi- ness). LDW systems are active automatically at the point of achieved vehicle spread upward 60 km per second.

The last line of warning is accordance with [9] ISO 17361:2007 given by the value 0.3 m within external boundary of the lane on road for personal vehicles or by value of distance 1 m for trucks and buses.

ADAS in the event of occurrence of dangerous situation automatically to generate and transmit alert message to all vehicle which is in the frequency range through C2C and C2I communications. Communications between particular elements of C-ITS running in VANET network which on the present consider to being the one from cardinal wireless communications technology in the area of ITS [10].

In the last decade in the whole world, a number of projects and consortiums come into existence, e.g. [11, 12]. An important position in the Europe has consortium Car2Car Communication Consortium (C2C-CC) [13] founded by six European automobile factories (Audi, BMW, Daimler Chrysler, Fiat, Renault, and Volkswagen).

Messages transmitted between vehicles are activated by the concrete event. Formats and types of messages depend on use applications. The messages for safety critical applications have the highest level of assurance. Every entity of C-ITS periodically to send safety-relevant messages (speed, the dimension of the vehicle) to all nodes within VANET with frequency of 1–10 Hz - Cooperative Awareness Message (CAM).

Father on the dependence of specific event (vehicle accident, danger in the way) node is able to transmit father types of messages - Decentralized Environmental Notification Message (DENM), which warns surrounding vehicles. The overall delay between verified messages should be bellow like the 100 ms.

At the present, the most critical safety-related applications within C-ITS is considered: Pre-Crash Sensing, Blind Spot Warning, and Lane Change Warning.

Authors in the contribution in the detail present the LDW system from the view of determinations of the optimal setting of input parameters for assurance of reliable detection and its quantifications on the base of SW simulation.

2 Principle of Advanced Driver Assistant System

Advanced Driver Assistant Systems use a variety of sensors on physical data collection to from vehicle and its environments. The principle of ADAS systems operation is presented on Fig. 1. After data gathering ADAS systems use object classification and detection, recognition and tracking processing techniques for evaluation safety risks and threats. Data, such as nearby vehicles, lane markings, road signs, obstacles in roads, etc. are captured by various environment sensors like radar, lidar, infrared, camera (Fig. 1). Detailed study and modeling of sensors characteristics are required to increase their applicability in a wide range of vehicles and applications. Camera, the radar sensor and infrared sensor enable realization of functions like Adaptive Cruise Control & Emergency Brake assist. Additional functionalities like Lane Detection or Lane Keep Assist, Traffic Sign Recognition, Pedestrian Detection, etc. In the vehicle there are several types of sensors (Fig. 1) and various restrictions that bring sensors, can be removed using sensor fusion [14, 15].

The step of processing presented on Fig. 1 is based on image analysis in the three different stages of data processing.

The first stage is data capture, second stage is pre-processing and third stage is post-processing. During pre-processing stage, various functions are applied to the full image captured - functions like classification of the image, stabilization, feature and signal enhancements, noise reduction, colour conversion and motion analysis. During post-processing stage, various features like tracking, image environment interpretation, system control, analysis and decision making are performed.

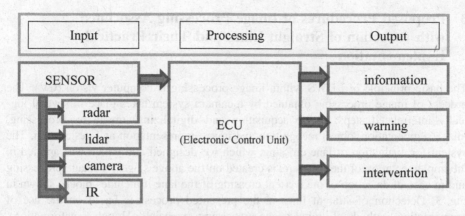

Fig. 1. Principle of ADAS systems [own study]

From the analysis of present state results from that the view of methods and principles of ADAS is ADAS development oriented into two primary areas:

- Lane Departure Warning systems which warn driver at the point of traffic line deviation via visual attention (on display), acoustic warning (audio signal spreads from reproducing unit on equivalent side) and/or haptic alert (by stimulating of human sense most often in form of the seat vibes or propriety part of wheel).
- Lane Keeping system, which warns car driver and if the driver is unresponsive application automatic give response (automatic deceleration or deviation of wheel) with the aim to keep the vehicle in the traffic line.

The authors in the contribution deal with problem of the image processing and its applying within searching of horizontal road markings. Some parts of this problem were to be already mentioned in the ours works, e.g. [16, 17]. The sense of ADAS system rests in the consequences reducing within inattention of driver during driving the vehicle which belong to the first place within reason of accident. On the Fig. 2 the statistic of the fatal road traffic accidents is illustrated published by automobile factory Volkswagen [18].

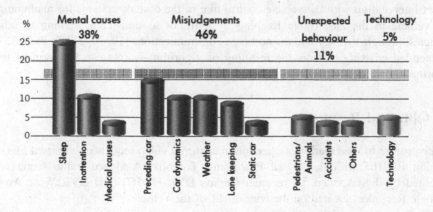

Fig. 2. Reasons of road traffic accident [18]

3   Proposed Procedures of Image Processing Associated with Detection of Straight Lines and Their Practical Implementation

The basic principle of LDWS within image processing is computer vision (CV). The process of image processing obtained by a camera system takes place in several logically interrelated steps: image acquisition and digitalizing, image pre-processing, image segmentation, object recognition and image representation and description. The system for evaluation of lane crossing, which we designed and which was applied in the implementation of the software, is created on the above stages of image processing and in case of detection of inadvertent crossing of the lane it includes blocks shown in Fig. 3. Detection of straight lines in the presented process is based on the use of segmentation methods of digital image processing (especially Hough transformation).

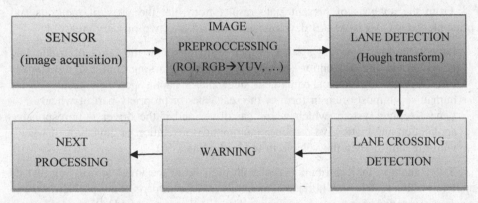

Fig. 3. The sequence of steps in the implementation of the detection of exceeding the lane and its links to block "Next processing" [21]

The "Next processing" block may represent, for example the system interface to a direct intervention with the vehicle control like in the case of systems for maintaining the vehicle in the lane (Lane Keeping System), or to transfer a warning to other vehicles through VANET and the on-board communications [19, 20, 21]. In the case of unintended crossing of lane, the resulting of algorithm warning in the form of text information which is shown on the display.

4   Obtained Results

In the process of road markings detection in captured video sequences was used Matlab version R2010a with the Signal Processing Toolbox. Analyzed traffic scene was obtained by digital video, hd recorder Genius DVR-HD560 (HD DVR Wide Angle Vehicle Recorder) located on the windshield of the vehicle.

In the process of image segmentation and the search of lines can be used many techniques such as segmentation [22–24] etc. Our proposed and implemented system for evaluating the line crossing that has been applied in a software implementation is based on using Hough transform [17]. In general, the Hough transform is a method for finding geometry structures in images. This work uses the Hough transform for straight line detection. Extracted line segments can be used to construct rectangles or generally any other polygons. Lines can be mathematically represented in many ways. For this module, the following representation was chosen. For reasons of simplification of lines detection, we have chosen linear model of lines found. The line is represented in polar coordinates:

$$x \cdot cos\theta + y \cdot sin\theta = r, \tag{1}$$

where r represents the smallest distance of line (in point X [x, y]) from the beginning of coordinate system, Θ is the size of oriented angle from the positive x-axis to the half-line guided from the beginning of the coordinate system upright to the searched line. When applying Hough transformation for straight lines, the point in the original image leads to a sinusoidal curve in the transformed image. Points in the original image belonging to one line result in sinusoids intersecting at one point in the transformed image (Hough space). The coordinates of this point describe the parameters r, Θ of the line and its value represents the number of points of the line.

Points from the original image with x and y-axes are transformed into the Hough space with r and Θ axes. We can say that the resulting (transformed) image is the Hough image. One point (pixel) in the original image is represented by a curve in the Hough image and one point in the Hough image defines a line in the original image by r and Θ parameters. One line in the original image is represented by an intersection of curves. More points (pixels) in one line lead to a higher value in the point of curves' intersection. For further evaluation, it is necessary to extract local maximum values from the Hough image which correspond to significant lines of the original image. This way we obtain the desired lines in form of r, Θ coordinate pairs. However, we do not obtain the end coordinates of the original line segment (r, Θ are line parameters).

$$y = \frac{x \cdot cos\theta}{sin\theta} + \frac{r}{sin\theta}. \tag{2}$$

The use of the normal form of line equation has the advantage that it is independent on the orientation of axes.

In our next text contribution, we will focus primarily on the reliability of detection straight lines in the evaluation process overruns line.

The algorithms for computer vision are characterized by considerable complexity to calculation because the processed image information is in the form of video record. At solving problems related to image analysis there was a need to find ways to speed up the computations. The area of image processed data, we reduced in the preprocessing phase with a suitable setting the size of the region of interest (Region of Interest). ROI was defined on the base of application of the road geometric model in a traffic scene. In segmentation of traffic scene we started from the fact that we know the approximate

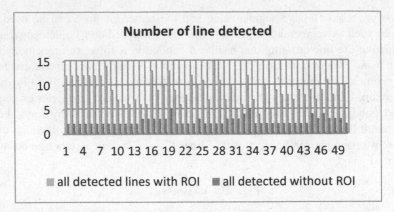

Fig. 4. Number of found straight lines without using ROI (240 × 480 pixels), and using the ROI (65 × 412 pixels) [own study]

location of the path relative to the camera and the road starts at the bottom on the picture and towards the horizon, it narrows. Thus, we can estimate the proportion of the image area which includes road with road markings. We also expect the same sensing system with a fixed position against the surface scan and then size and position of the ROI with respect to the overall image will still be the same. The size of the region of interest was determined empirically based on testing of captured video files [8, 25, 26].

Figure 4 shows the total number of found straight lines without using ROI (240 × 480 pixels) and using the ROI (65 × 412 pixels) for one possible set of parameters supplied to the algorithm of detection lines. It is evident that the number of total lines being compared to the same image several times fell.

4.1 Way to Increase the Reliability of Detection Lines

The basis for successful detection of the inadvertent crossing of the line is the reliable detection lines. When using HT in the process of detecting the lines we had to consider also possible detection of false of straight lines in the image. These are the objects in the image, such as e.g. crash barriers located on the side of the road, dirt on the road, the brake lines of cars, transverse bars, and the like. Thus they are places in the image where there is a step change of brightness and subsequent detection of false lines. Therefore, we are following the process of line detection in the image categorized all found lines on the right lines, respectively wrong lines. The Fig. 5 shows a simplified block diagram showing a categorization lines on the incorrect (wrong) and correct (right). The categorization of lines is based on comparing the obtained angle for each found line (obtained after applying HT) in 1 frame with the tolerance setting angles at the beginning of the program. In the process of segmentation, traffic scenes assume that we use the same sensing system with a fixed position and then size and position of the ROI with respect to the overall image will still be the same. Found lines, located near the line that has bound the lanes have approximately the same angle. Therefore, the line found with the same or similar angle (in selected tolerance) than the angle of the line on the way can be

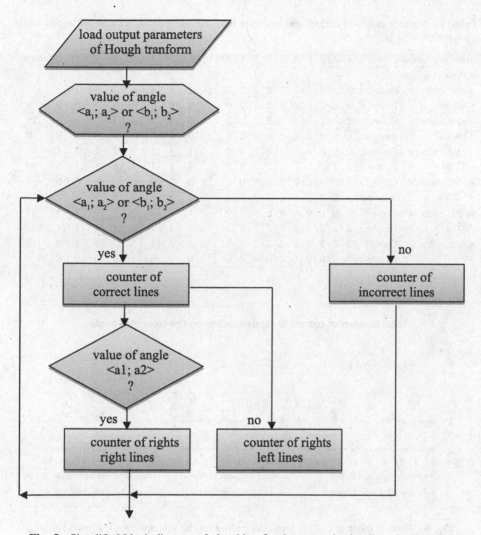

Fig. 5. Simplified block diagram of algorithm for the categorization lines [own study]

seen as finding the correct line. In the line detection algorithm of software application, we considered the different tolerances of angles that can be set at the beginning of the program. Based on the simulation and testing of the various angles of setting tolerances we have received various numbers of total correct and incorrect lines listed in the Table 1 and visualization of simulation results of different settings is on Fig. 6.

Table 1 shows that the algorithm to categorize all found lines that we have implemented in the process of detection of road signs is an important part of the algorithm by which significantly reduces the number of found lines that are relevant for further processing. On based of obtained results, as seems the most optimal setting tolerance angles to interval values $<-50°; -80°>$ to detect the correct horizontal straight lines and interval values $<50°; 80°>$ to detection of the correct horizontal left straight lines.

Table 1. Number of found correct and incorrect lines for different tolerances set angles [own study]

Number of frames		1	2	3	4	5	6	7	8	9	10		43	44	45	46	47	48	49	50	Sum
Tolerance of angles																					
<−50; −80 >or <50; 80>	Correct	2	2	2	2	2	2	2	2	2	2	...	2	4	3	4	3	3	3	2	125
	Incorrect	0	0	0	0	0	0	0	0	0	0	...	0	0	0	0	0	0	0	0	0
<−70; −50> or <50; 70>	Correct	2	2	2	2	2	2	2	2	2	2	...	2	4	3	4	3	3	3	2	125
	Incorrect	0	0	0	0	0	0	0	0	0	0	...	0	0	0	0	0	0	0	0	0
<−50; −65> or <50; 65>	Correct	0	0	0	0	0	0	0	0	0	0	...	1	3	2	3	2	2	2	1	45
	Incorrect	2	2	2	2	2	2	2	2	2	2	...	1	1	1	1	1	1	1	1	80
<−50; −60> or <50; 60>	Correct	0	0	0	0	0	0	0	0	0	0	...	0	0	0	0	0	0	0	0	1
	Incorrect	2	2	2	2	2	2	2	2	2	2	...	2	4	3	4	3	3	3	2	124
<−55; −65> or <55; 65>	Correct	0	0	0	0	0	0	0	0	0	0	...	1	3	2	3	2	2	2	1	45
	Incorrect	2	2	2	2	2	2	2	2	2	2	...	1	1	1	1	1	1	1	1	80
<−60; −70> or <60; 70>	Correct	2	2	2	2	2	2	2	2	2	2	...	2	3	3	3	3	3	3	2	120
	Incorrect	0	0	0	0	0	0	0	0	0	0		0	1	0	1	0	0	0	0	5

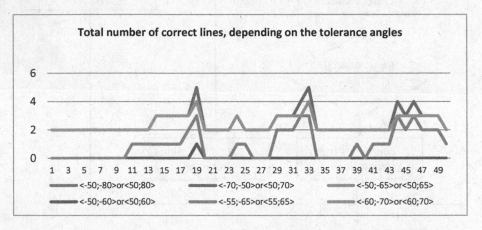

Fig. 6. Total number of correct lines, depending on the tolerance angles [own study]

Phase detection an inadvertent crossing of lane follows the phase detection lines (Fig. 2). The line crossing detection in software implementation works on the principle of searching intersections of lines. For each frame, the position of the intersection of the nearest lines found to the left and right to the vehicle and the fictional vehicle area is compared [16]. To guarantee reliable detection of any overrun lanes was necessary to ensure that in each processed frame will be detected by at least one line that corresponds to the left the road markings and at least one line that corresponds to the right road markings.

In a real recording can be in your captured dotted line is a gap of marking lines, which at some stage image analysis may occur to the fact that in the frame is missing one of the lines corresponding to any of the road markings. We solved this problem by using so-called floating window, whose size can be changed. Floating window before starting the work itself lines detection algorithm searches for the selected number of frames (given the size of the window W) frame with the maximum intensity, which is then used for further processing. Figure 7 illustrates a situation where the analyzed frame left missing traffic road signs and frame status before applying the line detection algorithm (Fig. 7a), without application window (Fig. 7b), after application of the floating window with size W = 10 (Fig. 7c) and with the size W = 15 (Fig. 7d).

From Fig. 8 it shows that increasing the reliability of the detection lines can impact appropriate setting of the size of the floating window at the beginning of the algorithm. As optimal in the processed video footage seems set to W = 15. That is confirmed by a graph in which the x-axis shows the serial number of the 50 frames of video the examined and the y-axis shows the result of a logical conjunction, based on the consideration: "It exists for each analyzed frame detected by at least one correct left

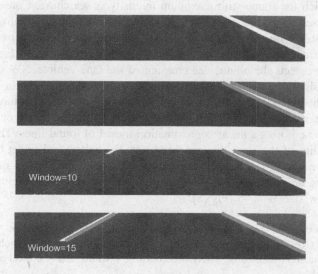

Fig. 7. Illustration processed images without (top figure) and using floating windows of different sizes - from above down (a), (b), (c), (d) [own study]

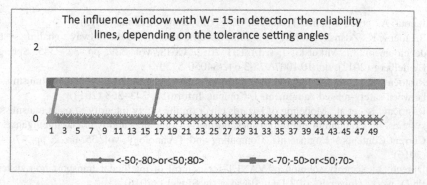

Fig. 8. Detection the reliability lines, depending on the tolerance setting angles [own study]

straight line together at least one correct right straight line?" Logic value 1 for all the images we have obtained at the window setting to W = 15 and at simultaneously setting the tolerance angles of <−50°, −80°> or <50°; 80°> or <−70°, −50°> or <50°; 70°> (for assessment angles). For W < *15* decreased the reliability of detection.

5   Conclusion

The aim of the contribution is describe the applying of modern techniques of computer vision in applications of digital image processing with the intention of transport processes. Practical implementation and verification of the results are focused on increasing the reliability of the detection lines. SW realization is possible to divide into two steps. In the first step we applied to the processed images so-called the floating window in which the frame with maximum intensity is searching. Consequently this frame is processed by the line detection algorithm. In the second step realized algorithm performs the categorization of all found lines in the region of interest to two categories: correct or incorrect lines. This categorization of lines is necessary for aspects of next processing that detects the unintended crossing of the lane vehicle. SW simulation of line detection algorithm and verification of the results was conducted in Matlab software tool in the toolbox Image Processing with using real data captured from real driving on the road. Line detection in realized algorithm is based on the Hough transform because it uses a linear approximation model of found lines. The data show that the reliability of the algorithm proposed was the better, and to set the tolerance angle is in the range of <−80°, −50°> or <50°; 80°> or <−70°, −50°> or <50°; 70°> (for assessment angles) (Fig. 6). We can improve the detection accuracy of used floating window size W = 15 (Fig. 8).

Acknowledgement. This paper has been supported by the Educational Grant Agency of the Slovak Republic (KEGA) Number 008ŽU-4/2015: Innovation of HW and SW tools and methods of laboratory education focused on safety aspects of ICT within safety critical applications of processes control.

References

1. Janota, A., et al.: Applied Telematic, EDIS University of Žilina (2015)
2. Rástočný, K., Ždánsky, J., Nagy, P.: Some Specific activities at the railway signalling system development. In: Mikulski, J. (ed.) TST 2012. CCIS, vol. 329, pp. 349–355. Springer, Heidelberg (2012). doi:10.1007/978-3-642-34050-5_39
3. Rástočný, K., et al.: Quantitative assessment of safety integrity level of message transmission between safety-related equipment. J. Comput. Inform. **2**, 343–368 (2014)
4. Rástočný, K., et al.: Modeling of hazards effect on the safety integrity of open transmission systems. In: Computing and Informatics, Formerly Computers and Artificial Intelligence. ISI Current Contents – Engineering, Computing and Technology, vol. 35, no. 2, pp. 470–496 (2016)
5. Ďurech, J.: Security solution of VANET for control of intelligent transportation systems. Ph.D. work, University of Žilina, Slovakia (in Slovak) (2016)

6. Ďurech, J., Franeková, M., Holečko, P., Bubeníková, E.: Performance Analysis of authentication protocols used within cooperative-intelligent transportation systems with focus on security. In: Mikulski, J. (ed.) TST 2015. CCIS, vol. 531, pp. 220–229. Springer, Cham (2015). doi:10.1007/978-3-319-24577-5_22
7. Roy, R.R.: Handbook of Mobile Ad Hoc Networks for Mobility Models. Springer, London (2011)
8. Bubeníková, E.: Detection of lines in applications of control within intelligent transport. Dissertation work. In: Slovak, University of Žilina, Slovakia (2014)
9. ISO 17361: 2007, Intelligent transport systems—lane departure warning systems – performance requirements and test procedures
10. Ďurech, J., et al.: Implementation of data from the mobile measurement platform to VANET application. In: Proceedings of the 10th International Conference ELEKTRO 2014, Rajecké Teplice, 19–20 May, pp. 430–434 (2014)
11. eSafety Initiatives. https://www.e-safetysupport.com/. Accessed 15 Jan 2017
12. EVITA project. http://www.evita-project.org/. Accessed 23 Jan 2016
13. Communication consortium Car2Car. https://www.car-2-car.org. Accessed 24 Jan 2016
14. Pirník, J., et al.: Integration of inertial sensor data into control of the mobile platform. In: Advances in Intelligent Systems and Computing, pp. 271–282 (2016)
15. Hruboš, M., et al.: Searching for collisions between mobile robot and environment. Int. J. Adv. Robot. Syst. 13(5), 1–11 (2016)
16. Bubeníková, E., Franeková, M., Holečko, P.: Evaluation of unwanted road marking crossing detection using real-traffic data for intelligent transportation systems. In: Mikulski, J. (ed.) TST 2014. CCIS, vol. 471, pp. 137–145. Springer, Heidelberg (2014). doi:10.1007/978-3-662-45317-9_15
17. Bubeníková, E., Franeková, M., Holečko, P.: Conceptual design of driving lane-crossing alarm threshold in C-ITS applications and its implementation. In: Cybernetics and Informatics (K&I) 2016, 2–5 February 2016, Levoča, Slovakia, pp. 1–6. IEEE, SCOPUS (2016)
18. Lienkamp, M.: Intelligent, Connected Cars - Volkswagen's Vision of the Future, Electronics and Vehicle Research, VOLKSWAGEN. http://www.itu.int/dms_pub/itu-t/oth/06/1B/T061B0000020056PDFE.pdf. Accessed 25 Jan 2016
19. Ďurech, J., et al.: Modelling of security principles within car-to-car communications in modern cooperative intelligent transportation systems. Adv. Electr. Electron. Eng. 14(1), 49–58 (2016)
20. Bubeníková, E., Franeková, M., Holečko, P.: Security increasing trends in intelligent transportation systems utilising modern image processing methods. In: Mikulski, J. (ed.) TST 2013. CCIS, vol. 395, pp. 353–360. Springer, Heidelberg (2013). doi:10.1007/978-3-642-41647-7_43
21. Bubeníková, E., Franeková, M., Holečko, P.: Secure solution of collision warning system integration with use of vehicular communications within intelligent transportation systems. In: 12th IFAC Conference on Programmable Devices and Embedded Systems, Veľké Karlovice Czech Republic, 25–27 September 2013, pp. 78–83
22. Huh, K., et al.: Development of vision-based lane detection system considering configuration aspects. Opt. Lasers Eng. 43(11), 1193–1213 (2005)
23. Kowsari, T., Beauchemin, S.S., Cho, J.: Real-time vehicle detection and tracking using stereo vision and multi-view AdaBoost. In: 14th International IEEE Conference on Intelligent Transportation Systems (ITSC), pp. 1255–1260 (2011)
24. Sotelo, A.M., et al.: A color vision based lane tracking system for autonomous driving on unmarked roads. Auton. Robot. 16(1), 95–116 (2004)

25. Bubeníková, E., et al.: The ways of streamlining digital image processing algorithms used for detection of lines in transport scenes video recording, In: PDES 2015: 13th IFAC and IEEE Conference on Programmable Devices and Embedded Systems. Cracow, Poland, 13–15 May 2015, [S. l.: Elsevier], SCOPUS, pp. 174–179 (2015)
26. Bubeníková, E., Franeková, M., Ďurech, J.: Security solutions of intelligent transportation's applications with using VANET networks, In.: ICCC, 2014 15th International Carpathian Control Conference (ICCC), Veľké Karlovice, Czech Republic, pp. 424–429 (2014)

Functional Configuration of ITS for Urban Agglomeration

Grzegorz Karoń[1(✉)] and Jerzy Mikulski[2]

[1] Silesian University of Technology, Akademicka 2A, 44-100 Gliwice, Poland
grzegorz.karon@polsl.pl
[2] Katowice School of Technology, Rolna 43, 40-001 Katowice, Poland
mikulski.jurek@gmail.com

Abstract. This article presents selected problems design of intelligent transportation systems (ITS) in aspect of functional and technical configuration of ITS. The configuration of ITS system may includes several subsystems described taking into account following aspects and components: main objectives, specific objectives, functionalities, integration with external systems, main installations and technical components, technical components location. The structure of an example ITS configuration based on ITS for urban agglomeration – case study for ITS KZK GOP (in Poland) – has been presented.

Keywords: Intelligent transportation systems (ITS) · Configuration of ITS · Urban agglomeration · ITS stakeholders aspirations · Aspirations-implementations matrix of ITS

1 Introduction

The variety of technical and functional solutions of ITS systems can include taking in the description of the system the functional and operational configuration of ITS which constitute a properly designed and cooperating structures [1, 2, 4, 6, 9, 14–16, 20]: logical architecture (functional) and physical architecture. The configuration of ITS system, in high-level design of systems engineering [18, 19], can be described taking into account the following aspects and components [2]:

- stakeholders aspirations,
- user needs,
- included transport modes,
- ITS architecture, taking into account:
  - ITS users,
  - ITS services,
  - ITS data storages,
  - ITS functions, taking into account: allocation of functions, decomposition of functions, synthesis of functions,
  - ITS data flows diagrams, taking into account following processes: functional processes, physical processes, communication processes, organizational processes,

© Springer International Publishing AG 2017
J. Mikulski (Ed.): TST 2017, CCIS 715, pp. 55–69, 2017.
DOI: 10.1007/978-3-319-66251-0_5

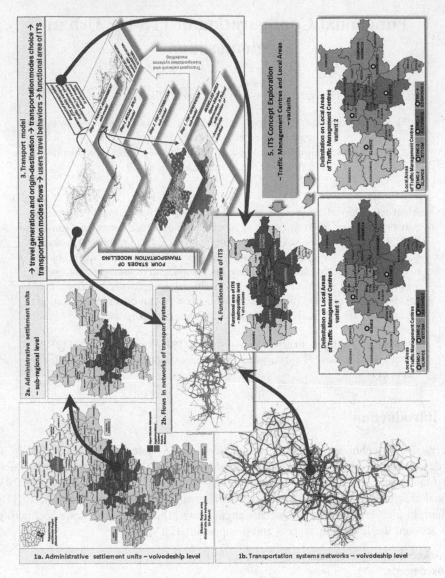

Fig. 1. Traffic Management Centres TMCs– concept exploration with use of transport model; result is in two variants of delimitation on local area: variant 1 – three TMCs, variant 2 – four TMCs. (own study based on [7, 12, 13, 17]).

- ITS viewpoints, taking into account the following views: functional, physical, communication, organizational,
- component specifications,
- ITS standards,
- communication requirements,
- organizational issues.

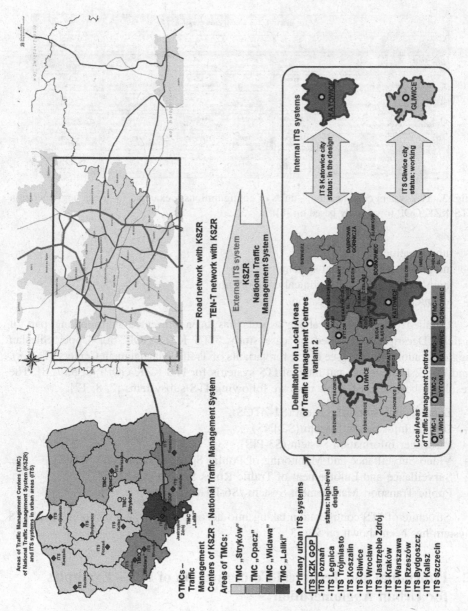

Fig. 2. External and internal ITS systems for urban agglomeration – ITS KZK GOP (own study based on [7, 10, 11, 17]).

In this paper, the system configuration has been presented, taking into account the following aspects and components of ITS system (see Sect. 2):

– main objectives,
– specific objectives,

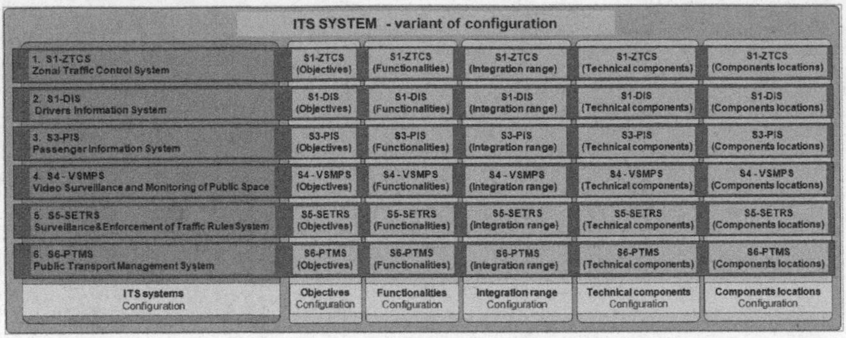

Fig. 3. ITS system configuration – fields of configurations – example for urban agglomeration ITS KZK GOP (own study based on [8]).

- functionalities,
- integration with external systems,
- main installations and technical components,
- technical components location.

Detailed description of above aspects has been presented in following part of article. Decsription is based on case study "ITS KZK GOP" for Upper Silesian agglomeration i Poland (see Fig. 1 for variants of Traffic Management Centers (TMCs), and Fig. 2 for external and internal ITS systems for ITS KZK GOP) [3, 5, 7, 8]. The result of the concept for this case are following ITS subsystems [7, 8, 17]:

- Zonal Traffic Control System (S1-ZTCS),
- Drivers Information System (S2-DIS),
- Passenger Information System (S3-PIS),
- Video Surveillance and Monitoring of Public Space System (S4-VSMPS),
- Surveillance and Enforcement of Traffic Rules System (S5-SETRS),
- Public Transport Management System (S6-PTMS).

Structure of ITS configuration taking into account aspects and components of ITS system has been shown on Fig. 3.

2 Functional and Technical Configuration of ITS – Examples for Urban Agglomeration

2.1 Zonal Traffic Control System (S1-ZTCS)

Main objectives:

- Reducing congestion in the transport network,
- Increasing smoothness traffic flows,
- Increasing level of service.

Specific objectives:

- Two-level optimization of traffic control – the central level (parent) to the local level,
- Delimitation of the transport network into several zones of traffic control,
- Grouping of signal-controlled intersections in three sets:
  - set I – includes the most important signal-controlled intersections – intersections into the main transport corridors with highest traffic flows between the cities of agglomeration; especially intersections on public transport lines (tram lines and bus lines) to apply priority.
  - set B – includes signal-controlled intersections into transport corridors with traffic flows less then set A,
  - set C – least important signal-controlled intersections.
- Grouping of signal-controlled intersections from sets I nad II **into coordinated control zones**.

Functionalities:

- Increasing the traffic handling capacity of roads,
- Reducing collisions and waiting time for both vehicles and pedestrians,
- Encouraging travel within the speed limit to meet green lights,
- Increasing smothness of traffic flows,
- Reducing unnecessary stopping and starting of traffic - this in turn reduces fuel consumption, air and noise pollution, and vehicle wear,
- Reducing travel time especially by public transport vehicles priority,
- Reducing driver frustration and road rage

Integration with external systems:

- ITS in Gliwice City – transmission of information about traffic conditions and road accidents, and data transfer enables coordination of signaling and priority for public transport vehicles.
- National ITS (KSZR) – transmission of information about traffic conditions and road accidents,
- traffic surveillance system in the tunnel under the roundabout in Katowicach city – transmission of information about traffic conditions and road accidents,
- urban information service with SMS in Katowice city – transmission of information about traffic conditions and road accidents,

Main installations and technical components:

- development of technical documentation,
- exchange or adaptation of existing signal controllers and signaling devices,
- installation of induction loops,
- installation of systems software for: area traffic control, management software of signal controllers, design solutions in the field of traffic engineering, macro and micro traffic simulation, PLC programming,
- providing workstations,

- adaptation of premises for the operators of traffic control center,
- construction of the adaptation intersections signaling,
- installation of fiber optics.

Technical components location:

- Traffic Management Centers,
- signal controllers and signaling devices on the intersections.

2.2   Drivers Information System (S2-DIS)

Main objectives:

- Information for drivers about traffic conditions in transportation systems of urban agglomeration.
- Information for drivers about recommended actions while driving due to difficulties – e.g. recommended routes.
- Information for drivers about vacant parking spaces and recommended routes.

Specific objectives:

- Information for drivers about traffic conditions (level of service) on main roads of agglomeration and about recommended alternative routes,
- Information for drivers about difficulties related to:
  - traffic incidents such as accidents, road works, detours etc.,
  - mass event (demonstration),
  - dangerous weather phenomena and weather conditions on the road (air temperature, surface temperature, information on rainfall, snow and the possible emergence of dangerous weather phenomena like black ice, fog, gale, hail, etc.),
- Information about the availability of parking spaces in parking lots, indicating the type of car park facilities, payment system, and recommended routes – to reduce search time of vacant parking spaces, which in turn reduces congestion on the surrounding roads for other traffic.

Functionalities:

- Information about traffic incidents – options:
  - full information about traffic incidents with identification of their type – dedicated equipment for the detection incidents,
  - traffic incidents detected as congestion by the traffic lights controllers,
  - level of service estimation – traffic measurements (volume, density and speed of traffic flows) and recognition of the structure of vehicles carried out by a dedicated devices and the traffic lights controllers,
- Weather information system – options:
  - information from the dedicated measuring devices of weather stations,
  - information from the external servers or a weather station,

- Parking guidance and information system – options:
  - detection of each individual parking space in the parking lot and on the street (the system supports the areas of paid and free parking lots), with software to manage the car parks and VMS (variable message signs) with information about the vacancies at several car parks,
  - the system counts the vehicles at the entrance and exit of the parking lots (the system does not support zones of paid parking); there is no communication between the VMS and the server system, so VMS inform about vacancies at the parking lot only; there is no exact information about the number of parking vacancies in parking on the streets; the system is not equipped with software for parking management.

Integration with external systems:

- ITS in Gliwice City,
- National ITS (KSZR),
- traffic surveillance system in the tunnel under the roundabout in Katowicach city,
- urban information service with SMS in Katowice city,
- transmission of information.

Main installations and technical components:

- installations with equipment (hardware and software) for:
  - measurement of traffic flows and structure of vehicles,
  - detection of traffic incidents and their recognition,
  - detection of vacant parking spaces,
  - weather conditions,
- VMS (variable message signs) for information about:
  - traffic condition (level of service) and recommended routes,
  - traffic incidents and recommended routes,
  - vacant parking spaces and recommended routes,
  - weather.

Technical components location:

- Software in Traffic Control Centres (TCC),
- VMS (variable message signs):
  - main roads and selected urban streets of urban agglomeration,
  - intersections of main roads and intersections of selected urban streets,
  - at the entrances to the car parks and park zones.

2.3   Passenger Information System (S3-PIS)

Main objectives:

- better able to conduct journey by travelers, including taking any necessary steps in the event of delays, by providing real time information; this helps to encourage greater use of public transport,

Specific objectives:

- provide real-time passenger information: arrival and departure times, nature and causes of disruptions,
- personalised channels (web service, mobile application, SMS service) will be set up to mimic the view from a station or stop and may in addition be linked to journey planners.

Functionalities:

- include both predictions about arrival and departure times, as well as information about the nature and causes of disruptions,
- use both physically within a transportation nodes and remotely using a web browser and mobile device,
- information at a **stations and stops** – provide up to date predictions of:
  - arrival times, lines/routes and destinations of the following few arrivals of public transport vehicles,
  - how closely vehicles are running to timetable,
  - general advice on current travel disruptions that may be useful to the passenger in understanding the implications for their travel plans,
- information on **vehicles** – provide up to date predictions of:
  - the next station or stop,
  - arrival time,
  - how closely it is running to timetable,
  - advice on connecting services,
- apart from **visual** information also **voice** information,
- information on **web service** and **mobile application**:
  - public transport network diagram,
  - location of public transport vehicles on the network diagram in real-time,
  - location of currently registered traffic incidents (accidents, road works, closures etc.),
  - timetables all lines with up to date departure times for the following few arrivals for each line serving a specific stop or station,
  - actual information about all transport modes in urban agglomeration,
  - travel multimodal planner,
  - information about the nearest destinations: park&ride, multimodal nodes, POIs etc.
  - location of parking lots with the number of free parking spaces,
  - traffic conditions on road network,
  - scanning the QR-code placed at bus stops, which allow the identification of the bus stop,
  - information provided via SMS,

Integration with external systems:

- integration with a similar system functioning at selected main stops,
- adaptation of public service vehicles to work with the system in the field of on-board computer, GPS, GSM, LCD or LED displays,

– integration with other urban information systems and web portals – exchange the information between systems about traffic incidents, mass events and traffic disruptions.

Main installations and technical components:

– software and hardware with management information system,
– LCD monitors and/or VMS on stops, stations and on-board in vehicles,
– on-board audio system for voice information in vehicels,
– web application and mobile application,
– main software with management information system.

Technical components location:

– software and hardware in public transport management center,
– LCD monitors and/or VMS on stops and stations,
– LCD monitors and audio system with voice information on-board in vehicels,
– mobile application installed on mobile devices (smartphones, tablets etc.).

2.4   Video Surveillance and Monitoring of Public Space System (S4-VSMPS)

Main objectives:

– increase security based on the possibility of rapid detection of road accidents and incidents with armed robbery, theft etc.,
– support the activities of the Voivodeships Crisis Management Centre in Katowice.

Specific objectives:

– automatic notifications and alerts (via email, SMS, by phone) appropriate authorities and services about incidents.

Functionalities:

– monitoring of roads, streets, public buildings and other public spaces,
– monitoring of public transport vehicles and stops, stations and multimodal transport nodes,
– provide video information for: police, city guards, regional and local rescue and crisis centers.

Integration with external systems:

– extension of existing systems with new camera points; maximum use of existing cameras and equipment or replacement of old devices that do not provide sufficient image quality.

Main installations and technical components:

– IP cameras and CCTV cameras
– video recorders, videoservers, storage devices – in centres and vehicles,
– workstations with software and joysticks/manipulators,
– video walls.

Technical components location:

- traffic management centres, public transport operator centre,
- terminals in the following institutions: police, city guards, regional and local rescue and crisis centers.

2.5 Surveillance and Enforcement of Traffic Rules System (S5-SETRS)

Main objectives:

- traffic monitoring and recording of road accidents,
- enforcement of traffic rules: speed and red light violation enforcement, stop sign and bus lane violation enforcement,
- commercial vehicles weight enforcement,
- hazardous materials management.

Specific objectives:

- video vehicle detection
- vehicle identification based on automatic number plate recognition (ANPR) cameras,

Functionalities:

- identification of speed violation vehicles – speed violation at fixed points and average speed violation between two fixed points,
- identification of stolen or wanted vehicles,
- identification of red light violation vehicles,
- identification of commercial vehicles weight violation – weight-in-motion system.

Integration with external systems:

- functioning speed cameras and radars,
- functioning red light violation cameras,
- functioning weight violation system.

Main installations and technical components:

- automatic number plate recognition (ANPR) cameras,
- speed and red light violation devices,
- stop sign and bus lane violation devices,
- weight-in-motion devices.

Technical components location:

- ANPR cameras and devices on selected points of transport Network,
- Hardware and software in Traffic Control Centers.

2.6   Public Transport Management System (S6-PTMS)

Main objectives:

- planning organization and management of public transport in urban agglomeration,
- management of transport infrastructure and vehicles,
- monitoring and evaluation of drivers.

Specific objectives:

- implementation of specialized software supporting the management of public urban transport and rail transport,
- integration of public transport services by introducing correlated timetables and common tariffs.

Functionalities:

- support for the development of strategic document – transport plan,
- analysis and settlement of transport work between municipalities and operators on the basis of information about the number of passengers,
- timetabling fir public transport services,
- management of companies and operators database,
- maintaining a register of stops and vehicles,
- monitoring and evaluation of service work.

Integration with external systems:

- integration with actual software used by operators of public urban transport and rail transport,
- data flows from other subsystems.

Main installations and technical components:

- hardware and specialized software for transport management.

Technical components location:

- hardware and specialized software in headquarter of urban public transport operator,
- hardware and specialized software in headquarter of rail transport.

3   The Configuration Matrix of Aspirations-Implementations – Examples for Urban Agglomeration

The most important and functional-utility aspirations A1÷A56 of The ITS KZK GOP case study are as follows. The implementation of following aspirations by the subsystems of ITS configuration – aspirations-implementations matrix of ITS – has been show in Table 1.

Table 1. The aspirations-implementations matrix of ITS [7, 17]

Implementation of aspirations by the subsystems of ITS configuration

No aspiration	S1-ZTC	S2-DIS	S3-PIS	S4-VSMPS	S5-SETRS	S6-PTMS	No aspiration	S1-ZTC	S2-DIS	S3-PIS	S4-VSMPS	S5-SETRS	S6-PTMS
A1	+	+	+	+	+	+	A29			+			
A2	+	+	+	+	+	+	A30			+			+
A3	+	+	+	+	+	+	A31		+	+			
A4	+	+	+	+	+	+	A32		+	+			
A5	+	+	+	+	+	+	A33		+	+			
A6	+	+					A34		+	+			
A7	+	+					A35		+	+			
A8	+						A36	+	+	+	+	+	+
A9	+						A37	+	+				
A10	+						A38	+	+				
A11	+						A39			+			+
A12	+						A40		+	+			
A13	+	+					A41		+	+			
A14		+	+				A42		+				
A15			+				A43	+					
A16		+	+				A44					+	
A17	+	+	+	+	+	+	A45					+	
A18			+			+	A46	+	+	+	+	+	+
A19				+			A47				+		
A20				+			A48		+				
A21	+						A49			+			+
A22	+						A50	+	+				
A23	+						A51	+		+			+
A24		+	+				A52						+
A25		+	+				A53						+
A26		+	+				A54	+					
A27		+	+				A55			+			
A28		+	+				A56	+	+	+	+	+	+

List of takeholders aspirations – example for ITS KZK GOP case study [7, 17]

A1. Technical documentation of the system should include a precise description of the data types and protocols for their transmission,

A2. System should be technologically open and flexible to allow for its continuous expansion,

A3. System should operate according to the principle "collect data once, use it many times,"

A4. System should have mechanisms to ensure the security of the collected and transmitted data,

A5. System should use open protocols for communication between devices, so that when the expansion was not necessary to rebuy/expansion license, a new device can be supplied by different manufacturers,

A6. System should collect data on traffic flows in the road network,

A7. System should recognize the different types of vehicles,

A8. System should calculate signal timing for traffic lights based on traffic data,

A9. System should allow the simulation of new solutions and strategies for road traffic control,

A10. System should include transport models both in macrosimulation and microsimulation,

A11. System should measure the traffic flows parameters and evaluate the effectiveness of control strategies,

A12. System should store sequence control (values of stages, phases, signal timing and detectors) for all intersections,

A13. System should detect traffic incidents,

A14. System should inform users about traffic incidents,

A15. System should support people with disabilities,

A16. System should inform users about the traffic situation in the network,

A17. System should diagnose own functioning and automatically report faults to operators (self-diagnosis),

A18. System should carry out electronic payments for transport services,

A19. System should detect stolen cars,

A20. System should detect vehicles properly labeled,

A21. System should support the priorities for public transportnej,

A22. System should handle the priorities for privileged vehicles,

A23. System should coordinate the work of signal controllers to create a 'green wave' for vehicles,

A24. System should provide free of charge traffic data for users,

A25. System should support users in the travel planning,

A26. System should support users in the plan a trip using various means of transport (multimodality),

A27. System should provide information about cultural events, sports, etc.,

A28. System should allow the location of the POI on the map,

A29. System should provide a road map for users,

A30. System should provide a map of public transport network,

A31. System should collect and provide information on available parking spaces,

A32. System should provide information using variable message signs (VMS),

A33. System should provide the information using the Internet,

A34. System should provide information using applications on smartphones,

A35. System should provide the information in a graphical, textual, audio,

A36. System should provide software tools for management and diagnostics of system,

A37. System should be able to exchange information about events with other telematics systems,

A38. System should be able to exchange information about the traffic on the roads with other telematic systems,

A39. System should provide information about the actual departure times of public transport vehicles,

A40. System should provide information on the traffic conditions in the network traffic both before and during the trip,

A41. System should indicate an alternative route,

A42. System should provide information about current weather conditions and related dangers,

A43. System should manage traffic to ensure travelers shorten travel time,

A44. System should monitor the situation on the roads, intersections, neighborhoods, parks, etc., to increase the security of citizens,

A45. System should monitor situations in public transport vehicles and in their vicinity,

A46. System should allow operators to remotely control and service equipment,

A47. System should detect vehicles exceeding the speed at a given point or on a given road section,

A48. System should allow the speed limit using variable message signs,

A49. System should record the routes, which passengers are traveling by public transport,

A50. System should collect data on events and make them available in the form of statistics,

A51. System should increase the efficiency and attractiveness of public transport,

A52. System should support the transport organizer in the development of timetables and manage transport services,

A53. System should allow the assessment of the quality of the operators of public transport,

A54. System should allow management of priorities for public service vehicles and privilaged vehicles,

A55. System should locate public transport vehicles on the map,

A56. System should use the data collected by other telematics systems.

4 Conclusion

The ITS configuration for urban agglomeration may be evaluated [17] using SWOT analysis. The results of this assessment for ITS KZK GOP case study are as follows. **Strengths:** open system architecture, technological neutrality, support for public transport, increase the safety of citizens, especially road users, support integration between urban public transport and rail. **Opportunities:** integration with other telematics systems, ability to finance this system with use of European funds. **Weaknesses:** high cost of building and maintenance of this system. **Threats:** large area of the system, need to agree of the architecture and functionality of the system with many institutions, long time to build the system, problems of interaction this system with other external systems, e.g. ITS system in Gliwice city.

References

1. Bělinová, Z., Bureš, P., Jesty, P.: Intelligent transport system architecture different approaches and future trends. In: Düh, J. (ed.) Data and Mobility, AISC 81, pp. 115–125. Springer, Berlin (2010). doi:10.1007/978-3-642-15503-1_11

2. European Intelligent Transport System (ITS) Framework Architecture (E-FRAME). http://frame-online.eu/. Accessed 12 Feb 2017

3. Feasibility study of National Traffic Management System, Kraków, General Directorate for National Roads and Motorways (GDDKiA in Poland) (2012). http://www.kszr.gddkia.gov.pl/. Accessed 12 Feb 2017

4. Gubbia, J., et al.: Internet of Things (IoT): a vision, architectural elements, and future directions. Future Gener. Comput. Syst. **29**(7), 1645–1660 (2013)

5. Karoń, G.: Analysis of the current status in the development of pre-design documentation for the project entitled Intelligent Transport Management System for Katowice City. Report of Research Work for City Hall of Katowice, Poland (2016). (in Poland)
6. Karoń, G., Janecki, R.: Development of various scenarios of ITS systems for urban area. In: Sierpiński, G. (ed.) Intelligent Transport Systems and Travel Behaviour. AISC, vol. 505, pp. 3–12. Springer, Cham (2017). doi:10.1007/978-3-319-43991-4_1
7. Karoń, G., Mikulski, J.: Expertise of the concept and architecture of the intelligent traffic management system (ITS) for the territory of the public transport operator in upper Silesian industrial region. Report of Research Work for Municipal Transport Union of the Upper Silesian Industrial District (KZK GOP), Katowice, Poland (2015)
8. Karoń, G., Mikulski, J.: Expertise of the functional analysis for intelligent transportation management system (ITS) for Katowice city. Report of Research Work for City Hall of Katowice, Katowice, Poland (2011)
9. Karoń, G., Mikulski, J.: Forecasts for technical variants of its projects: example of Upper-Silesian conurbation. In: Mikulski, J. (ed.) TST 2013. CCIS, vol. 395, pp. 67–74. Springer, Heidelberg (2013). doi:10.1007/978-3-642-41647-7_9
10. Karoń, G., Mikulski, J.: Problems of ITS architecture development and its implementation in upper-silesian conurbation in Poland. In: Mikulski, J. (ed.) TST 2012. CCIS, vol. 329, pp. 183–198. Springer, Heidelberg (2012). doi:10.1007/978-3-642-34050-5_22
11. Karoń, G., Mikulski, J.: Problems of systems engineering for ITS in large agglomeration – upper-silesian agglomeration in Poland. In: Mikulski, J. (ed.) TST 2014. CCIS, vol. 471, pp. 242–251. Springer, Heidelberg (2014). doi:10.1007/978-3-662-45317-9_26
12. Karoń, G., et al.: The methodology and detailed concept of traffic surveys and measurements and transportation model development for the territory of the public transport operator in Upper Silesian industrial region. Report of Research Work for Municipal Transport Union of the Upper Silesian Industrial District (KZK GOP), Katowice, Poland (2016). (in Poland)
13. Karoń, G., Żochowska, R., Sobota, A.: The methodology and detailed concept of traffic surveys and measurements and transportation model development for transportation study of central subregion territory. Report of Research Work for Union of Municipalities and Counties of Central Subregion of Silesian Region, Katowice, Poland (2016). (in Poland)
14. Małecki, K., Iwan, S., Kijewska, K.: Influence of intelligent transportation systems on reduction of the environmental negative impact of urban freight transport based on Szczecin example In: Procedia Social and Behavioral Sciences, vol. 151, pp. 215–229. Elsevier, London (2014)
15. National ITS Architecture 7.1. http://local.iteris.com/itsarch/. Accessed 12 Feb 2017
16. Sun, L., Li, Y., Gao, J.: Architecture and application research of cooperative intelligent transport systems. Proc. Eng. 137, 747–753 (2016)
17. Szarata, A., et al.: The concept and architecture of the intelligent traffic management system (ITS) for the territory of the public transport operator in Upper Silesian industrial region. Report of Research Work for Municipal Transport Union of the Upper Silesian Industrial District (KZK GOP), Katowice, Poland (2015)
18. Systems Engineering Guidebook for Intelligent Transportation Systems. U.S. Department of Transportation (2009)
19. Systems Engineering for Intelligent Transportation Systems. U.S. Department of Trans-portation. FHWA-HOP-07–069 (2007)
20. Żochowska, R., Karoń, G.: ITS Services Packages as a Tool for Managing Traffic Congestion in Cities. In: Sładkowski, A., Pamuła, W. (eds.) Intelligent Transportation Systems – Problems and Perspectives. SSDC, vol. 32, pp. 81–103. Springer, Cham (2016). doi:10.1007/978-3-319-19150-8_3

New Telematic Solutions for Improving Safety in Inland Navigation

Tomasz Perzyński$^{(\boxtimes)}$ and Andrzej Lewiński

Faculty of Transport and Electrical Engineering,
University of Technology and Humanities in Radom,
Malczewskiego 29, 26-600 Radom, Poland
{t.perzynski,a.lewinski}@uthrad.pl

Abstract. The article presents practical applications of telematics in improving safety of transport in inland waterway. Presented in the paper new telematic solutions, autonomous system of information about the event on the waters as well as mobile application, enable emergency services to react quickly. This is important particularly in situation where human life is in danger. For the solutions presented in the work it has been proposed a mathematical analysis on the basis of Markov processes.

Keywords: Telematic in inland navigation · Water tourism · Systems modeling

1 Introduction

The modern railway, road and water inland telematic infrastructure allows obtaining high levels of safety, environmental protection, better transport effectiveness and better control over transport means and the transport process. The execution of the functions related to management and control in transport imposes the necessity to ensure reliability indexes of a sufficient degree. The actions leading to establishing them regard a complex analysis of a single element and the entire system. The damage intensity is in this case the primary index. The use of the mathematical apparatus tools and simulations methods additionally allows for evaluating the transport telematics systems and modelling of various damage scenarios and also data transmission interference [1–3]. The transport telematics systems execute various functions and operate in various conditions. Each case of unreliability can generate side effects, including economic or environmental effects. The unreliability of the transport telematics systems can constitute a hazard for human life, which is of the highest value. In such cases, we are dealing with the unreliability of safety of the transport telematics systems.

One of the main tasks of telematics systems related to safety and management in transport is to obtain and process information. This is possible e.g. through the use of transmission standards and processing of information. Obtaining information is possible thanks to two layers, [1]:

- ground layer (IT networks),
- satellite layer (satellite navigation systems).

© Springer International Publishing AG 2017
J. Mikulski (Ed.): TST 2017, CCIS 715, pp. 70–81, 2017.
DOI: 10.1007/978-3-319-66251-0_6

The element influencing on the safety of the inland navigation is, apart from the skills of water equipment users, the use of telematics systems to inform about the possibility of a dangerous situation. An example of such a solution is running in 2011 warning system against dangerous atmospheric phenomena (Mazurian Lake District). Before the possibility of danger weather, users of yachts are informed by flashing lights on special masts. 40 flashes/min means that it is expected storm or strong wind, 90 flashes/min alerts to direct danger of the storm and strong wind, [4]. The list of the location of each mast of the system is shown in Table 1.

Table 1. Location data of the masts [1]

No.	The list of masts at the Mazurian Lakes District	Geographical coordinates
1.	Lake Mamry, Skłodowo	21 43 26.55″E 54 09 23.39 N
2.	Lake Mamry, Węgorzewo	21 43 10.08″E 54 11 51.8454 N
3.	Lake Święcajty, Ogonki	21 48 24.05″E 54 10 58.94 N
4.	Lake Ryńskie, Ryn City	21 31 51.83″E 53 55 49.75 N
5.	Lake Roś, Łupki	21 51 34.70″E 53 39 08.10 N
6.	Lake Mikołajskie, Mikołajki	21 37 13.90″E 53 45 53.60 N
7.	Lake Niegocin, Kępa Grajewska (island)	21 47 54.18″E 54 00 20.61 N
8.	Lake Kisajno	21 43 55.36″E 54 02 10.61 N
9.	Lake Mamry, Wysoki Róg	21 40 34.14″E 54 06 07.55 N
10.	Lake Jagodne	21 43 35.43″E 53 56 01.33 N
11.	Lake Boczne, Bogaczewo	21 45 00.05″E 53 57 48.69 N
12.	Lake Łuknajno	21 40 41.04″E 53 47 09.92 N
13.	Lake Śniardwy, Nowe Guty	21 49 11.56″E 53 45 49.40 N
14.	Lake Śniardwy, Niedźwiedzi Róg	21 41 45.05″E 53 43 06.51 N
15.	Lake Bełdany, Kamień	21 33 42.63″E 53 43 21.04 N
16.	Lake Seksty, Jegliński Canal	21 46 52.56″E 53 41 17.70 N
17.	Lake Tałty, Tałcki Canal	21 32 33.66″E 53 51 50.70 N

Component of the system, the mast with lamp, is shown in Fig. 1.

Fig. 1. The mast of warning system [own study]

Besides the above-mentioned solution it is also very important to have tools to quickly reach people in need of help. With regard to inland water transport the time of passing on the information to the emergency services equipped with technical support in the form of rescue boats and coordination with other emergency services are the most important. The notification without specifying the location of the event can significantly delay the time of arrival of water rescuers.

At the Faculty of Transport and Electrical Engineering University of Technology and Humanities in Radom it was designed and built a system dedicated to tourist boat on inland waters, especially for sailing yachts. The system is a part of the telematic solutions which improve the safety. Additionally it has been prepared mobile application for Android system which permit to send messages with the position of need of help.

The solutions presented in this paper are based on the data obtained from the navigation system, and the ground transmission is carried out via cellular network. In the solutions presented in the work, the cellular network is required to send information in the form of the *sms* text message. In the proposed solution dedicated to yachts, transmission is realized by typical GSM/GPRS module integrated with GPS module.

The GPS NAVSTAR (Global Positioning System) is currently the most popular satellite navigation system. Therefore, to sailing yacht system, authors proposed GPS module. The authors also envisage the possibility of using modules receiving GPS, Glonass and Galileo signals (GNSS module). The work of the receiver in many modes will allow for more accurate positioning. The GPS system is owned and operated by United States of America (USA). The system was launched in 1978, but for civilian use it was made available only in 1983 (formally in 1996). The system consists of three segments [1]:

- satellite segment,
- ground segment,
- user segment.

Presented in the paper solutions are the real telematics systems and also an example of the practical use of transport telematics systems.

2 Mobile Application

According to IDC (International Data Corporation) in 2016' almost 87% of mobile devices in the world worked on Android system, [5]. Due to this fact, at the Faculty of Transport and Electrical Engineering of UTH in Radom, it was created mobile app. that gives the possibility to send an *sms* (Short Message Service) with the call for help, [1]. Initially the application was written for the deaf-mute people, where in case of danger the person may send a request for help, together with the location data. Created application is a tool so versatile that it can be used to inform any rescue services about need for help, also about the event on the waters. In order to write an application it has been prepared the appropriate development environment, which includes, [1]:

- development environment JAVA - Java Developer Kit (JDK). This is a package of tools that allows, among others, to compile programs.

- Android SDK - this is an environment for application development for Android. It also contains useful tools to create or manage virtual devices, in which user-created applications can be checked.
- IDE Eclipse application (Integrated Development Environment), which was used for the mobile app.

The algorithm of the mobile app. is shown in Fig. 2.

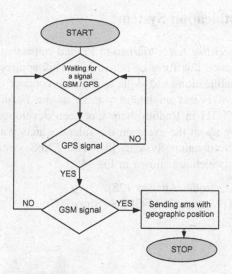

Fig. 2. The algorithm of the mobile app. [own study]

Figure 3 shows the result of mobile app test.

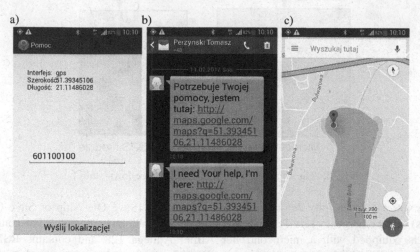

Fig. 3. (a) mobile app., (b) window with sms, (c) Google map with position [own study]

In the application it is possible to enter any phone number. The official phone number to water rescue service in Poland is 601 100 100. In the current version of the application (for testing) in a received text message (Fig. 3b) is automatically entered text: I need Your help, I'm here (pl. - *Potrzebuje Twojej pomocy, jestem tutaj*) and is added the link to the map. In Fig. 3c there is shown the view of Google Maps in open mobile device.

3   Emergency Notification System

Liberal regulations concerning water tourism in Poland cause that there is a real threat to the people who practice this form of relaxation. This applies primarily to people without appropriate qualifications and skills in the safe conduct of yachts. In order to counteract the effects of events on inland waters, at the Faculty of Transport and Electrical Engineering UTH in Radom there has been developed and built the electronic system informing about the event on the inland waters with the working name ENoS – Emergency Notification System, [6]. The ENoS system uses three main electronic components, which are shown in Fig. 4:

- control unit (microcontroller Atmega 128),
- integrated module GPS with GSM-GPRS (Sim908C),
- tilt sensor module.

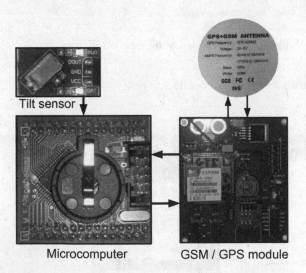

Fig. 4. The components of the ENoS device [own study]

ENoS system is fully autonomic. Sending an SOS (Save Our Ship or Save Our Souls) to the emergency services with a geographical position is automatic. The control unit is equipped with a microcontroller Atmel Atmega 128 and contains RS232 interface. GPS/GSM module also has an RS232 interface. Module control by the control unit is done with using standard commands for modems (AT commands) sent

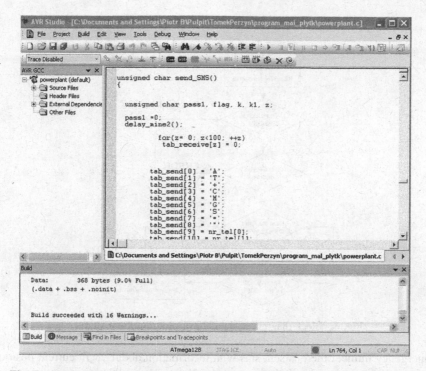

Fig. 5. Screenshot of the control program in the environment AVRStudio v.4.0 [1]

via RS232 interface. Information about the inclination of the yacht is delivered to the control unit by a tilt sensor via a standard digital input.

Software for the control unit is written in "C" for the microcontroller Atmega 128. For this purpose there is used AVRStudio v4.0 environment with compiler "C" for microcontrollers Atmega (GnuC compiler). Figure 5 shows a screenshot of the control environment AVRStudio v.4.0

After turning the device in the first phase it is followed an initialization of the RS232 interface for microcontroller Atmega 128. In this phase there are sets the transmission parameters, such as:

- speed transmission 9600 bit/s,
- stopbit 1,
- no parity,
- sent word of 8 bits.

The next step is conducted to initialize the GPS/GSM module. Initialization is conducted separately for part of the GPS and separately for part of the GSM. After the initialization GPS/GSM module is tested. It consists of sending a test message with the current location of a yacht on the number of recipient. In the case of an error in the operation of the GPS/GSM module the procedure is conducted again starting from reset and initialize the GPS/GSM module. The next step is reading the information about the tilting of yacht from the tilt sensor. If the yacht is tilted for a period longer than 30 s,

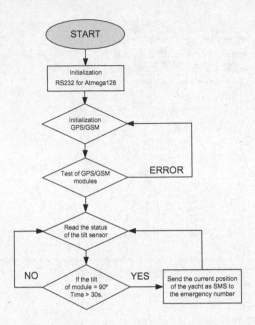

Fig. 6. Block diagram of the control program [own study]

on emergency number there is sent sms with information about yacht current position. In the case of the permanent tilt there are sent only three *sms* with the current position at intervals of 60 s. Figure 6 shows a block diagram of the control program.

Used in the device microcontroller system is programmable and completely autonomous microprocessor system. The device uses 8-bit microcontroller. The microcontroller is managing the device, which is adapted to direct cooperation with various outer devices and sensors without additional peripheral systems. In the future, this will allow to expand the function of the device. The GPS/GSM module system was built on SIM908, which allows connectivity: GSM, GPRS, and a GPS receiver. Communication with a module is via a serial bus. Figure 7 shows the view of the testing stand installed on the pontoon, [1].

a)   b)

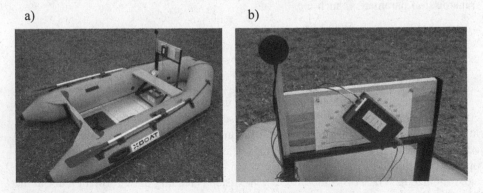

Fig. 7. The testing stand installed on the pontoon [1]

a)   b)

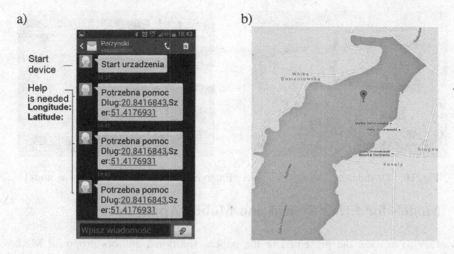

Fig. 8. (a) Received *sms*, (b) Google map with position [1]

In the testing stand, to allow tilting the ENoS device, there was used the rotary system (360°). Tilt of the device was initiated manually. After switching on and initialization of the GSM/GPS module the device worked in standby mode. The tests of the system was carried out on the waters area Domaniów near Radom. Figure 8 shows the test result (*sms* and a map of the location).

The tests allowed it to check the correct work of the system. Install the ENoS system on the sailing yacht will allow finish the next phase of work on ENoS. Yacht for the test is presented in the Fig. 9.

Fig. 9. Yacht for test the ENoS on Domaniów water area [own study]

In case of Masurian Lake District, the received information with the position of the yacht can be displayed on the TV monitor's map in the headquarters of Aquatic Volunteer Emergency Corps in Giżycko, as shown in Fig. 10. Currently, on the monitor screen are displayed only information with data of rescue boats.

Fig. 10. TV monitor - Aquatic Volunteer Emergency Corps in Giżycko [own study]

4 Models for ENoS System and Mobile App

In order to analyze the presented in the papers solutions, authors proposed Markov processes. In the Fig. 11 there is presented the system model, which includes the possibility of a dangerous situation on the waters. In case of the appearance of dangerous event, it is possible to notify the emergency services using autonomous system installed on the boat (ENoS) or using the mobile app.

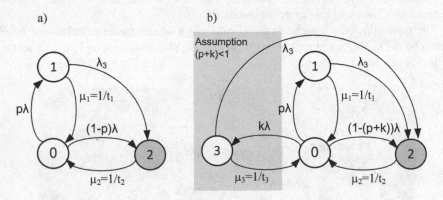

Fig. 11. Markov models, a) only ENoS, b) ENoS and mobile app. [own study]

In the models from Fig. 11, we can distinguish:

0 - state of correct work. No threat.

1 - state of danger. The yacht equipped with a system ENoS,

2 - state of critical danger,

3 - state of danger. The yacht not equipped with the system. Help requested by mobile app.,

p - the probability of equipment yacht in the system,

k - the probability of having the mobile app.

For models in Fig. 11a (1) and b (2) we can write the equation in the form of operators:

$$\begin{cases} s \cdot \widetilde{P}_0 - 1 = -p \cdot \lambda \cdot \widetilde{P}_0 + \mu_1 \cdot \widetilde{P}_1 - (1-p) \cdot \lambda \cdot \widetilde{P}_0 + \mu_2 \cdot \widetilde{P}_2 \\ s \cdot \widetilde{P}_1 = p \cdot \lambda \cdot \widetilde{P}_0 - \mu_1 \cdot \widetilde{P}_1 - \lambda_3 \cdot \widetilde{P}_1 \\ s \cdot \widetilde{P}_2 = (1-p) \cdot \lambda \cdot \widetilde{P}_0 - \mu_2 \cdot \widetilde{P}_2 + \lambda_3 \cdot \widetilde{P}_1 \end{cases} \quad (1)$$

$$\begin{cases} s \cdot \widetilde{P}_0 - 1 = -(p+k) \cdot \lambda \cdot \widetilde{P}_0 + \mu_1 \cdot \widetilde{P}_1 - (1-(p+k)) \cdot \lambda \cdot \widetilde{P}_0 + \mu_2 \cdot \widetilde{P}_2 + \mu_3 \cdot \widetilde{P}_3 \\ s \cdot \widetilde{P}_1 = p \cdot \lambda \cdot \widetilde{P}_0 - \mu_1 \cdot \widetilde{P}_1 - \lambda_3 \cdot \widetilde{P}_1 \\ s \cdot \widetilde{P}_2 = (1-(p+k)) \cdot \lambda \cdot \widetilde{P}_0 - \mu_2 \cdot \widetilde{P}_2 + \lambda_3 \cdot (\widetilde{P}_1 + \widetilde{P}_3) \\ s \cdot \widetilde{P}_3 = k \cdot \lambda \cdot \widetilde{P}_0 - \mu_3 \cdot \widetilde{P}_3 - k \cdot \lambda \cdot \widetilde{P}_3 \end{cases} \quad (2)$$

Using the properties of Laplace transform it was solved Eqs. (1) and (2) and it was calculated the probability of critical $P_2(t)$ for models from Figs. 11a (3), b (4):

$$A_{[Fig.9a]} = 1 - \lim_{t \to \infty} P_2(t) = \frac{(p\lambda + \lambda_3 + \mu_1)\mu_2}{(\lambda_3 + \mu_1)\mu_2 + \lambda(\lambda_3 + \mu_1 - p\mu_1 + p\mu_2)} \quad (3)$$

$$A_{[Fig.9b]} = 1 - \lim_{t \to \infty} P_2(t) = \frac{\mu_2(k\lambda(\lambda_3 + \mu_1) + (p\lambda + \lambda_3 + \mu_1)(\lambda_3 + \mu_3))}{(\lambda_3 + \mu_1)\mu_2(\lambda_3 + \mu_3) +}$$

$$\frac{1}{\lambda(\lambda_3^2 + k\mu_1(\mu_2 - \mu_3) + (\mu_1 - p\mu_1 + p\mu_2)\mu_3 + \lambda_3(\mu_1 - p\mu_1 + (k+p)\mu_2 + \mu_3 - k\mu_3))} \quad (4)$$

Assuming the values of rates:

- $\lambda = 0.00685$ h^{-1} (60 dangerous situation/year),
- $\lambda_3 = 0.000685$ h^{-1} (10% from λ lead to a dangerous situation),
- $\mu_1 = 6$ h^{-1}, $\mu_2 = 1$ h^{-1}, $\mu_3 = 6$ h^1, for model with additional application (Fig. 9b),
- $k = 0.4$,

it was given the results shown in Table 2.

Table 2. The value of availability [own study]

The value of probability p	Availability A: Model with ENoS and mobile app.	Availability A: Model with ENoS and without mobile app.
p = 0.1	0.996588242	0.99387339
p = 0.2	0.997268885	0.99455094
p = 0.3	0.997950303	0.99522927
p = 0.4	0.998632497	0.995908373
p = 0.5	0.999315468	0.996588242

5 Conclusion

The solutions presented in the article are a part of the telematics systems, which let to improve a safety. Currently operating in Poland system of warning against dangerous atmospheric phenomena with its coverage covers only Masurian Lake District. However, there are no solutions that can inform about the event on any water. The implemented solutions allow it to send an information to the emergency services, which allows for faster help. At this stage, there is also the possibility of making a voice call at the mobile application. The mobile application after modification can realize tasks similar to ENoS. For this purpose the mobile app will be complemented with the ability to taking an information from the internal accelerometer of the mobile device.

Presented in the paper system ENoS may become standard equipment on yachts on inland waterways. ENoS also can be complemented with additional functions, including the emergency button.

The conclusions of the tests the ENoS device:

- it was tested correct operation of the device during inclination of less than 30 s and greater than 30 s,
- the device does not send a message during inclination of less than 70° and a long tilting,
- during inclination in the range of 70°–80° and time <30 s the system does not send messages,
- an inclination close to 90° and the time <30 s system does not send messages,
- an inclination close to 90° and the time >30 s system worked and sent messages.

The conducted tests confirmed the assumptions from the stage of the concept and construction, [6]. The last phase of the ENoS test will be install system on a sailing yacht to carry out the test in real conditions during sailing.

The paper also conducted a mathematical analysis based on Markov processes, which are one of a lot methods of risk analysis, [7]. Result of the analysis indicates that in the event of a threat on the water using ENoS or mobile app allows for quickly reach water rescuers.

References

1. Perzyński, T.: Selected telematic systems in safety and management in land and inland transport. University of Technology and Humanities in Radom, Monographs Series, no. 201. Radom (2016). (in Polish)
2. Sumiła, M.: Risk analysis of interference railway GSM-R system in polish conditions. In: Zamojski, W., Mazurkiewicz, J., Sugier, J., Walkowiak, T., Kacprzyk, J. (eds.) Dependability Engineering and Complex Systems. AISC, vol. 470, pp. 469–478. Springer, Cham (2016). doi:10.1007/978-3-319-39639-2_41
3. Siergiejczyk, M., Pas, J., Rosinski, A.: Modeling of process of maintenance of transport systems telematics with regard to electromagnetic interferences. In: Mikulski, J. (ed.) TST 2015. CCIS, vol. 531, pp. 99–107. Springer, Cham (2015). doi:10.1007/978-3-319-24577-5_10

4. Łukasik, Z., Perzyński, T.: Telematic systems to aid in safety in Inland water tourism. In: Mikulski, J. (ed.) TST 2013. CCIS, vol. 395, pp. 89–96. Springer, Heidelberg (2013). doi:10.1007/978-3-642-41647-7_12
5. http://www.idc.com/promo/smartphone-market-share/os. Accessed 02 Feb 2017
6. Perzyński, T., Lewiński, A., Łukasik, Z.: The concept of emergency notification system for Inland navigation. In: Weintrit, A., Neumann, T. (eds.) Information, Communication and Environment. Marine Navigation and Safety of Sea Transportation, pp. 173–177. CRC Press, Boca Raton (2015)
7. Nowakowski, W., Łukasik, Z., Bojarczak, P.: Technical safety in the process of globalization. In: Proceedings of the 16th International Scientific Conference on Globalization and its Socio-Economic Consequences, Part IV, Rajecke Teplice, Slovakia, pp. 1571–1578 (2016)

Approaches to Quality Assessment of Traffic Information Services

Petr Bures[✉]

Faculty of Transportation Sciences, Czech Technical University in Prague,
Konviktská 20, 110 00 Praha 1, Czech Republic
bures@fd.cvut.cz

Abstract. The paper discusses different approaches of assessing quality of traffic data and related services. The quality is considered as all of the following, complying with formal data requirements like full documentation according to the specification, technical data requirements like complying to the specified data structure, content accuracy like how the description fit the real problem and finally the service quality like availability, timelines and other parameters. The article describes key features of published approaches to quality assessment i.e. Quantis project, TIH, ISO 21707, Qbench and QKZ and on one use case from Czechia shows the importance of quality assessment.

Keywords: Quality · Parameters · Traffic information · RDS-TMC · Quantis

1 Introduction

Delivering good quality ITS services starts with ensuring sufficient quality of underlying data and ends with delivering those data in a best possible way to the end user. For these purposes, a unified methodology for quality assessment of traffic information services is needed.

The methodology should be based on data quality standard ISO 21707 [1], outcomes from the research project Quantis [2, 3], studies of quality for US DoT [4] and current project EIP ITS [5]. The methodology and good practice documents will then serve for elaboration of a national manuals for accreditation of certification bodies that will subsequently issue the certificates approving data quality level.

This paper analyses several important steps and areas of the traffic information value chain where quality could and should be measured. It does not focus only on quality measures defined in ISO standard and further refined by European projects, but also on so call formal quality and delivery channel quality aspects.

– Real world data quality measures – standardized (ISO), tested and enhanced by number of projects. These measures are very hard to measure since the evaluator needs in most cases ground truth data to compare measured data to.
– Formal service quality measures – used for a certification procedures, these measures do not relate in any sense to real dynamic content of the service but only to how it is described, what metadata are provided.

© Springer International Publishing AG 2017
J. Mikulski (Ed.): TST 2017, CCIS 715, pp. 82–95, 2017.
DOI: 10.1007/978-3-319-66251-0_7

- Delivery channel quality measures – each delivery channel has specific technical and logical characteristics that affect quality of distributed information, we are aiming at the Internet and RDS-TMC delivery channels.

Even though all measures add up to the overall perceived quality by the user, so far just to the first two has been sufficiently researched. Therefore we in this paper focus more on the delivery channel quality aspects.

1.1 Quality Obligations

The European Union will regulate the market with traffic information [6, 7]. Obligations result from this not only for Member States, but also for entities creating traffic information contents and for service providers. Regulation will take place on the level of audits of provided services quality and certificates will be issued from the level of national state-authorized accreditation body.

However there is no legal decision on how to proceed in the quality of the data/services monitoring and what quality will entail. The EU delegated regulation [7] states that the topic of quality is of significant importance and shall be further investigated in cooperation of all ITS stakeholders.

Apart from the political documents [6, 7] that are specifically vague, the more specific approach to the quality assessment was described in a study from 2014 [8], which stated that providers were reluctant to define minimum quality levels for services. But they agreed that common definitions and understanding of quality criteria and that QUANTIS' work should be taken into account.

However it still was not as specific as study from 2011 [9] that compares different quality assessment methods and defines the Quantis and ISO approach as mandatory for all EU member states (as an obligation) as well as a rating for the Total Service Quality (TSQ) of the service.

To conclude, the more closer the service quality evaluation is and the more binding is the document, the less specific is the description what the quality is and how it shall be measured and less specific is the obligation to perform quality checks. We therefore perceive as more acceptable to start by relatively easy quality measurements like, formal and technical and logical delivery channel measurements and then to proceed to service content quality evaluation.

2 Real World Data Quality Measures

Real world quality measures relate to the message content, if it is precise and complete enough, in time, covers sufficient area, if the geographical accuracy is of accepted level etc. Most of these measures must be validated against true value and the reports and researches focus on how to obtain that true data (ground truth) and how to use it to validate service content.

2.1   ISO and US Approach

The ISO data quality standard [1] and US research [4] recommends six fundamental measures of traffic data quality. The measures are:

- **Accuracy** – The measure or degree of agreement between a data value or set of values and a source assumed to be correct. It is also defined as a qualitative assessment of freedom from error, with a high assessment corresponding to a small error.
- **Completeness** (also referred to as availability) – The degree to which data values are present in the attributes (e.g., volume and speed are attributes of traffic) that require them. Completeness is typically described in terms of percentages or number of data values. Completeness can refer to both the temporal and spatial aspect of data quality, in the sense that completeness measures how much data is available compared to how much data should be available.
- **Validity** – The degree to which data values satisfy acceptance requirements of the validation criteria or fall within the respective domain of acceptable values. Data validity can be expressed in numerous ways. One common way is to indicate the percentage of data values that either pass or fail data validity checks.
- **Timeliness** – The degree to which data values or a set of values are provided at the time required or specified. Timeliness can be expressed in absolute or relative terms.
- **Coverage** – The degree to which data values in a sample accurately represent the whole of that which is to be measured. As with other measures, coverage can be expressed in absolute or relative units.
- **Accessibility** (also referred to as usability) – The relative ease with which data can be retrieved and manipulated by data consumers to meet their needs. Accessibility can be expressed in qualitative or quantitative terms.

These measures focuses closely on the technical means of data generation (i.e. data sources like induction loops), the "service quality aspect" of the data provision in not taken into account. For example coverage relate to sensor used in data generation rather than dissemination coverage. These measures take into account that there are different types of traffic data and different customers and users and the data is used for different applications. Therefore, the needs and quality requirements are different for the different data customers and applications. The thresholds and computing methodology for the above defined measures have been experimentally evaluated by several case studies presented in [4].

2.2   EU Quality Measures

EU approach focuses more on the whole traffic information value chain. Projects like Quantis [3] and EIP+ [5] defining quality measures in EU uses the ISO and US approach as a baseline and adapt definition of the measures for assessment of a service quality. With the perspective of the service quality other measures had to be defined.

The aim of the Quantis project was to investigate the relationship between ITS service quality and benefits/costs, determine the optimum service quality in four European core services and to identify levels of data quality providing optimal services.

QUANTIS compiled and categorized the definitions of quality objects and their parameters according to the ISO standard [1]. Table 1 lists all objects and parameters which have been identified as relevant for the quality assessment of data and services.

Table 1. Quality objects and parameters for traffic data [10]

Q object	Parameter	Definition
Completeness	Geographic coverage	Dedicated road area including % of border numbers covered
	Physical coverage	Road network covered
	Percentage of physical	Coverage % of km covered according to physical coverage
	Percentage event coverage	Number of reported true events by number of total events
	Data types covered	Included relevant data types
	Depth of coverage	Density of measuring sites
Availability	Availability period	Period during which the availability is defined
	Up-time	Percentage uptime which can be expected during the availability period
Veracity	Error probability	Percentage of content provided outside stated quality boundaries
	Cross verified	Indicates whether data has been cross verified with one or more additional sources
Precision	Location accuracy	Accuracy in terms of location
	Forecast horizon	Time difference between forecast calculation and forecast date
	Duration accuracy	Deviations between reported start and end of event and actual start and end
	Content accuracy	Accuracy/weakness of information provided
Timeliness	Data latency	Definition provided for each service separately
	Data update mode	Event driven or periodic
	Data update interval	Event driven: delay between event and service provision Periodic: update interval as specified

Quantis methodology defines five quality objectives (objects) that are further decomposed to the "sub-measures". These measures could be in all quality evaluations, since it does not cover only data but also a service quality.

Some of the quality measures above could be only evaluated against a ground truth, these are mainly focused on provided traffic information compared to actual situation. There are several approaches to obtain the ground truth information, for example from induction loops [11], all of which are however ex post:

- Continuous monitoring of equipment performance and availability
- Manual verification of events or conditions
- Reference testing of data collected
- Time-space oriented reference test methods
- Monitoring of data completeness and latency
- Regular sampling of message or data content completeness and correctness
- Verification and calibration of traffic/weather conditions prognosis

- Surveys of perceived quality by users
- Collection of direct user feedback
- Monitoring of service use statistics

The commercial methods use "Time-space oriented reference test methods" like QKZ or QFCD [12, 13] for event occurrence validation and "Reference testing of data collected" like QBench [14] for travel time comparison. The QKZ method for example evaluates the real occurrence of an event obtained from ex-post processed time and space diagram against the event as it has been reported by a service.

The methods presented above are aimed to evaluate the real value of provided information to the user. There are, in most cases, difficult to obtain due to the fact that the ground truth information is needed for the evaluation. However, there are other aspects of provided traffic information that are of the comparable importance, the formal and technical quality. Even high quality information provided in unintelligible way or that is hard to obtain has in the end no impact on the traffic and the user.

3   Formal Service Quality Measures

With regard to ITS Directive [6] the EU member states are obliged to establish a national traffic data access point that serves as a central node for searching for traffic related data in a country. Any potential subscriber needs to know several information about the data and the service that will allow him/her to assess its usefulness. In 2011 the European Commission have had a detailed study elaborated on the issues related to the provision of free traffic information and the guaranteed access to it [9]. This study describes in detail the implementation of the services, obligation of member states and touches the topic of quality issues. All traffic information providers have the obligation to provide certain metadata about their service to the national access point, this could be perceived as the formal data quality. The metadata is being set up by coordinated activity AT, DE and NL in a common metadata catalogue [15] and also by individual access points (the Germany's MDM[1] see Fig. 1, Czech Register[2] etc.).

The formal requirements (metadata) are shown in Table 2, each service shall have at least, data published in a common format which, will contain the following information.

All of the measures are needed for assessing technical aspects of the data but one the most important is definitely data sample and validation schema. This allows potential subscriber to get practical experience with the data due to provided sample and to adapt the system for data reception (schema). Any service can be than evaluated against the criteria given in metadata, this we consider as fulfilling formal requirements, so reaching a minimum level of traffic information service quality.

---

[1] Mobility data Marketplace, Germany's traffic data access point http://service.mdm-portal.de/.

[2] Czech traffic data registry http://registr.dopravniinfo.cz/en/.

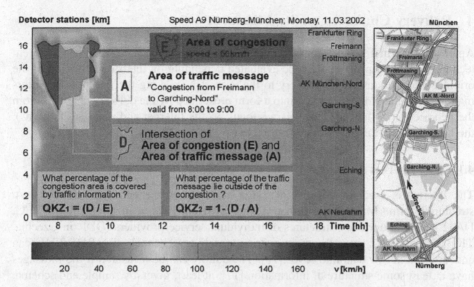

Fig. 1. Evaluation of quality by QKZ method [12]

Table 2. Formal quality measures for data sources [own study]

Measure	Description
Identification of the data owner	Contact details, phone, email
Description of the provided data service	Free text with data details
Data collection method	Technical and administrative means
Data specification and format	Format and its characteristic parts description, schema for validation of the data format and data samples
Supported mechanisms of data exchange	How can be data exchanged, ftp, html, email, with detailed set up information
The regime of data updating	Automated period or on occurrence
Physical coverage of the service	List of roads or a map with marked area
The guaranteed level of quality of the provided service	Only for such sources, where a common understanding what the quality is and how it is measured have been reached
The term of the guarantee availability of the service	What is the perceived availability of the service
The level of data validation	How are the data cross checked before releasing them to the public
Information about the license agreement on data reusing	Typically a blank contract, subscriber need to sign to get data
Terms of trade for using a data source	If it is for free or being paid for

4 Delivery Channel Quality Measures

Apart from formal requirements and real data content, each service has different technical quality aspects that can be evaluated depending on the transformation of the original information and its delivery to the user. This making the fine primary information coarser as it is being fitted into a form prescribed by the delivery channel. From the perspective of delivery channels we focus on the Internet and radio (RDS-TMC), the latter of which is much more restrictive.

4.1 The Internet

The Internet is used for delivery of traffic messages to the end users devices in a structured form. Most used format is the XML and its realizations like TPEG, DATEX II or proprietary structures of individual service providers (DDR in Czechia, TIC3 in Germany), even the more restrictive delivery channels like RDS-TMC can converted from raw data into XML [16]. The quality can be measured if the messages have at least some structure of information [17], at least, given by sample and schema, having:

– identification: message/event source and system id,
– time info: start time, stop time, time of reception,
– event description: described by a code (ALERT-C, TPEG or DATEX codes) and readable textual description,
– location description: described by a code or by coordinates of the location.

Quality of more restrictive delivery channels can evaluated against the XML message, assuming they use derived and coarser information. By comparing same information distributed via different services (channels) we can find out about its timelines and content, following:

– processing delay: start and stops of the event (as reported in XML) are (not) synchronized with reception in end user device,
– content deformation: event (as reported in XML) have same (or similar) content and though are describing situation in the same way.

4.2 RDS-TMC Technical Description

Radio Data System – Traffic Message Channel (RDS-TMC) uses radio broadcast for continuous delivery of traffic information suitable for reproduction or display in the language chosen by the user and without interrupting normal audio broadcast services. Received data are typically processed by a navigation system, which then offers the driver the proposal for alternative routes to avoid traffic incidents.

The RDS-TMC is relatively old technology with limited over the air capacity used for delivery of traffic information to navigation devices, which is being in recent years replaced by TPEG [18] or being enhanced [19]. It is however still being used and introduced in countries all around the world. Therefore it is meaningful to monitor end to end quality of RDS-TMC [20, 21]. The quality issues of the RDS-TMC could be separated to several groups, by the place in the value chain:

– Information encoding ("free form" to TMC) and service provider issues.
– RDS message formatting and delivery problems.
– End user device message reception and interpretation.

Information Encoding ("Free Form" to TMC) and Service Provider Issues

TMC defines data representation, introduces set of rules and tables (event codes and location codes table) that are together used by service provider to encode and by navigation device to decode traffic information. Each message consists of an event code and a location code in addition to expected incident duration, affected extent and other details, see Table 3.

Table 3. TMC message structure and its reconstruction according to location table [own study]

Event	Location	Direction	Extent	Duration
105 = "stationary traffic for 6 km"	LC = 1267	0 = Positive	3	2 = "30 min"

Message: D1 between exit 11 and exit 29 in direction to Brno, congestion for 12 km

Traffic situation location is a place between two locations from the look up table and is reconstructed from the location, direction and extent codes. Location codes, however usually describe intersections or important parts of the road network. The accuracy is therefore rather low, problem location described by precise start-end coordinates is being translated into a segment between two intersections.

Traffic event description can be as simple as shown in Table 1, but could get much more complex, using multiple event codes; quantifiers, making the information more specific; defining explicit start and stop time; proposing directions etc.

RDS Message Formatting and Delivery Problems

RDS allows distribution of digital data over analog radio channel and it is a part of every UHF FM broadcast. It has very limited capacity around 1 kb/s capable to convey a number of broadcast related data to the radio, like radio station name, programme type, traffic news indication etc. RDS data are transmitted in form of continuous stream of 11.4 RDS groups per second. From 32 possible group types, where each group type is used for different purpose, the TMC service use groups 3A (TMC service information) and 8A (actual traffic messages).

As RDS provides very limited capacity for amount of data, therefore TMC must be very space efficient. Just 4 bytes are enough to describe the simplest traffic situation.

Since RDS data are in real situation easily affected by external noise, so to prevent reception of mistaken data, each 8A RDS group (group containing traffic information) must be received at least twice with the same content to be accepted as correct.

Also to allow tuners to scan other frequencies and services, so called gaps are introduced. The service declares how many non-8A RDS groups are inserted in between two consecutive 8A groups. Tuner can then safely use the time of the gap to scan other frequencies without being in risk of losing an 8A RDS group.

End User Device Message Reception and Interpretation

The RDS-TMC standard does not specify in detail how to handle and interpret traffic information received over the RDS-TMC. Direct consequence of this is that every navigation system vendor had to decide themselves how to handle and interpret received traffic information. The result is however not satisfactory, sometimes vendors even overlook some crucial facts, because of limited understanding and almost no possibility of real testing.

Traffic information providers started to realize this problem and to mitigate it, for example Austria, started to issue conformity statements to navigation devices that properly interpret traffic information [22]. Also other providers started testing behavior of navigation devices under specified conditions more extensively [21], one of the approaches for enhancing quality of RDS-TMC value chain is proposed in [23]. On the other side the poor quality of TI could be caused by unsatisfactory implementation of RDS-TMC distribution system (technical and organizational) at the provider's side. So, it is necessary to focus on the analysis of whole RDS-TMC value chain and probable causes of bad quality of provided TI.

4.3 RDS-TMC Quality Measures

Baseline for all other quality measures in RDS-TMC delivery channel are parameters of RDS transmission and coding of messages. Service technical parameters directly influence the way how traffic messages are encoded on their processing and interpretation in navigation devices. In our previous research, reported in [23] we have defined 33 RDS-TMC service technical quality measures.

RDS Technical Quality Measures

Measures of overall quality of the service influence all other quality measures. The measures are inferred from parameters given by technical standards and handbooks (Table 4).

Table 4. Overall RDS quality measures [own study]

No	Measure	Description
1	RDS group error rate	Critical parameter. Message parts with errors are thrown out and message could not be reconstructed
2	TMC availability indication in RDS	TMC service shall be indicated by specific code CD46 in 3A group, not just inferred
3	Number of immediate repetition of RDS groups	Each TMC related group must be repeated 2 or 3 times by transmitter. 2 is optimal
4	Service parameters decoded from RDS groups	TMC service information: gap (3 groups), mode (basic), table number, service id
5	Gap between RDS groups with TMC messages	Real measured gap between TMC groups (shall be as indicated by service group)
6	Optimal use of RDS channel by TMC service	Standardized maximum is 25% of TMC in RDS, this is optimal setting

The measures sometimes just represent a quantity that is yet to be interpreted (i.e. error rate), in other cases there are expectations of their threshold value.

I.e. standard CEN ISO 14819-1 defines allowed gaps between 2 consecutive TMC groups. Minimal gap is 3 other RDS groups (25%). To achieve maximal throughput the minimal gap setting shall be used. However, lower gap setting than standardized one (like 1 or 2) may cause troubles while handling decoding of messages in navigation device. Also higher gap setting causes inefficient use of the channel. So the service provider shall maintain 8A groups at 25% rate.

TMC Channel and Message Quality Measures

TMC message quality measures are inferred from European standards and handbooks and best practices documents. They are also based on common sense of using the RDS channel efficiently and on getting maximum of actionable information, i.e. such information one can take an action upon.

Firstly, there are statistics common for whole TMC channel, measured without deep understanding and need for decoding of TMC messages. For example number of RDS groups per TMC message, even though the standard allows for 5 group (very detailed) message, the navigation devices usually discard details. Therefore it is inefficient to use long messages, it uses more of the channel capacity, is more error susceptible and brings no additional information to the user (is discarded or not implemented by the receiver) (Table 5).

Table 5. TMC channel quality measures [own study]

No	Measure	Description
7	Number of live TMC messages	Standard defines maximum 300 live messages, with 15 min repetition
8	TMC message sizes	% of 1 to 5 group long messages. Less groups the better
9	TMC message used classes	% of used classes. Class indicate a message type (accident, roadworks …)
10	TMC message with special attrs.	% of messages with problematic parameters. Vague messages, quantifiers, …

Apart from overall TMC quality measure each TMC message has its inner quality indicators. They are measured by decoding the message and tracing its lifecycle. The inner quality is content dependent, so it might be usage of too much event codes, inappropriate use of codes (not allowed by the standard), improper use of quantifier, location reconstructed out of the location table, too long time between repetitions etc. (Table 6).

Most of the measures above have indicative meaning and are used to flag possibly wrong messages and also to judge fitness of the service. For example, vague event codes, mean use of codes that are too broad and more specific code could have been used. Such phrases are without a meaning to the driver or to navigation device, they are not actionable: traffic problem, event code = 1, restrictions, event code = 493, major event, event code = 1501 etc.

Table 6. TMC message quality measures (per message) [own study]

No	Measure	Description
11	Number of repetitions	Number of message repetitions
12	Repetition period	Average repetition time. Time elapsed from last reception
13	Lifecycle info	First reception, last reception
14	Invalid content	Codes or part of the message with breaking the rules set up by the standard
15	Vague events	Use of event codes with unclear meaning
16	Superfluous information	Event codes used mean same effect, one is a subset of the other
17	Use of too many information	Too many event codes. Only effect is really necessary, cause is optional
18	Too long message	use of long message where exist shorter with exactly same meaning
19	Possibly wrong content	Use of quantifiers values 0 or 1
20	Invalid location	Reconstructed location does not exist in the table

TMC Interpretation Quality

Apart for technical problems on service provider side (wrong content, wrong encoding, etc.) there are also possible problems at the receiver side. Navigation system vendors are faced with difficult situation, due to limited experience with the standard, short implementation time and different implementation of TMC in every country. Therefore developers try to simplify received messages to small manageable subset.

Different developers experience and possibilities leads to different implementation of the TMC functionality, while service providers expect that all navigation systems will behave in the same way upon reception of the same message. Navigation devices therefore shall be tested if they respond in same manner to typical TMC situation like:

– How do they tune in a TMC service and what are the user possibilities?
– How do they present a TMC message with typical events and locations to the driver (different location types: point, segments, area; different event types: accidents, restrictions, travel times, weather, etc.)?
– How do they act on reception of certain important messages (ghost driver)?
– How do they filter received TMC information (by route, close area)?

Testing of navigation devices could be divided into two areas. The TMC user interface friendliness and precise processing and interpretation of individual TMC messages that do not degrade or even misinterpret received information.

5   Evaluation

The quality measures proposed in RDS-TMC delivery chain have been tested on data from national traffic information center in about year interval within the project "Monitoring of Quality of provided Traffic Information" CG944-051-120. Also synthetic test with several navigation devices have been performed.

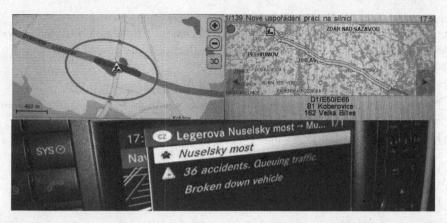

Fig. 2. Recommended rates for typical RDS groups (grey) and example of measured rates (black) [own study]

In testing, one of the basic RDS-TMC quality measure, the comparison between expected percentages of TMC groups and measured quantity has shown that only 11% instead of full potential 25% is used for the service.

Other indicators have shown excessive use of unnecessary level of detail, broadcasting information with no impact on the traffic flow, like the partial closure without any congestion. Use of unintelligible data in encoding the information, like use of implicit value of quantifiers or value of quantifiers equal to 0 (in TMC meaning 32, see following figure) due to bad implementation. In several cases we have been able by comparison of distribution channels to find out information being encoded to opposite traffic than it has really affected (again by bad implementation). In testing of navigation devices we found out problems with message reconstruction, for example for long locations where too long location was interpreted as being point (node) problem (Fig. 2).

One of the main problems of RDS-TMC has been non-existence of monitoring of the state of broadcasting and the guarantee of quality of the provided service.

5.1   Recommendation

Results from the testing show considerable options improving the existing technical quality of RDS-TMC through measures taken by data provider and data subscriber.

Recommendations for Producers of Navigation Devices
The "best practices" for reconstruction and interpretation of traffic information and for tuning the service should contain the following points:

– receive information ONLY from one service
– enable available services to be "scanned" and added (tuned) on request
– enable automatic tuning into TMC services
– enable classification of received information according to multiple criteria
– set the basic principles for the behavior of navigation devices with respect to certain types of traffic information

Recommendation for Service Providers

The current RDS-TMC broadcasting contains much information that navigation devices are not able to make use of. Therefore, data provider should make efforts to meet the conditions and:

- Do not broadcast messages with explicit duration time (from – to).
- Do not broadcast messages with explicit time of duration of an event (the situation will last at least for …).
- Eliminate broadcasting of supplementary info, use it only when necessary.
- Do not broadcast quantifiers with the value of 0 and 1 in messages.
- Do not broadcast messages with more than 2 events.
- Where it may be avoided, do not use information with a quantifier.
- Do not broadcast messages with are or a linear location.

The recommendations concurrently serve as the basic rules for assessing the quality of the TMC service from the point of view of its usability in a navigational device.

6   Conclusion

In this paper we have presented several measures how to evaluate traffic data and traffic service quality. We have defined three areas: real world data quality measures, formal service quality measures and delivery channel quality measures. Two areas were introduced in lower level of detail as being too complex, with number of possible solutions to the problem (first one), or too simple, with simple checking of yes/no criteria (second one). We have focused on the third area specifically on characteristic that can be evaluated with regard to RDS-TMC delivery channel. We have defined 3 subareas based on occurrence in a value chain, where we set up quality measures to be evaluated. We have performed several tests taking in defined criteria and found out a number of technical problems. In the evaluation we have presented some of the results and recommendations. In a future work we will focus on standardization of tests and form of quality certificates for traffic information services.

Acknowledgement. The authors acknowledge the financial support provided by the Technology Agency of the Czech Republic through project Kamelot (TA04031524) and through Centre for traffic research RODOS (TE01020155).

References

1. ISO 21707: Intelligent transport systems: integrated transport information, management and control—data quality in ITS systems. ISO, Switzerland (2008)
2. Kulmala, R., et al.: Quality criteria, requirements and assessment test plan. Deliverable of EIP+ 3.1 Testing and Validating of the Quality Recommendations and Results for Traffic Information from EIP (2015)
3. Kumala, R.: QUANTIS: how to utilise its results? Helsinki (2012)
4. Battelle: Traffic data quality measurement, Washington. https://ntl.bts.gov/lib/jpodocs/repts_te/14058_files/Traffic_Data_Quality_Measurement_Final.pdf (2004). Accessed 17 Feb 2017

5. Kulmala, R., et al.: Quality package for safety related and real-time traffic information services, EIP+ 3.1 Testing and Validating of the Quality Recommendations and Results for Traffic Information from EIP, EU (2017)

6. ITS Directive 2010/40/EU on the framework for the deployment of Intelligent Transport Systems in the field of road transport and for interfaces with other modes of transport. http://www.eltis.org/sites/eltis/files/celex-32010l0040-en-txt.pdf. Accessed 17 Feb 2017

7. Commission Delegated Regulation (EU) No. 886/2013 with regard to data and procedures for the provision, where possible, of road safety-related minimum universal traffic information free of charge to users. http://eur-lex.europa.eu/legal-content/EN/TXT/PDF/?uri=CELEX:32013R0886&from=EN. Accessed 17 Feb 2017

8. RAPPTRANS. ITS Action Plan D5 – Final Report: Action B - EU-wide real-time traffic information services. http://ec.europa.eu/transport/themes/its/studies/doc/2014-07-its-action-plan-d5-action-b.pdf (2014). Accessed 17 Feb 2017

9. Algoe, Rapp Trans: study regarding guaranteed access to traffic and travel data and free provision of universal traffic information: D8 – Final report, 1st edn, Brussels (2011). http://ec.europa.eu. Accessed 17 Feb 2017

10. QUANTIS deliverables. Quantis. http://www.quantis-project.eu/index.php/publications. Accessed 17 Feb 2017

11. Pribyl, O., Pribyl, P.: Mining data from induct loops. In: IEEE Intelligent Transportation Systems Conference (ITSC), pp. 514–518. IEEE, New York (2005)

12. Bogenberger, K., Weikl, S.: Quality Management Methods for Real-Time Traffic Information. Proc. Social Behav. Sci. **54**, 936–945 (2012)

13. Bogenberger, K., Hauschild, M.: QFCD: a microscopic model for measuring the individual quality of traffic information. In: Proceedings of the 16th ITS World Congress and Exhibition on Intelligent Transport Systems and Services, 21–29 September 2009

14. TISA: Guidelines for TISA QBench Calculations (2016). http://tisa.org/wp-content/uploads/SP16001_TISA_QBench_Calculations_v1.0.pdf. Accessed 17 Feb 2017

15. Coordinated Metadata Catalogue (2015). https://portal.easyway-its.eu/filedepot_download/1542/4389. Accessed 17 Feb 2017

16. Bures, P., Vlcinsky, J.: Converting recorded RDS-TMC services into XML to simplify following quality assessment. Trans. Transp. Sci. **6**(1) (2013)

17. Yearworth, M., et al.: Travel information highway, Tenth International Conference on Road Transport Information and Control, 2000. (Conf. Publ. No. 472), London, pp. 41–45 (2000)

18. Bures, P.: The architecture of traffic and travel information system based on protocol TPEG. In: Proceedings of the 2009 Euro American Conference on Telematics and Information Systems: New Opportunities to Increase Digital Citizenship, Prague, Czech Republic (2009)

19. Bures, P., Vlcinsky, J.: Implementing high density traffic information dissemination. In: Mikulski, J. (ed.) TST 2016. CCIS, vol. 640, pp. 243–256. Springer, Cham (2016). doi:10.1007/978-3-319-49646-7_21

20. Bures, P., Vlcinsky, J.: Monitoring of live traffic information in the Czech Republic. In: Mikulski, J. (ed.) TST 2011. CCIS, vol. 239, pp. 9–16. Springer, Heidelberg (2011). doi:10.1007/978-3-642-24660-9_2

21. Langebæk Hegner, J., Friis, H.: TMC messages in navigation systems: TMC navigation systems reproduction of TMC messages: Overall test results. Copenhagen: Danish Road Directorate, p. 35 (2008)

22. Asfinag. Free to air. ITS International. 2010, November/December, pp. 50–51. www.itsinternational.com. Accessed 17 Feb 2017

23. Bures, P.: Assessment of the quality of traffic information distributed by RDS-TMC service. In: Proceedings of the 11th European Transport Congress, pp. 29–39. CTU, Prague (2013)

Communication Systems' Safety and Security Challenges in Railway Environment

Marek Pawlik[✉]

Warsaw Railway Institute, Chłopickiego 50, 04-275 Warsaw, Poland
mpawlik@ikolej.pl

Abstract. Implementation of the communication based train control systems, is directly linked with safety of the train operation. The scope of the solutions as well as proposed way they can be classified have been presented by author at the Transport System Telematics conference in 2016. The communication based train control and management systems' safety and security impact reference model (CBTC SSIRM model), which was then defined and presented is an entrance point for the complementary communication systems' safety and security challenges analyse. Article shows why lack of safety and/or security of the communication systems in railway environment jeopardises safety of the whole railway transport system. It describes associated challenges distinguishing at the same time safety and security as well as sender and receiver, input and output, source and destination. Applied types of solutions from hardware and software redundancies via different addressing schemes to cryptographic technics are described and classified forming an important add-on to the SSIRM model in the cybersecurity domain.

Keywords: Transport · Railway · Safety · Security · Communication · Train operation · Control command and signalling

1 Introduction

In many languages the same words are used to describe safety and security. Although many publications link safety with technical and material characteristics of different solutions and security with adverse activities of unauthorised persons and entities both safety and security in some circumstances are used even as synonyms. For analyses, which are dealing with risk and its acceptance such ambiguity is not acceptable.

Since the very beginning of the railway transport mode engineers apply so called "fail-safe" principle, according which if system, which ensure operational safety, fails it has to switch to a predefined state in which safety is not jeopardized. That usually means shifting responsibility to other systems or/and to dedicated staff managing traffic disturbances. Usually such disturbances arise in such circumstances. Quoted operational safety meaning lack of unacceptable risk of accident is therefore ensured by both signalling equipment and operational rules applied by operational staff.

The "fail-safe" principle is adequate for mechanical, electromechanical, electrical and relay signalling solutions, but is not for modern electronic systems using data processing technics.

© Springer International Publishing AG 2017
J. Mikulski (Ed.): TST 2017, CCIS 715, pp. 96–109, 2017.
DOI: 10.1007/978-3-319-66251-0_8

For computer based systems different approach was elaborated and standardised in European norms [2]. This approach is known as 'safety integrity levels'. The range of solutions as well as risk acceptance criteria are subdivided into five levels from SIL 0 to SIL 4 and associated with random and systematic faults, which may occur during system life cycle. As signalling equipment is due to serve rolling stock movements on relatively long distances data processing is complemented by data communication. Safety in that respect is a challenge especially for signalling systems, which are based on track-train data transmission of electronic movement authorities. Such systems are known as control-command systems and are being implemented presently on all railway lines on which signalling equipment modernization takes place as part of publicly financed railway modernization projects.

The control command systems are subdivided into different classes, which have been presented by author at the Transport System Telematics conference TST 2016 [1] and shown as inadequate for present needs. The subdivision of the control command systems into Automatic Warning Systems AWS, Automatic Train Protection ATP, Automatic Train Control ATC as well as Automatic Train Operation ATO and Automatic Train Supervision ATS do not illustrate scope of safety related functions and do not illustrate levels of safety integrity [5]. In reality for nearly all such systems SIL 4 requirements are applied and assigning a system to ATP or ATC class does not support understanding their safety features.

Moreover data processing and data communication are presently also used by technical means supporting security. This makes the safety and security overall picture more complex. An overall proposal was defined by author and described as a reference model - the communication based train control and management systems' safety and security impact reference model (SSIRM model). This article is starting from SSIRM model and analyses communication safety in control command systems and security supporting systems from electrical interferences to cyber-crime.

2 Communication Based Train Control and Management Safety and Security Impact Reference Model: CBTC SSIRM Model

Presently used solutions require wider understanding of technical systems based on communication. Instead of talking about control command being subdivided into AWS, ATP, ATC and ATO belonging to signalling nowadays experts have to see communication based train control and management covering not only signalling functionalities but also different types of safety and security relevant processing of data transferred by communication systems. In wide understanding safety and security can be seen from different angles and provided and supported by different technical solutions. The CBTC SSIRM model distinguishes railway transport security and railway technical safety as two complementary areas with its' own technical and procedural means supporting overall railway operational safety.

All technical means, which are used to protect railway against incorrect preparation of the movement authority (MA), overpassing distance limits, overspeeding, and electrical hazard support railway technical safety. All technical, organizational and procedural means which are used to support passengers' health and protect railway against vandalism, against terrorism as well as against natural disasters support railway transport security. That forms four technical safety aspects associated with safety hazards and four security aspects associated with security hazards, which all support widely understood railway operational rules which are applied for ensuring overall railway transport operational safety. Complete overall picture, presented at Fig. 1., shows relationship between operational safety, technical safety and security.

Fig. 1. CBTC SSIRM model – overall picture [1]

Each of the four safety hazards as well as each of the four security hazards are then described in CBTC SSRIM model by four basic functionalities offered by technical means and procedures respectively for ensuring safety and security.

The safety domain (see Fig. 2.) covers safe lower level signalling composed by track occupancy, interlockings, block systems and level crossing protection systems constructed and operated in accordance with FS and SIL principles. It also covers two complementary control command parts usually seen as an upper signalling layer and referred to as control command - the trackside functionalities and the on-board functionalities. The key safety features at track side are the safe way to obtain data from lower level signalling, safe preparation and transmission of the movement authority. The key safety features on-board are safe receiving, safe processing, safe speed and distance measurement and safe brake application. The FS principle in that respect is applicable only partly, but SIL principle [2] has to be respected in full scale. The fourth segment covers two key features for manual train driving and two for automatic train driving which are applicable in case of advanced communication based train control systems providing automatic train operation. The first two are cab signalling and operational wireless communication, the second two are automatic driving (speeding up, cruise and braking) and automatic support for passengers access and egress (precise stopping, train doors and platform doors management).

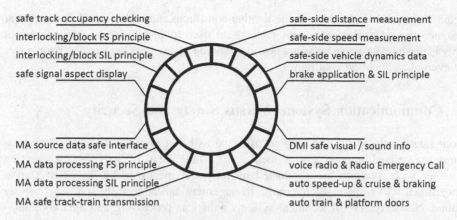

Fig. 2. CBTC SSIRM model – safety domain [1]

The security domain (see Fig. 3.) covers passenger personal security composed by providing passenger information at the stations and inside vehicles, ensuring fire protection at the stations and inside vehicles, ensuring electrical shock protection and protection against falling objects. It also covers two closely linked sets of functionalities ensuring protection against vandalism and against terrorism covering monitoring starting from station and vehicle interior monitoring ensuring protection against vandalism and thieves to video stream on-line analysing focused on automatic detection of left language and automatic detection of entering restricted areas as well as different types of public emergency call installations and medical emergency equipment up to language x-ray screening and magnetic security gates. Security related risks are presently shifting from vandalism towards terrorism and therefore such aspects have to be well defined and well understood.

The fourth segment of the security protection measures covers technical means against natural disasters. Transport systems can be and are in many cases protected

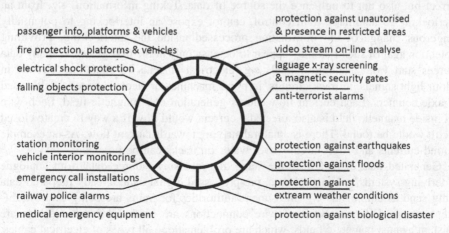

Fig. 3. CBTC SSIRM model – security domain [1]

against earthquakes, floods, extreme weather conditions and/or biological disasters. For instance special equipment (seismometers) are used to detect ground movements not only in seismically active areas but also e.g. where railway lines are running over active or even already closed coal mines.

3 Communication Systems Versus Safety and Security

Both safety and security can only be assumed as being provided if appropriate measures are defined, applied and verified along whole chain of activities disregarding whether such activities are conducted by technical means or by dedicated staff – operational staff responsible for traffic management and/or security staff e.g. railway police. Nowadays use of technical systems with data processing and data communication is growing significantly both in safety domain and in security domain. As a result not only procedural means used by staff but also technical means used by data communication and processing systems have to be protected along whole chain. Therefore cyber hazards from simple interferences to complex cyber-attacks have to be taken into account from the preparation of the communication system ideas to construction and exploitation.

3.1 Hazards in Communication Systems Relevant for Safety and/or Security

All technical systems supporting safety and/or security, which are using data processing and data communication can be affected by data corruption which can take place on any stage of widely understood data processing. It is necessary to distinguish data generation, data communication, data processing, data storage and data analyses. The risks against which protection is required as well as safety and security measures which can be applied are different on different stages.

Generating data requires appropriate means not only to produce values which are correct but also not to influence the source of data. Taking information, e.g. from an interlocking to automatic train control, cannot expose an interlocking to potentially dangerous change of the information processed inside the interlocking. Providing protection against such risk may appear to be easy only at the first glance as many data sources start from analogue signals e.g. electrical currents in signalling circuits in colour light signals, in track circuits, in point machines or electromagnetic fields used by axle counters. Each current flow causes generation of a magnetic field. Each wire put inside magnetic field is also affected – current would flow if a way to create closed circuit would be found. There is a natural striving towards current flow. As an example ground currents and short circuits over water on tracks can be pointed.

Generated data disregarding the stage of processing (source data from analogue underlying systems, raw digital data, pre-processed digital data, already prepared data being send as commands e.g. movement authorities for trains are being sent via different means. Different types of wire connections are used. Fibre optic cables are resistant against magnetic fields, which are problematic to all types of electrical cables.

In case of fibre optic cables it is necessary to minimise the amount of places where they are cut, otherwise expensive solutions are required to keep appropriate signal level. Generally affecting wired connection requires access to wires. Not necessarily direct electrical access. Protection becomes much more challenging in case of wireless transmission systems. Security measures which are used to protect communication systems like on-line verification of opening cabinets are useless in case of radio-communication. Technics which are used to detect interferences and cross-talks are not sufficient. It is required to provide solutions ensuring protection against different types of unauthorised attempts to change data being communicated.

3.2 Precise Describing of the Communication Systems in View of the Safety and Security Means: Senders and Receivers, Inputs and Outputs, Sources and Destinations

Different types of soft codes could be used to block or change data which is being generated, transmitted, processed, accumulated and analysed. Viruses, malware, fake firmware, spy-soft, etc. can affect sensitive data in safety and/or security relevant systems. Protection is possible and being introduced thanks to both software and hardware solutions. However each system requires preparation of a comprehensive and appropriate protection conception based on detail and precise description of a whole data management and complete identification of risks, which could arise on all processing stages. Therefore it is important to point that the same functions as well as the same pieces of equipment may be seen in different ways depending on the way they are seen by separate functions or pieces of equipment.

Safe communication can be provided over transmission systems which does not provide safety and have to be treated as non safety related ones. The overall concept is described in a dedicated standard [4] and shown on Fig. 4. Where different types of rectangles means:

- safety related equipment,

- non safety related equipment checked by safety related technics,

- non safety related equipment.

Standard [4] dedicated to railway specific solutions for signalling and control command can be applied not only for the safety related technical systems using data communication but also for the technical means which are used for security support. Precise and correct identification and assessment of risks requires looking on the same functions and the same pieces of equipment in different ways. In most cases the same piece of equipment can work as a sender and as a receiver. Sometimes its character changes over time in other solutions both roles are played at the same time. Sender and receiver are generally understood as pieces of equipment with inherent software. However in many analyses it is crucial to see functions.

Input and output create a pair of expressions which are used to describe hardware or data on the edge of analysed system but inside the system. Source and destination also

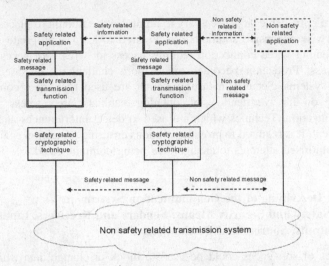

Fig. 4. Use of non safety related transmission system for safety related applications [4]

create a pair of expressions, which are used to describe hardware or data on the edge of analysed system, but which are outside the system. It is extremely important to be sure how we define the borders of the communication system.

Precise and consistent use of expressions: sender/receiver, input/output, source/ destination is a key for avoiding misunderstanding between different experts involved in planning, construction and maintenance. Usually separate experts sometimes from different companies, frequently working in different locations and speaking different languages are responsible for providing appropriate cybersecurity in different layers [3] of the transmission system from physical layer in which physical representation of '1' and '0' are distinguished via transaction, session, communication up to application layer in which an application dedicated for communication is using lower level functions defined individually and taken from libraries.

3.3   Protection Means in Communication Systems

The most common solution, which however cannot be used as the only one is providing data redundances – providing additional data which is generated on the basis of core data and data commonly known to senders and receivers and which is used to verify correctness of the core received data. Such approach would be adequate in case of protection against interferences and cross talks which take place due to random failures as well as against radical changes in external environment like e.g. sun storms generating coronal mass ejections significantly disturbing electrical equipment on earth. Basic principle in that respect is presented on Fig. 5, which is quoted according to standard [4].

Dedicated software creating data redundances and verifying correctness of core data on the basis of additional data does not provide protection against cyber-attacks. It is necessary to provide protection against intentional data repetitions, data deletions, data insertions, data re-sequencing, data corruption, data delay and masquerades. The range of

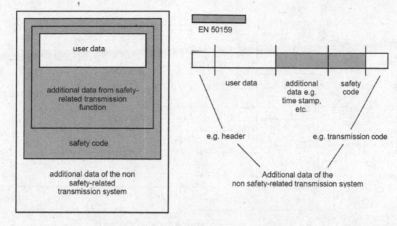

Fig. 5. Data redundances basic principle [4]

technics which are being used is wide from time stamping and double time stamping, via using key management systems based on private and public keys up to advanced cryptographic technics. It is however always very important to define and describe communication system in a way that allows complete identification, and precise description of risks to enable creation of adequate protection means providing complete protection.

Hazardous events against which transmission has to be protected include systematic failures like for instance: cabling errors, antenna misalignments, maintenance errors as well as random failures like for instance: hardware random failure and ageing, thermal noise, cross-talk, fading effects, overloading of transmission system, etc. as well as deliberately-caused events such as: wire-tapping, damage or unauthorised change to hardware, unauthorised change to software, monitoring of channels, transmission of unauthorised messages.

3.4 Safety Layer in CBTC Systems

Presently all types of data computation based systems have to communicate with external systems. For enabling such communication an already mentioned Open Systems Interconnections OSI basic reference model [3] has been defined and is maintained commonly by all international standards organizations. OSI basic model is defining interconnections between open systems using concept of seven layers: physical layer, data link layer, network layer, transport layer, session layer, presentation layer and application layer. Each open system from this point of view is seen as being logically composed by a set of pre-ordered subsystems (layers) usually represented in vertical sequence. The three lower layers are associated with media and therefore many systems, including railway specific applications, use communication schemes involving so called rely open systems working as routers. The upper layers are associated with hosts and, in case of railway specific applications, are seen as the ones which have to be precisely understood by experts developing individual applications. The overall model is shown on Fig. 6.

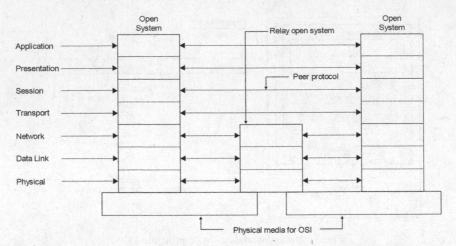

Fig. 6. Communication between layers of the OSI basic reference model [3]

Physical layer is probably the easiest to be imagined. Hoverer to be able to fully understand its nature it is important not only to know that the data can be physically transferred thanks to different kinds of physical-connections for bit transmissions, which are only physical media, but also to accept that such transfer requires physical and procedural means which are activating, maintaining and de-activating connections between so called data-link-entities. Presently railway applications are using all kinds of physical media from signal and point mechanical wires via different kinds of electrical cables, fibre optic cables, to different types of electromagnetic waves including spot, loop, radio and satellite transmissions. Classic railway applications were using them directly assuming that safety requires using only closed communication systems. This is no longer true since many years. However railway experts still usually see application layer as the one using physical transmission media. The five layers in between are either not seen and not understood or assumed to be hazard free which is of course not true. ISO model standard states that:

"The Data Link Layer provides functional and procedural means for connectionless-mode among network-entities, and for connection-mode for the establishment, maintenance, and release data-link-connections among network-entities and for the transfer of data-link-service-data-units. A data-link-connection is built upon one or several physical-connections.";

"The Network Layer provides the functional and procedural means for connectionless-mode or connection-mode transmission among transport-entities and, therefore, provides to the transport-entities independence of routing and relay considerations. The Network Layer provides the means to establish, maintain, and terminate network-connections between open systems containing communicating application-entities and the functional and procedural means to exchange network-service-data-units between transport-entities over network-connections.";

"The Transport Layer optimizes the use of the available network-service to provide the performance required by each session-entity at minimum cost. This optimization is achieved within the constraints imposed by the overall demands of all concurrent session-entities and the overall quality and capacity of the network-service available to the Transport Layer.";

"The Session Layer provides services to establish a session-connection between two presentation-entities, to support orderly data exchange interactions, and to release the connection in an orderly manner. For connectionless-mode communication Session Layer is providing mapping of transport-addresses to session-addresses.";

"The Presentation Layer provides for the representation of information that application-entities either communicate or refer to in their communication. The Presentation Layer ensures that the information content of the Application Layer data is preserved during transfer. Cooperating application-entities are responsible for determining the set of abstract syntaxes they employ in their communication. The Presentation Layer is informed of the abstract syntaxes that are to be employed. Knowing the set of abstract syntaxes to be used by the application-entities, the Presentation Layer is responsible for selecting mutually acceptable transfer syntaxes.".

This shows, that presently widely used open communication systems are on one side well organised, although not fully unified as many mutually noncompliant technical solutions exist, but on the other side not well suited for railway specific safety related applications. Safety critical influence on data can take place at different communication levels in different locations also outside the railway premises.

As an example we can point already widely implemented in Europe GSM-R (Global System for Mobile communication – Rail). This radio system is not only ensuring voice communication between signalmen and train drivers but also data communication used for transmission of the movement authorities. The GSM-R is a cellular centralised system using fibre optic backbone data transmission network. It is an open communication system not only because everyone can use a radio transmitter in any location to send data into the system, but also because all data being exchanged are being transferred between communication layers and are flowing via different locations over the backbone which are not fully strictly pre-defined and as a result can be easily accessible for many persons and applications. This requires firm complete defence strategy and powerful tools for ensuring safety. Namely cyber-security for maintaining technical safety of the safety critical control command functionalities which are used for communication of the movement authorities being curtail for signalling active operational safety protection.

Such tool forming so called safety layer is already defined for GSM-R and named Euroradio. The four Euroradio specifications, which belong to public domain as being pointed as legally binding in EU legislation, precisely define applied protection technics. This however does not mean, that everyone who can read and understand these documents can influence GSM-R radio or underlying communication. The key information is in the hands of GSM-R operators, which are the railway infrastructure managers. The challenge is that the railway experts have to learn how to leave with cyber-crime, and how to cope with the quick development of the digital world as new solutions are being continuously developed in all communication layers. If railway experts fail to follow hazardous events like the one shown on Fig. 7 may appear.

Fig. 7. Presently theoretical hazard which may appear in the future [own study]

This is a challenge not only for GSM-R and not only for control command. This is a challenge for all communication systems which are and will be used by all different kinds of systems with data transmission including voice and video streams, which are and will be used for supporting all different safety and security functionalities.

4   CBTC E-SSIRM Communication Based Train Control and Management Enhanced Safety and Security Impact Reference Model

It is possible to add cyber-security to CBTC SSIRM model as an extra dimension, which is required for all safety and/or security relevant systems using data communication. The cyber-security does not form an independent set of functionalities. It is rather an extra perspective on safety and security functionalities already included in the CBTC SSIRM model.

Adding cyber-security aspect in the security domain is visualising a set of cyber-security relevant transmission systems. The sixteen functionalities, which are naturally linked into four segments (passenger personal security, means against vandalism, means against terrorism, and protection against natural hazards) can be based on four different transmission systems, which have to be verified from cyber-security point of view as shown on Fig. 8. However use of transmission systems does not need to be organised according such segments. It is possible to use a separate transmission system for station monitoring, a separate for emergency call installations, a separate for oxygen masks and a separate for defibrillators, and a separate for informing railway police about pickpockets. Disregarding the amount of cyber-security relevant transmission systems each system has to be verified taking into account all internal risks, which may affect safety and/or security being provided.

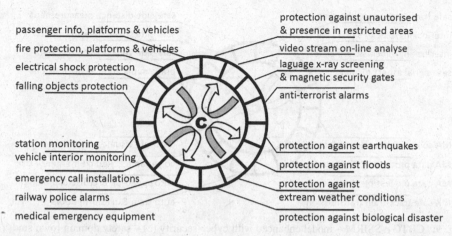

passenger info, platforms & vehicles

fire protection, platforms & vehicles

electrical shock protection

falling objects protection

protection against unautorised & presence in restricted areas

video stream on-line analyse

laguage x-ray screening & magnetic security gates

anti-terrorist alarms

station monitoring

vehicle interior monitoring

emergency call installations

railway police alarms

medical emergency equipment

protection against earthquakes

protection against floods

protection against extream weather conditions

protection against biological disaster

Fig. 8. CBTC e-SSIRM – model enhanced with cyber-security (c) – security domain [own study]

Adding cyber-security aspect in the safety domain shows, that two segments of the upper signalling layer namely the trackside preparation of the movement authorities and the on-board consumption of the authorities have to be linked by common transmission system. Such transmission may be based on more than one physical transmission media. For instance unified European automatic train control system (the European Train Control System ETCS) [5] in most cases is using two types of transmission media – balises and digital radio channel. This is because balises offer precise reference points for distance measurements while radio offers high data capacity and easy covering of tracks with radio signal. That shows correlation which cannot be omitted in cyber-security proving - due to direct dependency between safety functionalities in those two segments all risks have to be taken into account in a single multilevel cyber-security analyse.

Separate transmission systems may be used for other functionalities e.g. for track occupancy checking (e.g. network of wires between axle counters, aside electronic units and comparator defining track segments occupancy) and would require separate cyber-security verification.

Introducing cyber-security in CBTC SSIRM safety domain is shown on Fig. 9 below. The straight double arrow may cover also some lower level signalling functionalities and may cover automatic driving and/or emergency calls. The automatic driving is frequently using separate transmission system as such approach correlates with assigning inherent safety features to train control and ensuring safety for automatic driving by filtering ATO functionalities by ATC modules.

In all presently used systems radio voice communication is functionally totally separated. Even in case of unified European radio communication (the Global System for Mobile communication – Rail GSM-R) [5] subdivision between data transmission over radio and voice communication, which is also based on data transmission is not

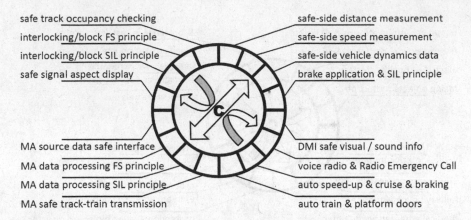

safe track occupancy checking

interlocking/block FS principle

interlocking/block SIL principle

safe signal aspect display

safe-side distance measurement

safe-side speed measurement

safe-side vehicle dynamics data

brake application & SIL principle

MA source data safe interface

MA data processing FS principle

MA data processing SIL principle

MA safe track-train transmission

DMI safe visual / sound info

voice radio & Radio Emergency Call

auto speed-up & cruise & braking

auto train & platform doors

Fig. 9. CBTC e-SSIRM – model enhanced with cyber-security (c) – safety domain [own study]

only logical or functional but physical. The two radio channels are being separately established for each traction vehicle between on-board voice radio and dispatcher terminal and between on-board data modem (ETCS Data Only Radio EDOR) and the Radio Block Centre.

5   Conclusion

The enhanced safety and security impact reference model for communication based train control and management systems - the CBTC e-SSIRM model is a powerful tool for safety, security and cyber-security analyses covering all technical aspects influencing finally provided operational safety, which is keeping visibility of the binding safety and security related operational rules.

As a result complex and large railway transport systems can be successfully designed, constructed, commissioned, maintained and operated commonly by engineers working for different entities, in different places on complementary functionalities, focusing on different layers of the communication systems [3], ensuring requested level of safety, security and cyber-security. In Poland such approach could be applied for construction of the high speed railway line and/or for the construction of the Warsaw Railway Ring, which would require shortening of the train headways approximately to 150 s for commuter trains.

References

1. Pawlik, M.: Communication based train control and management systems safety and security impact reference model. In: Mikulski, J. (ed.) TST 2016. CCIS, vol. 640, pp. 388–400. Springer, Cham (2016). doi:10.1007/978-3-319-49646-7_33
2. European Standard EN 50129:2003/AC:2010, Railway applications: communication, signalling and processing systems—safety related electronic systems for signalling

3. International Standard ISO/IEC 7498-1:1994, Information technology: open systems interconnection-basic reference model—The basic model
4. European Standard EN 50159:2010: Railway applications: communication, signalling and processing systems—safety-related communication in transmission systems
5. Pawlik, M.: European Rail Traffic Management System, functions and technical solutions overview – from an idea to implementation and exploitation (Europejski System Zarządzania Ruchem Kolejowym, przegląd funkcji i rozwiązań technicznych - od idei do wdrożeń i eksploatacji). KOW, Warsaw (2015)

Missing Data Problem in the Event Logs of Transport Processes

Mariusz Dramski[✉]

Maritime University of Szczecin, Wały Chrobrego 1-2, 70-500 Szczecin, Poland
m.dramski@am.szczecin.pl

Abstract. Data and process mining techniques are very helpful in analyzing transport problems. The model of the process can be built using the available data. It leads to make possible the operational support which improves the process. Very important task is to record data in the proper way. Unfortunately some errors may occur. In this kind of situation some data lacks can be observed. On the other hand the data may be complete but having very high error coefficient. The model of the process should have as minimum error as possible and has to be reliable. Although some missing data can occur, there are some ways to do some data recovery. In this paper the problem of missing numerical data in the event log is described. Different solutions and conclusions are presented.

Keywords: Data mining · Process mining · Missing data · Data lack · Event log

1 Introduction

A model is a simplified description of the system. In real world it is very rare that the system is given in an analytical form such mathematical equations etc. Initially for every researcher the system is a kind of "black box" (Fig. 1) and it's structure needs to be identified basing on input and output data. This step is impossible if the proper data doesn't exist. It is said that the model of the system should be built. The model makes possible carrying out some investigations in system's behavior without the full knowledge of it's structure. First task is always to have a correct and complete data which will be the base for further investigations. The data is delivered by the experts in specific domains, measurements etc.

The essential feature of every measurement process is the occurrence of errors. The errors can be systematic or casual. Generally there is no possibility of having data perfectly correct and complete. This situation occurs only when the exact analytical form of the system is given.

In this paper the problem of missing numerical data in the event log is raised. If the data is incomplete it doesn't mean that there is no solution of the problem. In some situations some methods of data recovery can be applied.

The simplest way of numerical data recovery is regression. Regression is a set of methods which allow for the analysis of different relationships occurring in the investigated system. There are two stages in regression: creating a model and scoring (applying the solution).

© Springer International Publishing AG 2017
J. Mikulski (Ed.): TST 2017, CCIS 715, pp. 110–120, 2017.
DOI: 10.1007/978-3-319-66251-0_9

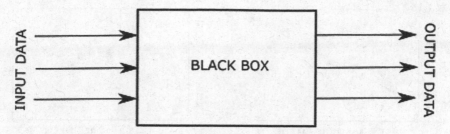

Fig. 1. Black box – unknown structure of the system [own study] .

2   Regression

According to the definition that can be found in encyclopedia, regression is a statistical technique for estimating the relationships among variables [1]. Of course currently, due to developing of new data or process mining techniques, regression not always must be related with statistics. This approach is very popular in the case of any missing numerical data. If we have a simple linear function the regression is done almost automatically. The only problem is to determine the mathematical equation which describes the system. This is an ideal situation when we know that the system is linear, and there are no errors in measurements.

In real world there's always some level of error which comes from the measurement methodology, tools used etc. Even the original system can have unexpected behavior which makes the identification process very difficult.

There are many solutions of the problem of missing or incorrect data, and further researches are carried out. One of the ways to find a solution of this problem is regression.

2.1   Anscombe's Quartet

Anscombe's quartet comprises four datasets that have nearly the same statistical descriptive characteristics but they show that regression can be quite complicated problem. Anscombe showed that the graphical analysis is also important but it is necessary to mention that it is not always possible. Anscombe's sets are considered only in two dimensions, so the illustration is very easy to understand. It is difficult to imagine the high dimensional system in a form of graphical chart.

Anyway, this simple quartet of sets shows one important thing. There are two very important criteria two evaluate the regression quality. First is of course the average error. The second is reliability.

Figure 2 presents the Anscombe's quartet. The regression line is always the same independently from the data samples. In first case (top left) we see that the regression line is reliable. The second and fourth cases are unreliable although the same error and standard deviation are observed. An interesting feature can be observed in the third case. There is one sample which clearly deviates from the main set of data. We can suppose that this sample is probably result of an error during measurement. Of course in this case the number of data samples is low, so making conclusions is not so easy.

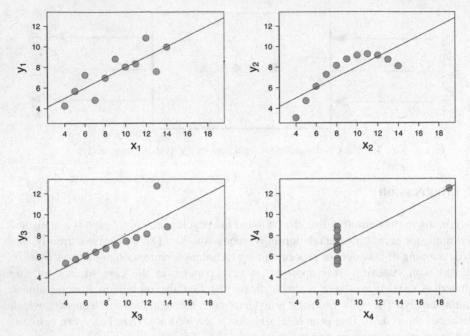

Fig. 2. Anscombe's quartet [2]

2.2 Types of Regression

In this paper three types of regression were chosen:

- linear regression
- ridge regression
- lasso regression

Linear regression fits a linear model with coefficients $w = (w_1, \ldots, w_n)$ to minimize the sum of squares between the observed responses in the dataset. Mathematically it can be expressed as: $min \, \| Xw - y \|^2$. This is the most often used method of regression and it's applied widely. Anyway, has one significant defect. It's very sensitive to measurement errors what will be proven in the experimental part.

Ridge regression solves partially the problem of linear expression by using an additional coefficient α. Now the formula to solve has a form: $min \, \| Xw - y \|^2 + \alpha \| w \|$ where $\alpha \geq 0$.

In lasso regression the objective to minimize is: $min \frac{1}{2n} \| Xw - y \|^2 + \alpha \| w \|$.

Of course, there are other methods mentioned in literature such multi-task lasso, elastic net or LARS models and more, but the three chosen were enough in the considered problem.

3   Event Log

The event log is a special data structure designed especially for process mining techniques. Table 1 illustrates a short example of the event log and the main data types are visible. The most popular file formats are: CSV and XES [3]. CSV is well known spreadsheet supported e.g. by MS Excel. XES is an official standard for process mining techniques based on XML structure. Each event log has different types of data. The most important is the timestamp data column, because the process always runs in a period of time. The source of the event log is the analysis of the process running. Every event, every activity is recorded and the obtained data is the base for the event log. If we're not able to record the data directly in XES format, there will be a need to convert it, especially when we're going to use tools such ProM (the most popular free of charge software designed at Technical University of Eindhoven) or some commercial applications such Disco (by Fluxicon).

Table 1. An event log example [3]

Case id	Event id	Properties			
		Timestamp	Activity	Resource	Cost ...
1	35654423	30-12-2010:11.02	Register request	Pete	50 ...
	35654424	31-12-2010:10.06	Examine thoroughly	Sue	400 ...
	35654425	05-01-2011:15.12	Check ticket	Mike	100 ...
	35654426	06-01-2011:11.18	Decide	Sara	200 ...
	35654427	07-01-2011:14.24	Reject request	Pete	200 ...
2	35654483	30-12-2010:11.32	Register request	Mike	50 ...
	35654485	30-12-2010:12.12	Check ticket	Mike	100 ...
	35654487	30-12-2010:14.16	Examine casually	Pete	400 ...
	35654488	05-01-2011:11.22	Decide	Sara	200 ...
	35654489	08-01-2011:12.05	Pay compensation	Ellen	200 ...
3	35654521	30-12-2010:14.32	Register request	Pete	50 ...
	35654522	30-12-2010:15.06	Examine casually	Mike	400 ...
	35654524	30-12-2010:16.34	Check ticket	Ellen	100 ...
	35654525	06-01-2011:09.18	Decide	Sara	200 ...
	35654526	06-01-2011:12.18	Reinitiate request	Sara	200 ...
	35654527	06-01-2011:13.06	Examine thoroughly	Sean	400 ...
	35654530	08-01-2011:11.43	Check ticket	Pete	100 ...
	35654531	09-01-2011:09.55	Decide	Sara	200 ...
	35654533	15-01-2011:10.45	Pay compensation	Ellen	200 ...
4	35654641	06-01-2011:15.02	Register request	Pete	50 ...
	35654643	07-01-2011:12.06	Check ticket	Mike	100 ...
	35654644	08-01-2011:14.43	Examine thoroughly	Sean	400 ...
	35654645	09-01-2011:12.02	Decide	Sara	200 ...
	35654647	12-01-2011:15.44	Reject request	Ellen	200 ...

The definition of the event log is given in [3]. It consists of traces and the traces consist of events. A trace is a finite sequence of events $\sigma \in \varepsilon^*$ such that each event appears only once. An event log is a set of cases $L \subseteq \mathbb{C}$ such that each event appears at most once in the entire log. The typical event log contains: case id, event names, timestamp, resources etc. The example of the event log can be given as follows:

$$L = [\langle a,b,c \rangle, \ \langle a,b,d \rangle, \ \langle b,b,c \rangle] \tag{1}$$

The equation above describes the event log L with three traces. The first trace consists of the events a, b and c. The second one are the events a, b and d. The last are b, b and c. The letters a, b and c are the shortcuts used in the place of full names of the events. Thanks to this simple procedure, the event log is easier to read and analyze.

In the event log given, some observations can be made. Two process instances begin with the event a. The third instance begins with the event b repeated twice (the loop is detected). Only the event b is present in all the traces. This event log is short and simple. It can be easily observed that the event a is probably the beginning action in the process. The final action is c. Some deviations are observed too. One trace begins with the event c, and one trace ends with the event d. There are four classes of the events – a, b, c and d. Three traces are present.

When the event log is given, then the process model can be constructed. Various techniques may be applied such induction mining, alpha algorithm (Petri nets), BPMN models etc. Most of them are described in [3].

3.1 Recording Data

As mentioned above the most important part of the event log are timestamps which are necessary for the process model construction. So it is very important to remember about this feature. If the timestamps are not present, then process discovery fails. Of course some data recovery techniques can be used, but the reliability of the model would be difficult to predict.

Data recording process is not always possible or easy to proceed. Every data may contain errors or lacks. Besides, the type of the data is significant. It forces to use the suitable data recovery techniques.

If the timestamps are given in the event log, then the problem of any data lacks or errors is less complicated. In some situations the conclusions can be directly taken from the characteristic of the process e.g. - if the student passes an exam, he will be evaluated, if the ship arrives to the port, it will leave it in some time.

Such easy predictions described above are not always possible. Then some other techniques should be used e.g. data mining, statistical methods etc. All these approaches are the elements of process mining [3] and are widely applied in process discovery. Anyway, it is always important to remember about the reliability. Let's discuss shortly the extrapolation. It is very useful in the case when the external part of the process is under the investigation. Extrapolation is always possible, but reliable only when it's domain is close to the process data.

The clear structure of the event log is given e.g. in [3]. In this paper the numerical data is the question of interest.

In current times the amount of the information is huge. Every activity can be connected with the data processing. The sources of the information are everywhere. The typical data sources are:

- business processes,
- machines,
- people,
- organizations,
- software systems (messages, transactions etc.),
- and more.

Independently of the data source it needs to be considered that the sources are usually heterogenous. So the preparation of the data requires prior processing.

4 Experiments

First part of experiment was to choose some datasets basing on the known analytical form with given mathematical equations. A random error was added to each data sample. Then the regression was carried out and the result compared to the expected answer of the modeled system.

Table 2. Examples of the systems with the known analytical forms [own study]

No.	Function	X and Y
1	$f(x) = x$	$X = [1, 2, 3, 4, 5, 6, 7]$ $Y = [1.2, 1.9, 2.2, 4.1, 5.4, 5.8, 7]$
2	$f(x) = 2x + 3$	$X = [1, 2, 3, 4, 5, 6, 7]$, $Y = [5.3, 7.1, 9, 10.3, 13.8, 15.1, 17.4]$
3	$f(x) = x^2$	$X = [0, 1, 2, 3, 4, 5]$ $Y = [0.4, 1.1, 3.9, 9.4, 16.1, 24.9]$
4	$f(x) = \sqrt{x}$	$X = [0, 1, 2, 3, 4, 5]$ $Y = [0.2, 1.1, 1.5, 1.8, 2.1, 2.4]$
5	$f(x) = e^x$	$X = [0, 1, 2, 3, 4, 5]$ $Y = [0.9, 2.6, 7.6, 19.9, 57, 141]$
6	$f(x) = \ln(x)$	$X = [0.1, 1, 3, 4]$ $Y = [-2.3, 0.3, 1.2, 1.3]$
7	$f(x) = \sin(x)$	$X = [0, 0.1, 0.2, 0.3]$ $Y = [0.05, 0.12, 0.17, 0.33]$
8	$f(x) = x^x$	$X = [0, 1, 2, 3]$ $Y = [0.9, 1.2, 4.4, 21]$

In the Table 2, the datasets are given. Knowledge of the analytical form of the modeled system makes the verification of it's correctness very easy. As it can be seen only two of them are linear (the model will be very easy to create). The others are nonlinear but in some conditions every system can be modeled using linear models. From the other side the nonlinear systems can be almost always decompsed to smaller local linear models. Real systems usually are nonlinear, but if there is no knowledge about the analytical form, the system needs to be identified. The identification process can be as simple as very complicated, depending on the problem being investigated.

Anyway, if we know that the system is linear or can be represented by a linear model with low value of the errors, the model construction process can be done faster supporting the reliability.

Table 3. Linear regression experiment [own study]

Function	a	b	x	y	$\hat{y}$	e [%]
1	1.01	−0.11	1.7	1.7	1.61	5.29
2	2.04	2.99	1.7	6.4	6.45	0.82
3	4.94	−3.06	2.1	4.41	7.32	66.05
4	0.41	0.50	1.9	1.38	1.27	7.86
5	25.03	−24.4	2.2	9.03	30.66	239.70
6	0.82	−1.53	1.3	0.26	−0.47	278.33
7	0.89	0.03	0.15	0.15	0.17	11.67
8	6.35	−2.65	1.5	1.84	6.88	274.25

Table 2 illustrates the experiments carried out using datasets given above. The results aren't unexpected. Linear systems are modeled in the proper way and the error is quite low. Modeling of the nonlinear systems is complicated, even when we know the analytical form of it. It leads to the obvious conclusion that the experiment should always be well designed (e.g. taking into account the Anscombe's quartet). Anyway, there are some approaches e.g. decomposition of nonlinear systems into small linear subsystems but this is not the subject of this paper.

The same data was used in ridge and lasso regression experiments. The results are presented in Table 3. This time only the error in [%] is shown. It has been proven that in some situations a lower value of errors is observed. Functions (1) and (2) are linear and the error for them is similar like in linear regression. In function (1) it can be observed that the parameter α is very useful and can decrease the value of error. Interesting thing can be observed in the case of (5) and (6) functions. The error is very high of course, but the highest value of α significantly decreases the error. Functions (7) and (8) show that it is not important what kind of regression is carried out.

If we are able to find an answer if the system is linear or not, different ways can be chosen. There are several approaches to solve it, and it depends on the characteristics of the data. Modeling of nonlinear systems using linear models is possible, but there is a need to consider the linearization or making some smaller linear submodels. Then the whole model can be constructed by creating a set of submodels.

The event log has it's own specific properties [4]. The only independent variable is always the time (some exceptions can be observed, but the time is the most important). So using regression in data recovery means looking for the answer, if in the given time

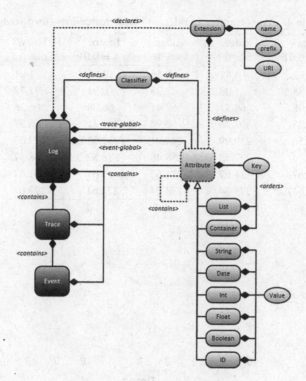

Fig. 3. The XES data format [3]

period the dependent variable changes in linear way. If yes, the only problem is the question of how many data samples should be used to create a model. If not, it should be considered, if there is a sense using linear regression. As mentioned above in some situations nonlinear systems can be modeled using linear algorithms and the same conclusion is applied to the event logs.

Let's consider the data $t = [0, 1, 2, 3, 4, 5, 6, 7, 8]$, $f(t) = [1, 2, 3, 3, 3, 4, 4, 3, 2]$. The variable t is the time and it's independent. The answer 4 for t = 4.5 is expected. Figure 3 illustrates the shape of the system. It can be seen that the system is nonlinear, but can be decomposed to even five small linear subsystems.

In Table 4 it is presented that the conclusions mentioned above are confirmed. The error is quite high because the data in the event log is not linear. Of course in some situations there is no knowledge if the data is correct e.g. the data may be linear but there were some errors during measurements. Therefore there is a need to remember that the good model never should have lower level of error that the real data, because some problems with the reliability may occur.

Table 4. Average errors in ridge and lasso regressions [own study]

Function	Ridge $\alpha = 0.01$	Ridge $\alpha = 0.1$	Ridge $\alpha = 1$	Lasso $\alpha = 0.01$	Lasso $\alpha = 0.1$	Lasso $\alpha = 1$
1	5.25	4.81	0.56	4.96	1.91	28.53
2	0.85	1.08	3.35	0.91	1.72	9.80
3	66.08	66.31	68.47	66.08	66.36	69.16
4	7.85	7.76	6.90	7.71	6.37	7.05
5	239.56	239.99	244.01	239.53	239.63	240.65
6	279.74	277.63	258.48	278.82	268.37	163.88
7	11.67	11.67	11.67	11.67	11.67	11.67
8	273.64	273.64	273.64	273.64	273.64	273.64

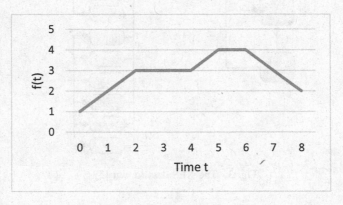

Fig. 4. The data from event log in the function of time f(t) [own study]

5   Conclusion

What is the aim of constructing of the process model? As it can be found in [3] the main reasons are:

- Insight – the modeller is triggered to view the process from various points of view;
- Discussion – important for stakeholders;
- Documentation – processes are documented;
- Verification – processes can be analysed;
- Performance analysis, animation, specification, configuration etc.

It has been proven that the data in the event log should be treated in the same way as in other data structures.

Figure 5 illustrates the problem of missing data. The data is extracted from the event log. The way of recording says that there is a need to choose the proper columns of the event log and the most important independent variable is the time. Besides, two types of errors can occur. Data lacks and wrong data. In the second type the researcher should consider why the error has occurred. It is also necessary to know that all the events are related to each other, so their order and appearance are important.

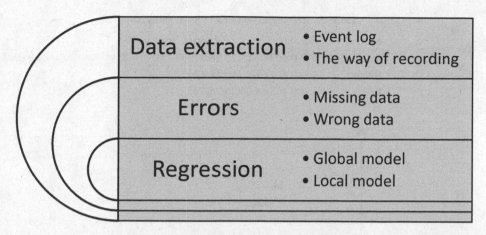

Fig. 5. Missing data problem [own study]

One of the ways of solving the problem of missing data is regression. It has been shown that there are different ways to this approach. Sometimes simple linear regression using least squares method can be carried out. Ridge and lasso regression add the additional coefficient marked α which makes the regression less sensitive to some errors. There is a need also to consider if the global model is possible to build. If not then some smaller local linear submodels may be created (Fig. 4).

The problem with the regression and other similar techniques is that it models very well especially linear systems. If the system is nonlinear then the problem is more complicated. The same thing happens when the multidimensional systems are under investigation. It is proven that the higher number of dimensions causes more problems in geometrical way. This problem is called the curse of dimensionality. There are several ways trying to solve it. One of them are significance indices (in some situations the regression can be used as this index too). It is assumed that in the whole system sometimes occur the situation that the data is not important, although it is present in the event log (or other datasheet). Then it can be omitted in further investigation. The same action occurs when the one input of the system is the function of the other one. In the case of more than one output, the system can be decomposed to more subsystems with one output (Table 5).

In this paper the regression was applied to solve the problem of missing data in the event logs focusing on the domain of transport. Of course the same technique can be used in other areas of science, engineering, economy etc.

All the regression method (linear, ridge and lasso) gave satisfactory results. It was shown that the linear regression is very sensitive to the single errors which are able to make the reliability of the model significantly lower.

It should be mentioned that the data recovery is not always possible. It depends strongly on the amount of the data lacks, measurement errors and many other factors. The distinction between such situations is not always easy and requires the experience from the researchers.

Table 5. The regression of f(t) [own study]

Method	Equation	Average error in [%]
Linear LSQ	$\hat{y} = 0.167x + 2.11$	28.47
Ridge	$\hat{y} = 0.167x + 2.11$	28.47
Lasso	$\hat{y} = 0.165x + 2.12$	28.49

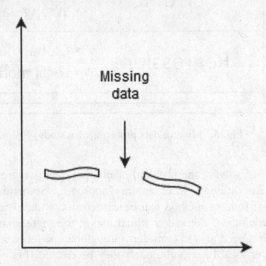

Fig. 6. Missing data problem [own study]

Figure 6 illustrates the simple system with some data missing. In this case the reconstruction of the model is easy because it can be seen that the model can be linear. Of course the system is nonlinear, but can be represented in the simplified form using e.g. regression methods. The reality is not always so clear. It confirms the fact discussed above.

Summing up, the regression can be applied to solve the problem of missing data in the transport event logs. The experiments showed that this approach was correct. It is also necessary to consider the other facts mentioned in this paper e.g. given above.

Numerical data is easy to recover, especially when the characteristic of the system is known or it can be assumed with the low error and high reliability. The further research will be carried out and the conclusions presented in the next papers.

References

1. Freedman, D.A.: Statistical Models: Theory and Practice. Cambridge University Press, Cambridge (2009)
2. Anscombe, F.J.: Graphs in statistical analysis. Am. Stat. **27**(1), 17–21 (1973)
3. van der Aalst, W.M.P.: Process Mining. Discovery, Conformance and Enhancement of Business Processes. Springer, Heidelberg (2011)
4. Dramski, M.: Extensible event stream format for navigational data. Sci. J. Marit. Univ. Szczecin **47**(119), 61–65 (2016)

Traffic Modelling on the Roundabout in the City of Žilina with Capacity Assessment, According to New Technical Conditions

Ján Palúch, Simona Kubíková, and Alica Kalašová[✉]

The Faculty of Operation and Economics of Transport and Communications,
University of Žilina, Univerzitná 1, 01026 Žilina, Slovakia
{jan.paluch,alica.kalasova,
simona.kubikova}@fpedas.uniza.sk

Abstract. Nowadays, road transport is on the rise. The number of vehicles are still increasing and road network in terms of capacity and permeability ceases to satisfy. The fundamental problem is overloaded communications, overcrowded junctions and big delays particularly in the cities and also significant economic losses are associated with this situation. Therefore, it is necessary to design junctions and communications with regard to future needs of population. In the Slovak republic is designing and capacity calculating of junctions based on technical conditions, "TC 04/2004" which is approved by Ministry of Transport and Construction of the Slovak Republic. At the end of 2015 the new document, "TC 16/2015" came into force and replaced the older one. In our paper, we will discuss the traffic modelling of roundabout in the city of Žilina, where in the peak hour, congestions periodically occur and the network is impermeable. The analysis and capacity assessment of this junction will be conducted according to both technical conditions TC 04/2004 and TC 16/2015. In conclusion, we design some possibilities how this situation can be solve.

Keywords: Technical conditions · Junctions · Transport survey · Capacity · Modelling

1 Introduction

People are moving according to their needs always from the origin to the destination of their travel. An origin of travel causes into productivity of the area and a destination causes into attractiveness of the area. It is very important to find out which mode of transport people prefer the most and assess the impact of transport modes on surrounding communications.

Shopping centers are very often one of the most common travel destination of citizens. In the city of Žilina, there is shopping center Dubeň, which contains lot of shops. An important link between this shopping center with parking and road communication leading around the resort is an adjacent roundabout. This roundabout should ensure qualitative, safe and smooth connection between road communication and parking slots. In the Slovak republic is designing and capacity calculating of junctions based on technical conditions "TC 04/2004" which is approved by Ministry

© Springer International Publishing AG 2017
J. Mikulski (Ed.): TST 2017, CCIS 715, pp. 121–130, 2017.
DOI: 10.1007/978-3-319-66251-0_10

of Transport and Construction of the Slovak Republic. This document was using for a quality assessment of traffic situation at the chosen junction. At the end of 2015 the new document "TC 16/2015" came into force and replaced the older one. In this paper, we compare these two documents and their applications in the transport modelling software Aimsun [1, 2].

2 Characteristic of the Chosen Roundabout

The chosen roundabout is situated at Obchodná and Vysokoškolákov streets, which are classified as local communications, in the city of Žilina. Shop OBI and a petrol station are located near this roundabout too. In Fig. 1, entrances to the intersection are shown. Entrances A and C are located on the Vysokoškolákov street, which connects the center of the city with settlement Vlčince. Entrances B and D are located on Obchodná street, which connects the shopping center also with the settlement Vlčince [3, 4]. On entrances B and C are situated also exits from parking slots near shop OBI and on the enter D from parking slots near the shopping center Dubeň.

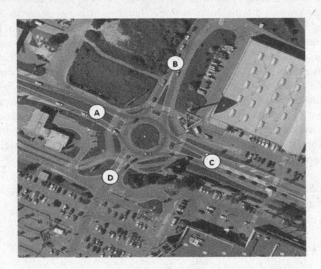

Fig. 1.   Marking of individual entrances of roundabout (own study based of [5])

3 Realization and Evaluation of Transport Survey

We have performed a profile transport survey on Wednesday 12th of October 2016. A temperature of air reached from 10 °C to 15 °C. The sky was cloudy and wind reached speed from 2 to 5 m/s. The survey was focused on traffic intensity on each enter to roundabout. We divided the transport survey into two parts regard to traffic variations. The first part we were performing in time from 6 a.m. to 11 a.m. and the second one from 1 p.m. to 6 p.m. [6]. Our task was to find out when during the day the intensity of traffic reaches the maximum value. The interval of counting we set on 15 min. Following table shows load of roundabout during the day.

Table 1. Intensities on the roundabout during transport survey with peak hour marking [own study]

Interval	Entrance				Sum
	A	B	C	D	
6:00–6:15	72	36	70	9	187
6:15–6:30	89	55	90	11	245
6:30–6:45	143	84	139	15	381
6:45–7:00	205	86	177	27	495
7:00–7:15	194	103	148	21	466
7:15–7:30	224	57	153	27	461
7:30–7:45	264	74	183	25	546
7:45–8:00	243	113	206	25	587
8:00–8:15	145	109	173	37	464
8:15–8:30	99	94	143	43	379
8:30–8:45	109	74	144	38	365
8:45–9:00	109	95	175	54	433
9:00–9:15	93	77	146	30	346
9:15–9:30	125	88	176	66	455
9:30–9:45	98	94	115	56	363
9:45–10:00	112	109	120	60	401
10:00–10:15	255	97	167	81	600
10:15–10:30	204	85	188	119	596
10:30–10:45	223	90	145	81	539
10:45–11:00	210	86	168	86	550
13:00–13:15	162	97	178	109	546
13:10–13:30	182	86	158	121	547
13:30–13:45	183	73	167	104	527
13:45–14:00	165	92	173	94	524
14:00–14:15	154	83	171	104	512
14:15–14:30	182	101	164	93	540
14:30–14:45	212	104	169	115	600
14:45–15:00	231	104	147	94	576
15:00–15:15	270	117	162	110	659
15:15–15:30	226	78	139	119	562
15:30–15:45	291	100	124	106	621
15:45–16:00	256	118	154	101	629
16:00–16:15	261	122	98	118	599
16:15–16:30	273	100	119	111	603
16:30–16:45	280	114	116	117	627
16:45–17:00	255	127	100	106	588
17:00–17:15	258	91	139	126	614
17:15–17:30	228	116	156	108	608
17:30–17:45	236	86	133	106	561
17:45–18:00	273	95	161	104	633
Sum	7794	3710	5954	3077	20535

The highest intensity of traffic during the day, the peak hour, is from 3 p.m. to 4 p.m. as you can see in Table 1. The whole intensity during the transport survey achieved value of 20 535 vehicles. During the peak hour 2 471 vehicles passed throughout the roundabout.

The next step was to perform another transport survey based on peak hour and determine vehicles routing. This transport survey was carried out on Thursday 20th of October. Only peak hour was recorded on the camera. We put the camera on tripod on the vehicle roof. We parked the vehicle on adjacent parking lot near the gas station. In this way, we ensured sufficient visibility of each entrance [3, 7, 8] (Fig. 2).

Fig. 2. Photography from the record of camera located on the vehicle roof [own study]

When the transport survey finished, we analysed the camera record. The main goal was to determine vehicles routing throughout the chosen roundabout. Table 2 shows a number of vehicles and their types.

Table 2. The composition of traffic flow in peak hour [own study]

Roundabout	Cars	Bus	Trucks	Cyclists	Sum
Entrance A	924	8	13	0	945
Entrance B	359	5	1	2	367
Entrance C	425	2	0	0	427
Entrance D	621	1	13	0	635
Sum	2329	16	27	2	2374

The highest intensity was recorded on the entrance A in direction from the city center. During an hour, the intensity has reached the value of 945 vehicles. In Fig. 3 there is the proportion of each types of vehicles occurred in traffic flow in peak hour.

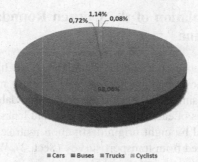

Fig. 3. The proportion of each types of vehicles occurred in traffic flow in peak hour [own study]

Routing of all vehicles from each origin to each destination points is shown in Table 3 and Fig. 4.

Table 3. Origin/destination matrice [own study]

Origin\destination	A	B	C	D	Sum
A		234	587	124	945
B	238		41	88	367
C	552	43		40	635
D	247	98	82		427
Sum	1037	375	710	252	2374

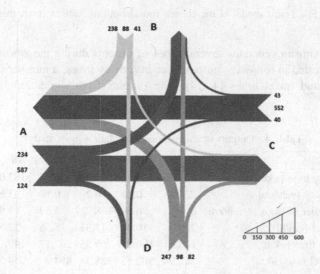

Fig. 4. Routing of vehicles throughout the chosen roundabout hour [own study]

4 Microscopic Simulation of the Chosen Roundabout in the City of Žilina

For microscopic simulation of the chosen roundabout, we have used software tool Aimsun. This software allows making microscopic and also macroscopic simulations [9, 10]. First step of our simulation was to draw the roundabout and create a traffic model. We defined a radius of each entrance and exit and lanes width (see Fig. 5). The traffic volume was defined by eight origin/destination matrices. We have used traffic volume, which was obtained from transport survey (Sect. 3). When everything was set up, we started the microscopic simulation [11–13].

Fig. 5. Traffic model of the chosen roundabout in Aimsun [own study]

Software Aimsun generates several types of outputs during the whole simulation. We were interested in following outputs: density, delay times, a number of stops, stop times, speed and travel times [4, 14]. The outputs from simulation are shown in Table 4.

Table 4. Outputs of microscopic simulation [own study]

	Cars	Trucks	Buses	Cyclists	All
Delay time [s/km]	181.26	228.28	250.31	125.37	182.09
Density [veh/km]	32.72	0.56	0.2	0.17	33.66
Number of stops [#/veh/km]	8	6.96	8.73	3.59	7.98
Speed [km/h]	22.62	19.41	17.31	8.36	22.51
Stop time [s/km]	148.61	190.2	209.35	262.17	149.76
Travel time [s/km]	252.94	302.53	328.34	484.2	254.56

In Table 4, there are average values of outputs. It is necessary to take into account a standard deviation when simulating, for each mode of transport (Table 5).

Table 5. Outputs of simulation with a standard deviation [own study]

Cars	Value	Standard deviation	Trucks	Value	Standard deviation
Delay time [s/km]	181.26	35.43	Delay time [s/km]	228.28	36.45
Density [veh/km]	32.72	4.9	Density [veh/km]	0.56	0.07
Number of stops [#/veh/km]	8	0.43	Number of stops [#/veh/km]	6.96	0.59
Speed [km/h]	22.62	0.99	Speed [km/h]	19.41	0.42
Stop time [s/km]	148.61	36.34	Stop time [s/km]	190.2	37.99
Travel time [s/km]	252.94	35.44	Travel time [s/km]	302.53	36.52
Buses	Value	Standard deviation	Cyclists	Value	Standard deviation
Delay time [s/km]	250.31	24.69	Delay time [s/km]	125.37	45
Density [veh/km]	0.2	0.05	Density [veh/km]	0.17	0.02
Number of stops [#/veh/km]	8.73	1.07	Number of stops [#/veh/km]	3.59	0.27
Speed [km/h]	17.31	0.77	Speed [km/h]	8.36	0.22
Stop time [s/km]	209.35	22.78	Stop time [s/km]	262.17	56.27
Travel time [s/km]	328.34	24.41	Travel time [s/km]	484.2	44.97

Table 5 shows that the buses and trucks have the highest average delay and cyclists have the lowest delay. It is mainly due to the fact that cyclists were recorded only 0.08% during whole survey, and they can manipulate their bikes easily [15]. They do not need as much space as bus or truck. On the other hand cycling shows the highest value of travel time 484.2 s/km. The Fig. 6 represents delay time running during the simulation [3, 7].

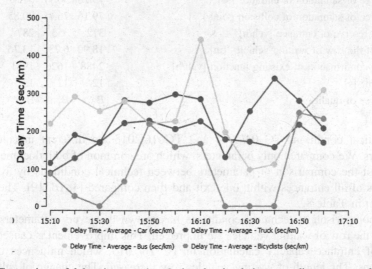

Fig. 6. Comparison of delay time running during the simulation according to mean of transport [own study]

Based on simulation outputs we can say that the roundabout does not have sufficient capacity reserve and congestions are occurred at the entrances and also at the exits
to/from junction. In this case, the total travel time is increasing. Therefore, it is
appropriate to make a capacity assessment of the chosen roundabout [8, 11].

5 The Capacity Assessment of the Chosen Roundabout and Comparison of Calculations According to Technical Conditions

We have used obtained information from transport survey to capacity calculations
according to technical conditions. Calculations were performed according to TC
04/2004 and TC 16/2015 and then they were compared. New technical conditions TC
16/2015 focused on roundabout exits, while the older one were more focused on
entrances. Table 6 shows the calculated parameters [18, 19] (Table 7).

Table 6. Calculated parameters according to TC 04/2004 (own study based on [1])

Parameter\entrance	A	B	C	D
Traffic intensity on exits [veh/h]	1037	375	710	252
Traffic intensity on entrances [veh/h]	945	367	635	427
Traffic intensity on round between exits and entrances [veh/h]	169	839	579	862
Distance between collision points [m]	23.04	23.04	23.04	20.16
α [°]	0.10	0.10	0.10	0.14
β [−]	1.00	1.00	1.00	1.00
γ [−]	1.00	1.00	1.00	1.00
Capacity of entrance [veh/h]	1257	720	922	702
The degree of saturation of entrance [%]	75.18	50.97	68.87	60.83
The degree of saturation of collision point [%]	79.16	76.41	80.85	81.64
Capacity reserve of entrance [veh/h]	312	353	287	275
Length of the row of waiting vehicles [m]	18.90	6.73	13.76	9.25
Capacity confirmation of existing junction [veh/h]	2488	1626	1981	1699
Delay time [s]	12	11	13	13
The degree of quality	B	B	B	B

Technical conditions TC 04/2004 and TC 16/2015 are different in calculation
parameters. We compared only parameters, which are common of both documents. We
carried out the comparison of parameters between technical conditions by averaging
the values of all entrances within one exit and then compared [4, 16, 19]. The results
are shown in Table 8.

The comparison of technical conditions has shown that every parameters except
length of the row of waiting vehicles are improved. This improvement is caused by the
change of entrance capacity calculation in TC 16/2015, which influences the next
calculations. The length of waiting vehicles row increased. The change of calculation
causes this negativity.

Table 7. Calculated parameters according to TC 16/2015 (own study based on [2])

Parameter\entrance	A	B	C	D
Traffic intensity on exits [veh/h]	1037	375	710	252
Traffic intensity on entrances [veh/h]	945	367	635	427
Traffic intensity on round between exits and entrances [veh/h]	169	839	579	862
Distance between collision points [m]	23.04	23.04	23.04	20.16
Critical time gap [s]	0.00	0.00	0.00	0.00
Average follow time gap [s]	0.00	0.00	0.00	0.00
Minimum time gap between vehicles on roundabout [s]	2.10	2.10	2.10	2.10
Basic capacity on entrance of roundabout [veh/h]	1433	789	1309	980
Capacity reserve	488	422	674	553
The degree of saturation [veh/h]	0.66	0.47	0.49	0.44
Assessment of congestion length [m]	34.02	15.44	16.79	13.76
Capacity of exit	1282	1500	1500	1349
Assessment of exit capacity [−]	0.81	0.25	0.47	0.19
Average delay time [s]	7.34	8.51	5.33	6.50
The degree of quality	A	A	A	A

Table 8. Comparison of technical conditions and outputs of Aimsun [own study]

Parameters\entrance	TC 04/2004	TC 16/2015	Aimsun
Capacity of entrance [veh/h]	900.25	1127.75	900.00
Capacity reserve of entrance [veh/h]	306.75	534.25	306.75
The degree of entrance saturation [%]	63.96	51.14	63.96
Length of the row of waiting vehicles [m]	12.16	20.00	–
Average delay time [s]	12.25	6.92	18.63
The degree of quality	B	A	B

6 Conclusion

The main goal of our paper was to compare outputs of microscopic simulation with capacity assessment of roundabout according to TC 04/2004 and TC 16/2015. Within the microscopic simulation, we created traffic model of the chosen roundabout. The next step was to process simulation outputs, and compared them with capacity assessment according to technical conditions. The basic parameters for comparison were entrance capacity, entrance capacity reserve and an average delay time. Entrance capacity increased due to changing the capacity calculation according to TC 16/2015. This increasing of entrance capacity influenced the degree of quality of movement on roundabout from level B to level A. However, the simulation outputs are more closely to calculated values according to TC 04/2004. We can attribute these deviations to the fact that we used fixed values of deceleration and acceleration of vehicles in microscopic simulation [10, 17, 18].

Acknowledgements. Centre of excellence for systems and services of intelligent transport II, ITMS 26220120050 supported by the Research & Development Operational Programme funded by the ERDF.

References

1. Technical conditions TC 04/2004: Designing roundabouts on road and local communications, Ministry of Transport and Construction of the Slovak Republic
2. Technical conditions TC 16/2015: Calculation of road communications capacity, Ministry of Transport and Construction of the Slovak Republic
3. Ondruš, J., Černický. Ľ.: Usage of Polcam device for parameter monitoring and traffic flow modelling. Commun. Sci. Lett. Univ. Žilina **18**(2), 118–123 (2016). ISSN 1335-4205
4. Černický, Ľ., Kalašová, A., Mikulski, J.: Simulation software as a calculation tool for traffic capacity assessment. Commun. Sci. Lett. Univ. Žilina **18**(2), 99–103 (2016). ISSN 1335-4205
5. www.mapy.cz. Accessed 20 Jan 2017
6. Gnap, J., Konečný, V.: The impact of a demographic trend on the demand for scheduled bus transport in the Slovak Republic. Commun. Sci. Lett. Univ. Žilina **10**(2), 55–59 (2008)
7. Iwan, S., Małecki, K., Stalmach, D.: Utilization of mobile applications for the improvement of traffic management systems. In: Mikulski, J. (ed.) TST 2014. CCIS, vol. 471, pp. 48–58. Springer, Heidelberg (2014). doi:10.1007/978-3-662-45317-9_6
8. Cyprich, O., Konečný, V., Killianová, K.: Short-term passenger demand forecasting using univariate time series theory. Promet Traffic Transp. **25**(6), 533–541 (2013)
9. http://www.aimsun.com/wp/. Accessed 20 Jan 2017
10. Barcelo, J., et al.: Microscopic traffic simulation: a tool for the design, analysis and evaluation of intelligent transport systems. J. Intell. Robot. Syst. **41**(2–3) (2005). doi:10.1007/s10846-005-3808-2
11. TSS-Transport Simulation Systems: Aimsun 8 adaptive control interfaces manual
12. TSS-Transport Simulation Systems: Aimsun 8 dynamic simulator users manual
13. Smith, J., Blewitt, R.: Traffic modelling guidelines. TfL traffic manager and network performance best practice. Version 3.0. http://www.tfl.gov.uk/assets/downloads/traffic-modelling-guidelines.pdf. Accessed 20 Jan 2017
14. Gnap Skrúcaný, T., Šarkan, B., Gnap, J.: Influence of aerodynamic trailer devices on drag reduction measured in wind tunnel. Ekspolatacja i niezawodność Maint. Reliab. **18**(1), 151–154 (2016)
15. Földes, D., Csiszár, C.: Conception of future integrated smart mobility. In: Koukol, M. (ed.) 2016 Smart Cities Symposium Prague (SCSP). Konferencia helye, ideje, Praha, Czech Republic, 26–27 May 2016, pp. 29–35. IEEE, New York (2016)
16. Novák, A., Mrazová, M.: Research of physiological factors affecting pilot performance in flight simulation training device. Commun. Sci. Lett. Univ. Žilina **17**(3), 103–107 (2015)
17. Turiak, M., Novák-Sedláčková, A., Novák, A.: Portable electronic devices on board of airplanes and their safety impact. In: Mikulski, J. (ed.) TST 2014. CCIS, vol. 471, pp. 29–37. Springer, Heidelberg (2014). doi:10.1007/978-3-662-45317-9_4
18. Poliak, M.: The relationship with reasonable profit and risk in public passenger transport in the Slovakia. J. Econ. **61**(2), 206–220 (2013)
19. Poliak, M., et al.: Defining the influence of the support of bus service on road safety. Commun. Sci. Lett. Univ. Žilina **18**(2), 83–87 (2016)

Telematics Systems in Supply Chains

Janusz Figura[⊠] and Karolina Lubieniecka-Kocoń

University of Economics, 1 Maja 50, 40-287 Katowice, Poland
{janusz.figura,lubieniecka}@ue.katowic.pl

Abstract. The goal of the hereby article has been to present and to identify chosen aspects of telematics system, as well as presenting them as support elements of contemporary supply chain both treated as a whole and in the situation of examining their parts. Moreover, the authors have shown the integrity of telematics systems, then proceeding to analyze the model of contemporary supply chain, which has been presented as the platform for applying telematics systems. In the final part of the article there has been presented the attempt to identify telematics systems in supply chains and to present chosen aspects of the performed research.

Keywords: Telematics systems · Supply chain

1 Introduction

When analyzing contemporary challenges supply chain faces, one needs to remember about the possibilities offered by various tools to deal with the challenge. As there are multiple problems, issues and questions supply chain ought to undertake and answer, there seem to be unlimited range of support ready to be used to solve potential difficulty or a problem. Even if contemporary perspective forbids referring to problems and chooses to face challenges instead, there seem to be plenty of notions world economy demands to focus on.

Supply chains used in world economy are of complex structure and consists of many tasks and operations. Each of the stages to perform bears the risk of inefficiency, thus limiting or slowing down the effect all chain focuses on. Therefore there are multiple danger or risk areas in such performance that are being identified and various questions concerning the supply chain effectiveness are asked in order to help improve work performance.

It seems prudent not to forget telematics from among the range of support tools offered to be used in supply chain planning and monitoring. Telematics may be the perfect answer for many of the challenges every-day work performed within supply chain offers, thus helping to facilitate the work and improve its effects, either by shortening the time needed for it or upgrading the smoothness of each tasks made.

Work performance ought to be monitored, analyzed and examined for many reasons, obviously financial benefits being one of them, however, other aspects, such as improved performance quality, raised safety level or upgraded green footprint seem to be equally important, even if they operate in longer time horizon.

© Springer International Publishing AG 2017
J. Mikulski (Ed.): TST 2017, CCIS 715, pp. 131–141, 2017.
DOI: 10.1007/978-3-319-66251-0_11

2 The Meaning and the Range of Telematics Systems as Support Elements of Contemporary Supply Chains

2.1 Telematics Systems

In order to describe the role telematics systems play in the performance of contemporary supply chains, one ought to declare the range of definition of telematics systems, as well as the meaning of supply chain. Thus, it seems important to refer to some of the explanations of telematics, especially in the context of supply management.

It is claimed [1] that telematics used in automotive sector is defined as the information-intensive applications that are being enabled for vehicles by a combination of telecommunications and computing technology. They highlight that telematics systems focus on processing the data, capturing it, storing and exchanging in order to achieve the goal which is obtaining remote services [1]. From such perspective, the most important part of the telematics system seem to be the information stored by IT applications with the help of proper infrastructure, which may be used in constantly updated version to create the picture of the situation in which potential user is, or in which the objects, such as the application user's goods may be.

On the other hand, it is said that *telematics works at the interface of road vehicles and the road network, helping to ensure the former make the most efficient and rational use of the latter* [2]. Such perspective seem to focus more on hardware, nonetheless keeping in mind the main goal of launching telematics systems into the "service" of its automotive use, gaining increased efficiency level of using technically enhanced tools.

From more general perspective, telematics is defined [3] as *the support of interaction between people and/or processes while bridging distance and/or time, through the integrated application of information and telecommunication technology*. That seems to be the widest of all applied definitions, allowing to perceive the use of telematics systems in broader light. Telematics obviously may be defined in many ways and from various perspectives, yet for the sake of the supply chain management it seems prudent to perceive it mainly as a support tool, extremely crucial one in contemporary economy demands, which help to realize supply processes for great distance in a manner and quality of the service that would not be possible if such tool were not introduced.

Thus, with the focus on its use in case of managing the transport of the supply chain, one may claim telematics systems connect vehicle technology and fleet management together with computer science through navigation systems, vehicle applications, automotive navigation systems and the road infrastructure, combining constantly updated knowledge on the vehicle in move situation with the experience on the most advisable method of managing the fleet involved in the supply chain.

Thanks to telematics systems, the efficiency of managing the information acquired in constant flow of data would not be possible to such a great extend and would definitely not be so useful in managing supply chain transport. telematics seem to be one of the most extensively used equipment to ensure proper work of vehicles.

2.2   Supply Chain

How to understand the notion of supply chain? There are numerous definitions, both concerning the supply chain itself as well as its management. One of the most general dictionary definitions [4] declare that supply chain is:

The movement of materials as they flow from their source to the end customer. Supply Chain includes purchasing, manufacturing, warehousing, transportation, customer service, demand planning, supply planning and Supply Chain management. It is made up of the people, activities, information and resources involved in moving a product from its supplier to customer.

Moreover, the Canadian Study Supply Chain Sector Council [5] underline that

Definitions of a "supply chain" virtually universally encompass the following three functions: supply of materials to a manufacturer; the manufacturing process; and, the distribution of finished goods through a network of distributors and retailers to a final customer. Companies involved in various stages of this process are linked to each other through a supply chain.

The Council [5] highlights the meaning of the information in supply chain maintenance:

To facilitate the flow of products, information is shared up and down the supply chain, i.e. with suppliers and clients. This sharing of information enables all parties to plan appropriately to meet current and future needs. Numerous goals can be achieved through successful supply chain management:

- inventory can be minimized
- costs can be reduced
- product time to market can be improved
- flexibility can be enhanced

The more the companies within a supply chain are able to integrate and coordinate their activities, the more likely they'll be to optimize the flow of goods from supplier to customer and to react efficiently to changes in demand.

Hence the importance of using properly adjusted tools to improve the effectiveness and efficiency of the supply chain. When supply chain is perceived from the perspective of cohesive activity, the value of tool enabling smooth action is even greater. It is declared [6] that:

Supply chain management is the integration of key business processes from end user through original suppliers that provide products, services, and information that add value for customers and other stakeholders.

There also appear consumer focused definitions [7], where the supply chain is defined as:

The management of upstream and downstream relationships with suppliers and customers to deliver superior customer value at less cost to the supply chain as a whole.

One ought to mention one more, highlights extremely important factor, that is that supply chain and its management is currently mostly based on IT technologies [8]:

As a result, SCM consultants and researchers are building models in one limited field, often without a deeper knowledge of established theory, practical usefulness, economic benefits, or the effects of their developments on the system as a whole. In today's world, businesses are shaped by complexity, fast-changing conditions, and constant development. (...) Companies are trying to respond to dynamic developments and complexity, striving to achieve stability and to carry out operations more efficiently. The goal of IT development, to a great extent, is to create a better (which often means simpler and easier) way to conduct business.

Thus, when connecting the goal of IT development from business perspective and the aim of supply chain management, one ought to consider the support telematics system could offer to supply chains.

2.3   Support Telematics Systems Offer to Supply Chains

There are many tools enhancing supply chain efficiency; however, telematics seem to be one of the most important ones. Obviously, the range and the area of usefulness of such tool may vary depending on the nature of the supplied goods. Having cold storage as example, portal SupplyChain247 explains [9]:

Telematics continue to help cut costs for fleets, but one segment of the industry is keenly feeling the impact: cold storage applications. In these environments, inefficiency of any kind creates tremendous expenses, and fleet management solutions are paying off.

Any way to improve the work with simultaneous lowering of the expenses seem to be desired one, cold storage may be a good example (followed below) as the expenses in such type of supply are of high level.

The help telematics offer is often described as the internet of things. Taking storing process as an example [10]:

The modern lift truck, for instance, is outfitted with a ton of sensors that collect and communicate operational data from the truck to other systems and people. As distribution centers become more automated and order fulfillment requirements become more complex, conveyors, sorters, automated storage systems and automatic vehicles are talking to one another as well as the warehouse control and management systems.

What can be reported as telematics supporting the supply chain? One may mention just the following [11]:

- improve warehousing work,
- increase safety in driving,
- increase safety in mining industry,
- enhance cargo management,
- control drivers' performance,
- improve fleet management,
- speed up the flow of stages in rescue operations.

There are just a few of potential area of improved efficiency of supply chain work thanks to the support offered by telematics.

3 The Identification of the Telematics Systems in the Supply Chains

It seems difficult to imagine contemporary supply chain functioning without support of any of telematics systems. From the most basic devices providing connection to the most developed satellite location systems, telematics systems determine the functioning of current supply chains. Identifying telematics systems in the supply chains is not an easy task. Implementing, maintaining and developing telematics systems is a complex, interdisciplinary area within technical, organizational and financial fields [12]. Identifying requires professional combining of supply chains telematics aspects within the area of lead supporting, managing, navigation and safety. The basic subjects of telematics there one may then include identifying, data transmission, applying sensors, following the database, software and the coupling systems. Usually, telematics systems are identified due to their technological and technical action possibilities based on devices and their tailored applications, such as the following:

- GSM mobile networks.
- Internet WLAN networks.
- GPS.
- Radio connection RDS-TMC.
- Threshold data.
- Road traffic monitoring (sensors, detectors, cameras, radars).
- Weather monitoring.
- Passing data to the transport system users (variable message signs).

However, scientific interests of the Authors have focused not on presenting specific telematics systems and their functional applications, as there are many such types of papers. Telematics systems have been perceived as a homogeneous object – an element extracted from the system as the one responsible for realizing telecommunication and IT solutions. In the process of the research the following five scientific aspects were focused on:

- Identifying the concentration of telematics systems being present in the subject structure of the supply chain,
- Identifying telematics systems within transport structure of the supply chain,
- Indentifying telematics systems in the functional structure of the supply chain,
- Indentifying telematics systems as the support element of the managing function,
- Identifying the application specialization of telematics systems in the supply chain.

The chosen results of the empiric studies concerning the allocation of the telematics systems together with the characteristics of the run examination are presented in the following part of the paper.

4 The Chosen Results of Empiric Studies Concerning the Allocation of the Telematics Systems in the Supply Chains

The aim of the conducted research was to indentify the method telematics systems appear as objects within the supply chains. Pilot research was conducted in December 2016 and January 2017 among the group of purposely chosen 59 entrepreneurs functioning in the conditions of Polish transport-freight forwarding – logistics market. Likert 5-grade scale (where 1 means minimum and 5 maximum) was used to assess telematics systems identification there.

The first of the examined areas was the issue of identifying the appearance of telematics systems within the structure of the supply chains. The chains that were chosen to the research were composed of five elements that is supplier – producer – distributor – retailer – customer. The result of the research show that the greatest degree of intensity of telematics systems appearance in the supply chains are among distributors: 32% and retailers: 29%. Both the distributors and the retailers are indirect parts of the supply chain and their combined intensity of telematics systems using is there as much as 61%. The degree of using telematics systems in the group of suppliers was at the level of 19% and in the group of producers: 11%. The smallest degree of intensity of telematics systems use was noted among the customers – 9%: see Fig. 1.

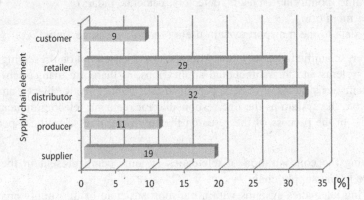

Fig. 1. Areas of telematics support intensity in the supply chain structure [own study]

The result of the research on the degree of intensity of the telematics systems use in the supply chain structrure show that telematics systems have been intensively used, especially by the distributors and retailers, as shown on the Fig. 1. It is therefore worth considering the reason of dominating presence of telematics systems among distributors and retailers in the examined supply chains. It is also possible to show the most probable aspects that determine high intensity of supply chains use in the given supply chain. Such considerable degree of telematics systems supprot in the distribution stage may be explained in several ways. Distribution and retail stages are characterized by the major dergree of the supply operationalization and include the activities connected with covering the space, time, quantity and assortment differences that take place among the production zone and consumption. Therefore, telematic systems in distribution and retail stages support crucial functions of coordinating and organizing the supply of goods.

Another factor that may influence so great degree of telematics systems presence is the type of realized model, especially when taking distribution into consideration. Intensive distribution that includes the supply of goods in as many packing centres as possible enforces considerable degree of coordination and organizing the delivery functions.

When taking into account selective distribution, it is admittedly characterized by lower number of packing centres where the product is available, nonetheless, the intensity degree of coordination and organizing functions and thus of the level of telematics systems support is conditioned by the means of goods supply, ways of packing, transport route optimizing, warehousing time, the complexity of handling operations etc. The level of telematics systems use in the model of exclusive distribution may result from the number of distributors and retailers that serve the given area, who have exclusive rights to sell given assortment of goods. Thus, the intensity level of telematics system support in the examined supply chain may be explained by the distribution model in use, which would be supported by the specific telematics system. It seems reasonable then, in separate, in-depth research, to examine the influence of the used telematics systems on the model of goods distribution within the supply chain.

Another of the examined areas of the research was the notion of telematics systems identification within the transport structure of the supply chain. Three elements were chosen to be examined: the means of transport, the cargo and the route of the transport, which create the transport structure of each of the supply chain. The results of the research show that using telematics systems in relation to the means of transport is of dominant: 40%, then in relation to the cargo: 32% and the route: 28%, as shown in the Fig. 2.

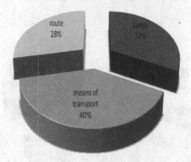

Fig. 2. The identification of telematics systems within transport structure of the supply chain [own study]

The identification of telematics system in the function structure of the supply chain was the following aspect of the research. There were five elements connected with the supply chain that were chosen for the examination: orders, packaging, transport, warehousing, outlets. The result of the examination show that two of the supply chain elements where telematics systems are of key importance are dominant: those are warehousing: 32% and outlets: 26%. Lower intensity of telematics systems support was noted in the rest of the supply chain elements, such as transport: 17%, packaging: 14% and orders: 11%, as presented in the Fig. 3.

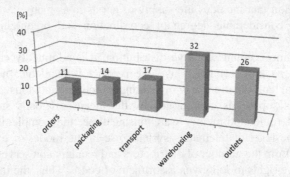

Fig. 3. The identification of telematics systems within function structure of the supply chain [own study]

The next of the examined aspects of the research was the issue of indentifying telematics systems as the support element for the managing function. Four functions which support managing were chosen for the examination: planning, organizing, controlling and leading.

The result of the study point to two main areas of managing function where telematics systems are of great importance, that is organizing: 36% and planning: 32%. It seems worth noticing the asymmetry of the intensification towards two other managing functions, where telematics systems appear to be of much smaller importance: in controlling: 21% and in leading: 11%, as shown in the Fig. 4.

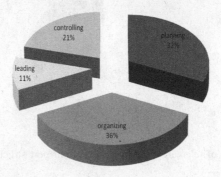

Fig. 4. The identification of telematics systems as support elements of the managing function in the supply chain [own study]

The last of the examined aspects of the research was the identification of the specialization of telematics systems implementation in the supply chain. It appears telematics systems specialization within the examined group of transport-freight forwarding – logistics market is mainly focused on improving the management of the transport means: 31%. The conduced research show that within telematics systems specialization area in the supply chains there appear to be relatively evened level, especially in three of the areas, that is improving health and safety: 20%, enhancing cargo management 19% and monitoring staff work 17%. The lowest level of the specialization of telematics systems was noted in the area of improving work operations in warehouses: 13%, as presented in the Fig. 5.

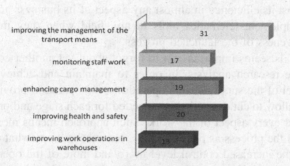

Fig. 5. The identification of telematics systems specialization in the supply chain [own study]

The identified aspects of telematics systems support in the supply chains are only some of the field that is to be analyzed. On the basis of the conducted research is appears challenging to form unambiguous conclusions, even more so, as the research methodology has covered only pilot study and the range of the indentified aspects of telematics systems support has concerned only the chosen notions.

Nonetheless, from among known studies on telematics systems supporting the supply chains there seem to be some of the remarks to be considered. Those are as follows:

- Telematics systems constitute the active element of shaping the structure and the functionalities of the supply chains,
- The allocation of telematics systems in the supply chains is asymmetric,
- Telematics systems are shaping positive effect of the goods transported via supply chains,
- The development of the telematics systems in the supply chain consitutes the active element of support.

5 Conclusion

Telematics systems are crucial in many aspects of contemporary economics. Help they provide and the approach to the methods of conducting various aspects of daily business activities are not to be overrated. With such tools not only is the work smoother, but there appear more advanced means of spread positive control over performed activities. Thanks to telematics systems support the work performance becomes facilitated, more developed, simply-easier to manage.

The same aspects of implementing telematics tools are valued when analyzing the performance in the supply chains work. One may say telematics system perform there a role of a facilitator, enabling more flawless passing from one stage to another, also speeding up the flow of operations and providing constantly updated stream of information, thus offering control and facilitation tools over each stage of the processes within the supply chains.

The areas pointed to as crucial be the participants of the conducted study seem to indicate that the most important for the entrepreneurs is the company performance and the means to boost its efficiency in almost any aspect of its business activity, therefore they seem to implement telematics tools in any field when they may promise the increased effectiveness of the launched process.

Nevertheless, is seems important not to forget about yet another conclusion which appears from the research analysis. In order to maintain and achieve the increased performance level of the supply chains operations, the operation providers implement the tools which allow to cut down the time needed for each stage and provide increased control level over every aspect of the work. Such approach means most often as great mechanization of the process as possible; thus, one reaches somewhat of a paradox: in order to spread the increased control level more and more of the operation aspects is taken from human hands into machine network, at the same time depriving humans not only from the possibility to perform the task (desired effect) but also from the possibility to stop the process immediately when emergency happens or to reflect some of the performance stages. Nobody forgets about control mechanism of the Internet of things performance, however, when there is nobody watching the watchers, the space for uncontrolled mechanisms and actions appear. Such potential risk area should also be taken into consideration when launching most effective, modern solutions, telematics systems being one of them.

References

1. Duri, S., et al.: Framework for security and privacy in automotive telematics. In: Proceedings of the 2nd International Workshop on Mobile Commerce, WMC02, pp. 25–32 (2002)
2. Walker, G., Manson, A.: Telematics, urban freight logistics and low carbon road networks. J. Transp. Geogr. **37**, 74–81 (2014)
3. Tettero, O., et al.: Information security embedded in the design of telematics systems. Comput. Secur. **6**(2), 145–164 (1998)
4. Supply chain definitions. http://www.supplychaindefinitions.com/. Accessed 2 Feb 2017
5. Canadian Supply Chain Sector Council: http://www.supplychaincanada.org/en/supply-chain. Accessed 2 Jan 2017
6. Lambert, D.M., Cooper, M.C., Pagh, J.D.: Supply chain management: implementation issues and research opportunities. Int. J. Logist. Manag. (1998). Enarsson, L. (ed) What do We Really Mean by Supply Chain Management? http://www.supplychainquarterly.com/topics/Logistics/scq200901book/. Accessed 2 Feb 2017

7. Christopher, M.: Logistics and supply chain management. Enarsson, L. (ed) What do We Really Mean by Supply Chain Management? Prentice Hall, London (1998). http://www.supplychainquarterly.com/topics/Logistics/scq200901book/. Accessed 2 Feb 2017
8. Enarsson, L.: Supply chain management: just a simple system, or a determining solution? Paper given at the 15th International Conference on Production Research. University of Limerick, Ireland. In: Enarsson, L. (ed.) What do We Really Mean by Supply Chain Management? (1999). http://www.supplychainquarterly.com/topics/Logistics/scq200901book/. Accessed 2 Feb 2017
9. Bond, J.: Telematics add degrees of efficiency to cold storage. http://www.supplychain247.com/article/telematics_add_degrees_of_efficiency_to_cold_storage/Telematics. Accessed 2 Jan 2017
10. Trebilcock, B.: Automation: what happens when machines talk to machines? http://www.supplychain247.com/article/automation_what_happens_when_machines_talk_to_machines/Telematics. Accessed 2 Jan 2017
11. Logis Advantech. http://advcloudfiles.advantech.com/ecatalog/2015/08111416.pdf. Accessed 2 Jan 2017
12. Mikulski, J.: Contemporary situation in transport systems telematics. In: Mikulski, J. (ed) Advances in Transport Systems Telematics, Vol. 2. Silesian University of Technology, Katowice (2007)

Impact of Intelligent Transport Systems Services on the Level of Safety and Improvement of Traffic Conditions

Jacek Oskarbski$^{(\boxtimes)}$, Tomasz Marcinkowski, and Marcin Zawisza

Gdansk University of Technology,
Ul. Narutowicza 11/12, 80-233 Gdańsk, Poland
joskar@pg.gda.pl

Abstract. The positive effects of the services of Intelligent Transport Systems (ITS) on the level of transport systems operation was confirmed by long-term studies conducted, inter alia, in the USA, Japan and Europe. Benefits resulting from the application of ITS services can be presented through performance indicators. The indicators represent in a numerical or qualitative manner to what extent ITS services can contribute to improving the safety of travellers and the quality of travel, upgrading the efficiency and reliability of the transport system, the efficiency of transport services providers, energy saving and environmental protection. The paper presents the impact of selected ITS services on the safety and efficiency of traffic, as well as indicators describing that impact. ITS services were specified according to the standard classification of services, prepared within the framework of the RID 4D research project "The impact of the use of Intelligent Transport Systems services on the road safety level," funded by the National Centre for Research and Development and the General Directorate for National Roads and Motorways (GDDKiA).

Keywords: Intelligent transportation systems · ITS service · Road safety

1 Introduction

The purpose of ITS services is to improve the efficiency, effectiveness, reliability and safety of the transport system. Effective implementation of ITS measures depends in part on the knowledge which of them most effectively solve problems related to road congestion and safety. Therefore, it is important to understand the real benefits of operation of existing and implementation of new technologies. Based on documented experiences gathered from available literature the paper presents the results of studies on the effects of selected ITS services in European countries, the USA and Japan. This paper presents the indicators that enable an assessment of the impact of the proposed solutions of Intelligent Transport Systems on road safety and improvement of traffic conditions. Additionally, the authors analysed - based on the diagnosis of the traffic status - the possibility of the impact of existing and planned ITS services on improving the safety and efficiency of traffic in the area of the National Road Traffic Management System (Krajowy System Zarządzania Ruchem, - KSZR). This impact is presented on

© Springer International Publishing AG 2017
J. Mikulski (Ed.): TST 2017, CCIS 715, pp. 142–154, 2017.
DOI: 10.1007/978-3-319-66251-0_12

the basis of a benchmark classification of ITS services developed within the framework of the RID 4D research project "The impact of the use of Intelligent Transport Systems services on the road safety level."

2   Effects of ITS Services Application

Advantages of using ITS systems are defined by various performance indicators. These indicators represent the way in which ITS systems can improve safety and mobility of the traveller, efficiency of the transport system, performance of providers of transport services, energy saving and environmental protection. These measures include:

- Safety: direct safety measures may include the number of accidents and changes in their number, the number of injuries and fatalities (absolute measurements). Direct relative measures can be analysed as micro indicators (automotive, transportation, severity) specifying the number of events in relation to the volume of traffic (miles travelled), the number of trips or proportion of involved fatalities or seriously injured persons in the total number of traffic incidents; and macro indicators (e.g. the density of accidents in the defined area or road section). Safety analyses also apply indirect measures which include traffic parameters and their dynamic range (e.g. vehicle speed, speed variation, changes in numbers of infringements of road safety traffic rules, rescue operation time, as well as drivers' and pedestrians' behaviour measures). Safety can also be examined by defining social or individual risks [1];
- Mobility: these measures may include travel time, variability of travel times (travel time reliability), variable speed (traffic flow), time to restore normal/typical traffic conditions (from the point of view of the user and efficiency of the transport system), miles travelled;
- Capacity: measured by the maximum number of people, goods or vehicles passing a point in the road (junction, perimeter or the reference point of the road network per unit of time), as well as traffic conditions (e.g. number of stops, number of vehicles/persons in the queues, time lost);
- Satisfaction of a service user (traveller, supplier of goods): related to the choice of means of transport and the quality of service measured by the level of satisfaction. Typical results of satisfaction with services provided include: assessment of professionalism of the service provided, meeting the traveller's expectations, quality of use, as well as the level of service efficiency and reliability;
- Productivity: activities related to sufficient operational performance and security of the service costs [2];
- Energy and environment: measures include changes in the level of pollutant emissions (carbon dioxide and carbon monoxide, nitrogen oxides, hydrocarbons and volatile organic compounds) and energy consumption [2].

The efficiency of road traffic is closely related to the aforementioned indicators. Traffic performance is upgraded by increasing safety, improving mobility (reduction of travel time), increasing road capacity and use of public transport vehicles, reducing the cost of freight, reducing the negative impact of road traffic on the environment, as well as satisfaction of the road user (ITS services are to make, among other things, the journey more pleasurable and less tiresome).

The result of the implementation of ITS services can be a balance between all the indicators, e.g. improvement of the level of safety in such a way as not to cause a significant increase in travel time which translates directly into user satisfaction. On the other hand, the capacity should improve along with shortened travel time (taking into account all travellers, including cyclists and pedestrians) - but not at the expense of safety. In conclusion, the implementation of a new ITS service should consider comprehensively all aspects of traffic, so that the overall balance of such implementation was favourable, which requires developing methods defining such balance.

Currently, in Poland there are not methods allowing for a comprehensive assessment of the implementation of individual ITS services which significantly hinders the development planning of ITS services on transport networks and makes it impossible to make optimal decisions take into account the costs of implementation and maintenance of these services. The planned development of multi-method within the RID-4D project can improve decision-making processes in the field of ITS services location and development, taking into account the potential benefits of their use.

3 Efficiency of Selected ITS Services

The following are examples of ITS services that can improve safety and efficiency of transport systems to the greatest extent.

3.1 Electronic Tolling Systems

The literature review indicates that in its early years the introduction of intelligent systems of road tolling resulted in an increase in the number of accidents caused by drivers' confusion and ignorance concerning these services. Currently, due to the greater availability of these systems and greater awareness of drivers, smart tolling leads to increased safety and reduced number of accidents in the areas of their application. However, the presence of road charges leads to migration of traffic onto lower standard roads, which in turn involves the risk of increasing the number of accidents on such alternative roads [3].

Implementation of the Open Road Tolling (ORT) may provide better performance with respect to charges collection than a conventional construction of a toll site. The technology of Automatic Vehicle Identification (AVI) for electronic toll collection system (ETC) is a concept that has revolutionised the way of tolling, and the ever increasing volume of traffic on the tolled roads results in more frequent application of such innovative methods as ORT. For example, following the introduction of ORT at the University Toll Plaza they observed [4]:

- reduction of the average time losses for cash paying customers by 49.8%, while in case of customers using the automatic toll collection - by 55.3%;
- increase in the average speed of vehicles by 57%;
- decrease in the number of accidents by 22% at the place of tolling, and by 26% in the impact area.

A high-occupancy vehicle lane (HOV) is a lane reserved for vehicles carrying a driver and one or more passengers. HOV lanes are designed to reduce road traffic when many persons travel in a single vehicle. But foreign experience (mostly from the USA) shows that HOV lines do not always meet their objectives and under-utilise their potential, which is why more and more often HOV lanes are converted into High-Occupancy Toll (HOT) lines, which are available both for vehicles with a large number of passengers and not tolled, as well as other vehicles with road charges paid [5]. The fees depend on the time of day and number of people in a vehicle. There are also differences in tariffs depending on the day of the week. Pricing strategies are selected in such a way as to improve traffic conditions in a given traffic corridor as much as possible.

Charges also apply for entry points to the central areas of cities. The introduction of fees for driving into the city centre in London in 2007 reduced the number of vehicles entering that zone by 14%; travel time was shortened by 14%; and the average speed increased by approximately 30%. The charges resulted in an increase in the use of public transport by about 40%, which in turn reduced traffic congestion in the centre. In London it was observed that a 10% increase in travel costs for a car user (without parking) reduced the number of car trips by 4–5% [6].

3.2 Speed Management

Variable Speed Limits (VSL) are displayed by Variable Message Signs (VMS). VSL systems have been implemented in many countries, particularly in Europe, as a method of improving traffic flow and safety. VSL systems through sensors collect data on current road and/or weather conditions, and then send the recommended speed limits that are dynamically updated in order to impact the drivers' behaviour. Passing onto drivers speed limits that are appropriate for the current conditions can reduce vehicle speed and speed variations in the overall traffic stream. Properly designed VSL systems result in reducing the number of accidents, travel time and exhaust emissions thanks to harmonised traffic speed [7].

In 2008, during the implementation of the VSL system on the I-495 road in Virginia, an analysis of some operating VSL systems was carried out. The results of that analysis are presented in the Table 1 [8].

VSL signs are useful in the implementation of various strategies, such as Hard Shoulder Running (HSR) which allows for periodic increase in the number of traffic lanes by means of ITS equipment. Implementation of variable obligatory speed limits on four lanes with the possibility of using shoulder to keep ongoing traffic led to a 55.7% decline in the number of collisions on the main motorway in England (M42) [9]. VSL signs are also useful in roadwork areas. The research in Lansing showed that VSL helped to reduce travel time and increase the average speed of vehicles in the area of such works [10].

Table 1. VSL applications throughout the world [8]

Real conditions	Germany, Autobahn 5	Reduction of accidents involving injury by 30%
	UK, M25	10–15% reduction of accidents
	The Netherlands	16% reduction of accidents, capacity increase by 3–5%
	Germany, Autobahn 9	Free traffic flow during increased congestion, 20–30% reduction of accidents
	Finland, Motorway E19	Increase of an average speed, reduction of speed variability, expected 8–25% reduction of accidents
	Utah, I-80	Reduction of speed variability in the area of VSL impact
	Minnesota, I-494	Capacity increase by 7% at rush hours, no changes at other hours, compliance with speed limits increased by 20%–60%
Simulation	North Virginia	VSL minimises the dangerous imbalances in speed and length of queues, it was less effective in case of long queues

3.3   Detection of Traffic Violations and Support for Law Enforcement

Systems for automatic recording of traffic violations are used both in urban and rural areas. In the UK in 2003 the research results were presented, aimed at determining the impact of such systems on direct and indirect traffic safety measures in each city covered by the study. The results obtained were compared to those obtained in the adjacent areas (with no enforcement systems), also, the long-term trends were fore-casted. A pilot study was conducted in the period from April 2000 to March 2002 at 599 locations in eight regions (Cleveland, Essex, Lincolnshire, Northamptonshire, Nottingham, South Wales, Strathclyde and Thames Valley). The following elements were researched: changes of speed, a change in the KSI (killed or seriously injured) indicator, and a change in the PIA (personal injury accidents) indicator. The KSI rate has been reduced by 31–67%, while the PIA – by about 14–64%. In most of the areas the researchers noted the decrease in the number of vehicles exceeding the speed limit (about 61–81% for fixed speed cameras and 24–46% for mobile speed cameras) [11].

In Scottsdale, Arizona, speed cameras registering speeding on the highway caused a reduction in the number of accidents by 44–54%, the number of injured by 28–48%, and accident costs by 46–56%. Whereas in Chicago there was a 65% reduction in the number of vehicles exceeding speed in three weeks after installation of the surveillance system. Implementation of the system to detect vehicles passing a red light in Texas contributed to reducing by 11% the number of road accidents occurring within the junctions [11, 12].

3.4   Providing Traffic Information to Drivers

Advanced communication technologies allow for fast transmission of information to travellers. Drivers can now receive important information about the current traffic situation, including specific road conditions, in many ways, both before travelling - via websites or smartphone applications, as well as during the journey by means of Variable Message Signs or information systems in the vehicle.

Providing road users with accurate and timely information about travel conditions is very important because it can affect the choice of route and in case of obtaining information before the trip - the choice of means of transport or the time of departure. Studies show that information about road events (e.g. the occurrence of queues or roadworks) and recommended alternative routes displayed on a VMS leads to traffic dispersion among routes covered by those systems. However, the percentage of drivers who change their route rarely exceeds 40%. In the Swedish study, from 6% to 41% of drivers chose the recommended alternative route in order to avoid traffic jams, while the research in Houston showed that as much as 85% of drivers changed their route following the information displayed on a VMS [13].

In 2012 in Minneapolis - St. Paul and Seattle-Tacoma they surveyed drivers' reaction to the information about the actual travel time displayed on variable message signs. The results indicate that drivers are more likely to alternate their route when the displayed travel time is almost two times longer than the typical travel time on a given route (sections with a regular travel time of 5–20 mins were studied) [14, 15].

Crash reductions resulting from VMS to be 28% for injury related crashes in the UK, 35% for all crash types in Switzerland, and 10–30% for property damage and injury crashes in Germany. Weather monitoring VMS system was estimated to reduce crashes by 30–40% in various European countries; fatalities and injuries were conservatively estimated to reduce by 1.1% and 2.0% respectively [16]. The USDOT implemented a fog detection and warning system in December 1990. The system incorporated a VMS that alerted road users to the fog and slower traffic speeds, and a variable speed limit during foggy conditions. Following the introduction of this system, no fog-related crashes have been reported, compared with the 200 crashes that occurred between 1973 and 1990.

Messaging services are used among others in traffic incident (adverse event) management systems interconnected with Traveller Information Systems [17, 18].

3.5 Ramp Metering

Ramp metering is an integral part of the Highway Management System, which outlines strategies to reduce congestion and increase safety on Highway. Traffic light together with a signal controller that regulates the flow of traffic entering freeways according to current traffic conditions. It is the use of traffic signals at freeway on-ramps to manage the rate of automobiles entering the freeway.

Kansas Department of Transportation (KDOT) and Missouri Department of Transportation (MoDOT) implemented a pilot ramp metering programme in Kansas City (KC) Scout in order to improve traffic safety and flow on the road I-435. The main objective of ramp metering is to reduce the number of sudden changes of lanes and braking in the vicinity of the motorway entry point, as well as to improve the flow of traffic on the main road by maintaining the stream of traffic in conditions of optimal traffic parameters. The conducted analyses indicated significant improvement in safety when entering the motorway (lower number of dangerous braking and too small spacing between vehicles) [19]. Preliminary conclusions of the evaluation were consistent with the data collected in Milwaukee, Portland, Detroit and Denver which show that installing ramp metering systems can reduce the number of accidents by 26–50%. The research carried out after one year of ramp metering systems introduction showed a decrease in the number of accidents by 64% in the study area, including an 81% reduction in accidents in the area of entering the main road. Before the implementation of the system, the average time to remove the effects of the incident on the I-435 was approx. 22 min, while after the implementation of the system - 18 min. Travel time and speed were maintained at the same level as before the implementation of the system, despite the increased traffic volume [19, 20].

In 2000, in response to drivers' dissatisfaction with the ramp metering system, Minnesota DOT conducted an experiment which excluded all 433 ramps in the area of Minneapolis-St. Paul, to test their effectiveness. The study was conducted by Cambridge Systematics and it showed that after disabling the system the motorway capacity decreased by 9%, travel time increased by 22%, the average speed on the motorway fell by 7% and the number of road accidents increased by 26% [21].

3.6 Incidents Management and Support for Rescue Operations

Studies conducted in USA in the 1990s showed that the use of priorities for emergency vehicles allowed to shorten their travel time to victims by 14–23% (surveys were carried out in the cities of, inter alia, Colorado and Texas) [22].

In 2007, the city of Savannah implemented the wireless GPS system for prioritising rescue vehicles along the Derenne Avenue corridor. The system was installed at key junctions with which rescue vehicles communicated. The travel time of emergency vehicles along Derenne Avenue was reduced from 7 min before the implementation of the system to about 1 min [23].

Integration of systems of dispatching emergency vehicles and the Traffic Management Center in Utah allowed for more effective incident management. Observations carried out in the first months of implementation indicated that the process of detection, verification and reporting of events was shortened by 35–105 s. Accuracy of events location was improved through the direct import of information from CAD systems to the incident management system in TMC [24].

In Georgia, the program for adverse events management, Navigator, reduced the average duration of the incident from 67 min to 21 min [24].

With the implementation of the Washington State DOT Incident Response Team, 98% of traffic disruptions caused by incidents are removed in less than an hour, and 75% in less than 15 min. Whereas the Bay Area Incident Response System (BAIRS) allows to shorten the duration of the incident by approx. 15%. In Utah, the service of event management in the area of Salt Lake Valley made it possible to shorten the duration of an incident by an average of 20 min. In 2009 the Miami-Dade Traffic Incident Management (TIM) allowed to reduce the average duration of an incident by 11% as compared with the previous year, while in Maryland - by 28.6%. TIM in Atlanta helped to reduce the average duration of an incident by 46 min and it lowered the number of secondary accidents by 69%.

4 Analysis of the Possible Impact of ITS Services on Road Safety and Traffic Efficiency

On the basis of obtained data a preliminary analysis was made of the possible impact on current and planned ITS services on the safety of traffic. The impact was analysed through surveys and comparison of selected measures of safety and efficiency of traffic.

Surveys were conducted in order to diagnose the status and manner of implementation of ITS services in Poland. In addition, relevant plans and needs were analysed. A database of services currently provided and planned was developed, as well as of services identified by the stakeholders as required. The analysis of surveys helped to define current and future participation of stakeholders in the services and identify cases of shared services (logical connections). The main conclusion of the surveys is that currently there is no sustainable plan for the implementation of ITS services and systems in Poland. The authorities that implement or plan to implement some ITS services operate independently of one another. Respondents highlighted the lack of a common ITS architecture for the whole country. Currently, each administrator performs its tasks for the city or the region and does not have a coherent policy of implementing new technologies. The exchange of data is also problematic. A good first step towards the development of Polish ITS architecture was to develop guidelines for the National Traffic Management System (KSZR).

Regarding the links between various institutions, such cooperation results mainly from statutory requirements or regulations. Road operators provide data concerning violations of traffic rules, as well as ensure exchange of information on road traffic safety. However, some of the respondents were willing to widen the scope of cooperation in the area of ITS services, *inter alia* those related with the management of road incidents (including rescue operations), Variable Message Signs and information displayed on those signs.

In the survey respondents were asked about the impact of ITS services on traffic safety and flow. Services were analysed according to the standard classification of ITS services developed under the RID-4D project. The impact was evaluated according to a five-point scale, where 1 equals strongly negative impact; 3 - no impact; 5 - definitely positive impact. Among ITS services regarded by respondents as having the most positive impact on the safety and efficiency of traffic are (in parentheses the percentage of responders who pointed to the positive impact of the ITS service on road safety and traffic flow):

- Speed management (87% on road safety, 84% on traffic flow);
- Lanes management (78% on road safety, 83% on traffic flow);
- Detection of potential traffic violations (86% on road safety, 49% on traffic flow);
- Traffic control strategies management (94% on road safety in cities and 70% on road safety in non-urban area, 73% on traffic flow);
- Providing traffic information to drivers (87% on road safety, 89% on traffic flow);
- Traffic control at entry points (ramp metering) (79% on road safety, 81% on traffic flow);
- Traffic management at bridges (84% on road safety, 75% on traffic flow) and tunnels (91% on road safety, 82% on traffic flow);
- Incident management services (95% on road safety, 90% on traffic flow);

– Monitoring and sharing environmental information (weather) (88% on road safety, 80% on traffic flow);
– Support for maintenance of roads in winter (87% on road safety, 80% on traffic flow);
– Rescue operations management services (85% on road safety, 76% on traffic flow).

The above selection was made based on a number of responses indicating a positive impact of a given service (4 or 5 points) in relation to the total number of answers to a given question.

Within the framework of the project an initial analysis of the possibility of impact of existing and planned ITS services on traffic safety and efficiency was made. In order to determine the effect of selected ITS services on the national road network in Poland, the analysis was performed taking into account the selected factors affecting traffic conditions on the roads. The following maps were developed:

– Traffic volumes on reference sections of national roads in 2010 and 2015 based on data from general traffic measurements GPR2010 and GPR2015;
– Changes in the level of road safety: the number of accidents, the number of injured, the number of seriously injured, the number of fatalities - in years 2010–2015 based on data on accidents on national roads in Poland received from GDDKiA;
– Changes in the level of road safety in relation to traffic volume: the number of accidents, the number of injured, the number of seriously injured, the number of fatalities - per 1 million vehicle kilometre consecutively for years 2010 and 2015 based on data on accidents on national roads in Poland in years 2010, 2015 received from GDDKiA;
– Accident severity – the number of fatalities per 1 accident and the number of fatalities and seriously injured per 1 accident;
– Level of service for 2010 and 2015 based on data on traffic volume, road class and vehicle structure [25].

Based on the aforementioned maps road sections were selected, for which an in-depth analysis were carried out. The table below (Table 2) presents sample indicators for a few key road sections of motorways and expressways.

Among the implemented modules of KSZR, the largest group are devices associated with traffic parameters (speed, traffic volume, detector occupancy – road data collection) and vehicle data collection (devices that identify the features of individual vehicles), monitoring weather and road surface conditions, video monitoring and providing information to drivers (via VMS). The location of the selected devices is presented in Fig. 1. We can also mention ITS services relating to the detection of incidents and the management of speed, however, such services require the extension of implementation and functional development. The number of devices and functional structure of ITS services on individual road sections differ from each other. Table 2 shows examples of road sections where the range of the implementation of ITS services and equipment is more comprehensive (S8, A8) and those for which ITS services have been implemented to a smaller extent (S6) or sections without ITS equipment (A1). The concentration ratio of accidents (number of accidents per million vehicle-kilometer) were compared, which is a measure of individual risk. Individual risk is defined as the probability of involvement of a single participant in the process of road traffic in collision/accident or the probability

Table 2. ITS equipment and accidents per mln vehicle-kilometres on selected roads in 2015 [own study]

Road no	Average daily traffic/weighted average	Section length	Lanes	Implemented KSZR modules	Accidents per mln veh-km
S6	29660 – 77141/ 61672	36.54	2 × 2	VMS (2) Road data collection (3) Weather stations (3) Video data collection (2)	11.73
S8	26789 – 29252/ 28003	37.39	2 × 2	VMS (27) Road data collection (28) Traffic lights (1) Weather stations (29) Video data collection (29) Vehicle data collection (15)	6.82
A1	19927 – 35683/ 23960	37.41	2 × 2	None	19.98
A8	38925–59528/ 50621	22.72	2 × 3	VMS (18) Road data collection (14) Traffic lights (1) Weather stations (11) Video data collection (30)	5.09

Fig. 1. Selected ITS devices on Polish national roads in 2015 [own study]

of material or physical loss in such incident [1]. The examples (Table 2, Fig. 2) show that on the road sections where the state of implementation of ITS equipment is at a higher level (A8, S8), the concentration is lower. However comparison of intensity of traffic on individual road sections do not give a definite answer on the impact of traffic flows on road safety. There is relatively less traffic on the road S8 than on S6, which may affect the level of road safety. Analyses carried out for selected sections of motorways has shown that despite a higher level of service of traffic on the highway A1 (less traffic per lane) than on the A8, the concentration ratio has a lower value, providing a higher level of road safety in terms of individual risk. Note, however, that the level of road safety is influenced by many factors, among others: class of road, road geometry, road surrounding, speed limits, interchange entries and exits areas (their number and geometry within the interchange, form of interchange). On the basis of the common indicators, it cannot be unambiguously determined that the implementation of the ITS services will improve the level of road safety, since a number of other factors could adversely affect this level.

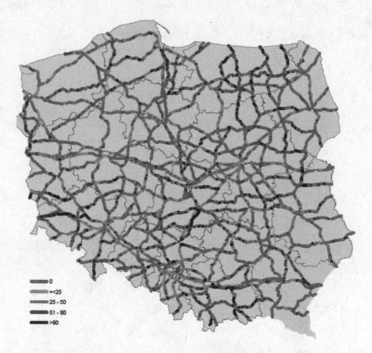

Fig. 2. Number of accidents per mln veh-km on Polish national roads in 2015 [own study]

5 Conclusion

In-depth study of literature have demonstrated that providing ITS services can significantly influence both the improvement of road safety and efficiency of the transportation system. Among the services, which have the greatest impact on increasing the level of road safety, while also have a positive impact on the efficiency of traffic on motorways and express roads are ramp metering, traffic incident management (including providing information on speed limits, alternative routes and warnings to drivers), providing traffic information to drivers about weather and road surface conditions with associated speed limits, electronic tolling systems as well as detection of traffic violations and support for law enforcement. Literature studies were confirmed by surveys carried out as a part of the project RID 4D. Respondents indicated the ITS services that they think positively affect the level of safety and efficiency of traffic. Most respondents indicated ITS services (for the rural roads), such as incident management, monitoring and providing information to drivers (especially about weather conditions), speed management and detection of traffic violations (speed and red light enforcement).

A preliminary analysis of the impact of the application of ITS services on the level of traffic safety based on accident data and the location of ITS equipment were also carried out. The number of devices and functional structure of ITS services on individual road sections differ from each other. Selected road sections were analyzed where the range of the implementation of ITS services and equipment is more comprehensive, and those for which ITS services have been implemented to a smaller extent or sections without ITS equipment. Generally on the road sections where the state of implementation of ITS equipment is at a higher level the individual risk to be involved in accident is lower. However the level of road safety is influenced by many factors e.g.: class of road, road geometry, road surrounding, speed limits, interchange entries and exits areas (their number and geometry within the interchange, form of interchange). On the basis of the common indicators, it cannot be unambiguously determined that the implementation of the ITS services will improve the level of road safety, since a number of other factors could adversely affect this level. Due to the above, it is necessary to carry out more detailed studies, taking into account a variety of factors, which may affect the road safety. Based on preliminary studies pilot road sections were selected, which will be analyzed in detail with the use of traffic models (macro, micro and mesoscopic depending on particular ITS service or groups of services). Models of the impact of various factors on road safety will be also developed for selected sections in order to isolate the effect of ITS services.

Acknowledgement. This paper describes realization effects of RID 4D (41) called "The impact of the usage of Intelligent Transport Systems services on the level of road safety" funded by National Research and Development Center and General Directorate for National Roads and Motorways (agreement no. DZP/RID-I-41/7/NCBR/2016 from 26.02.2016).

References

1. Jamroz, K.: Metoda zarządzania ryzykiem w inżynierii drogowej. Wydawnictwo Politechniki Gdańskiej, Gdańsk, Poland (2011)
2. Koonce, P.: Benefits of Intelligent Transportation Systems Technologies in Urban Areas: A literature Review. Portland State University, Portland (2005)
3. Hatcher, G.: Intelligent transportation systems benefits, costs, and lessons. U.S. Department of Transportation, Washington, USA (2014)

4. Klodzinski, J.: Evaluation of impacts from deployment of an open road tolling concept for a mainline toll plaza. Paper Presented at the 86th Annual Meeting of the Transportation Research Board, Orlando, USA (2007)
5. Bhatt, K.: Value Pricing Pilot Program: Lessons Learned. U.S. Department of Transportation, Federal Highway Administration, New Jersey, USA (2008)
6. Arnold, R.: Reducing Congestion and Funding Transportation Using Road Pricing in Europe and Singapore. Federal Highway Administration, U.S. DOT, New Jersey, USA (2010)
7. Randolph, L.: Texas variable speed limit pilot project. In: National Rural Intelligent Transportation Systems Conference, San Antonio, USA (2015)
8. Fudala, N.: Work Zone Variable Speed Limit Systems: Effectiveness and System Design Issues. Virginia DOT, Virginia (2010)
9. Arlow, A.: M42 MM Monitoring and Evaluation Three Year Safety Review. Highways Agency, Department for Transport, London, UK (2011)
10. Scriba, T.: Intelligent Transportation Systems in Work Zones: A Case Study - Real Time Work Zone Traffic Control System. Federal Highway Administration, U.S. DOT, Springfield, Illinois, USA (2004)
11. Gains, A.: A cost recovery system for speed and red-light cameras - two year pilot evaluation. Department for Transport, Road Safety Division, London, UK (2003)
12. Cheeks, J.: Speed Limit and Safety Nexus Studies for Automated Speed Enforcement for the District of Columbia. District Department of Transportation, Washington (2014)
13. Elvik, R.: The Handbook of Road Safety Measures, 2nd edn. Emerald Group Publishing, Bingley (2009)
14. Texas DOT: Travel Time Messaging on Dynamic Message Signs—Houston. www.ops.fhwa.dot.gov/publications/travel_time_study/houston/houston_ttm.htm. Accessed 13 Feb 2017
15. Athey Creek Consultants: Impacts of Traveler Information on the Overall Network: Final Report, Enterprise Transportation Pooled Fund TPF-5(231), Minneapolis-St. Paul, USA (2012)
16. Organization for Economic Co-operation and Development: Road Safety: Impact of new technologies, OECD PUBLICATIONS, Paris, France (2003)
17. Oskarbski, J., Zawisza, M., Miszewski, M.: Information system for drivers within the integrated traffic management system-TRISTAR. In: Mikulski, J. (ed.) Tools of Transport Telematics. Communications in Computer and Information Science, vol. 531, pp. 131–140. Springer Verlag, Berlin Heidelberg (2015). doi:10.1007/978-3-319-24577-5_13
18. Oskarbski, J., Zawisza, M., Żarski, K.: Automatic incident detection at intersections with use of telematics, Transport Research Arena TRA2016. Transportation Research Procedia, vol. 14, pp. 3466–3475 (2016)
19. Kansas and Missouri Departments of Transportation: Ramp Metering 2011 Evaluation Report. Kansas DOT, Missouri DOT, Missouri, USA (2012)
20. Kansas and Missouri Departments of Transportation: Ramp Metering 2010 Evaluation Report. Kansas DOT, Missouri DOT, Missouri, USA (2012)
21. Cambridge Systematics: Twin Cities Ramp Meter Evaluation. Minnesota DOT, Minneapolis, USA (2001)
22. Taylor, S.: Helping Americans. ITS World, Albuquerque, USA (1997)
23. Global Traffic Technologies: Traffic Signal Priority Control for emergency vehicle preemption, Global Traffic Technologies, Savannah, USA (2010)
24. Guin, A.: Benefits Analysis for the Georgia Department of Transportation NaviGAtor Program: Final Report, URS Corporation, Atlanta, USA (2006)
25. Oskarbski, J., et al.: Report RID-4D—Periodic Report on the Implementation of Tasks 2.3. Gdańsk University of Technology, Gdańsk, Poland (2016)

Application of BAN Network to Increase Security in Transport Systems

Malgorzata Gajewska[✉]

Faculty of Electronics, Telecommunications and Informatics,
Gdansk University of Technology, Gabriela Narutowicza 11/12,
80-233 Gdańsk, Poland
Malgorzata.Gajewska@eti.pg.gda.pl

Abstract. In the article general characteristics of the BAN network with M2M communications are presented. These are networks that enable the implementation of wireless transmission of signals using special sensors located on the body or implanted subcutaneously. These sensors allow monitoring of different type life parameters of a human. In the next part of work there is proposed the implementation of BAN networks to transport systems as a structural part of V2X systems. It allows the increasing of a functionality and working range of these systems. In addition, the analysis of the possibility of using the BAN network, to improve the safety of road users is presented. The use of such solutions allows, amongst others, permanent monitoring of the state of health i.e. drivers and rapid response in case of danger, which undoubtedly can prevent road accidents and/or increase the safety of potential victims.

Keywords: BAN · 5G · M2M · V2X

1 Introduction

Nowadays, the development of transport systems telematics are very intensive. We have continuous increase of the number of different road users (pedestrians, drivers, engine drivers, passengers). Unfortunately, if we have the more participants of transport systems than we will have the more of potential traffic risks. Therefore, it is necessary to search for such of technical solutions which use give us ability to reduce the number of threats.

Maybe proposition of the solution, which give us change for eliminate some of the risks, is application the modern methods of wireless communication, for example M2M (Machine-to-Machine) communication, in the transport systems [1].

In general, M2M communication is the technology of data transmission from a machine (e.g. a car) to a machine (e.g. a bus) or from one terminal to another terminal or a sensor to a sensor as well as a mobile to a machine. But also, it can be transmission of information between a man (e.g. pedestrian) and a machine (e.g. a car). This application is often characterized by the absence of a human decision maker directly. But we have some M2M applications in the transportation sector in which also interaction with drivers is required [1, 2].

© Springer International Publishing AG 2017
J. Mikulski (Ed.): TST 2017, CCIS 715, pp. 155–166, 2017.
DOI: 10.1007/978-3-319-66251-0_13

In this article, the new proposition of the solution, which consolidate two elements of M2M communication is presented. These elements are automotive V2X (Vehicle-to-Everything) [3–5] system and a short-range network of BAN type (Body Area Network), as a structural part of this V2X system.

The use of such V2X systems in transport systems telematics is not a new proposition. However, the proposition which consolidate V2X systems with BAN networks is a completely new solution. It appears that such solution allows significant reduction of the number of road accident victims and improves road safety.

2   Principles of BAN Network with M2M Communications

The BAN networks enable the implementation of wireless transmission of signals using special sensors which may be surface-mounted on a body in a fixed position (so-called on-body sensors), can be implanted subcutaneously inside the body (in-body sensors) or outside the body but close to it (off-body sensors). These sensors allow monitoring of different type life or health parameters of a human as well as can be used to monitoring of environmental parameters of environment surrounding a human [6, 7].

Example of the BAN network using wireless M2M communication is presented in Fig. 1. As we can see, the person has (for example) six sensors on its body which control its life parameters:

- sensor for blood pressure monitoring,
- sensor for pulse monitoring,
- ECG sensor,
- sensor for measuring the brain waves,
- sensor for monitoring the patient's breath,
- sensor for measuring the concentration of glucose,
- and others.

These sensors are designed to monitor the current state of this person health. So, they send, in continuous and real time, signals to the communication device which is located on its clothes or outside a body. Next, this device send signals, through the M2M gate implemented for communication, to some mobile network. The transmission in mobile network is made through radio access network, core network and then through internet to the M2M application platform using algorithms which enable health data analysis. Note that for transmission can be used different types of radio communication networks e.g. satellite networks or other types [1].

On application platform, signals are continuously analyzed and preferred are real-time or quasi real-time operations during transmission and analysis. But small delays can be accepted. The application platform can be located at server which is typically equipped with specialized software to make analysis, using artificial intelligence algorithms. Artificial intelligence algorithms can take decisions of sending emergency information to a doctor and/or to patient. Of course using artificial intelligence is not included in all applications of BAN-based medical networks. All information can be sent individually or in specified groups including other network users, in order to use them for different purposes [6].

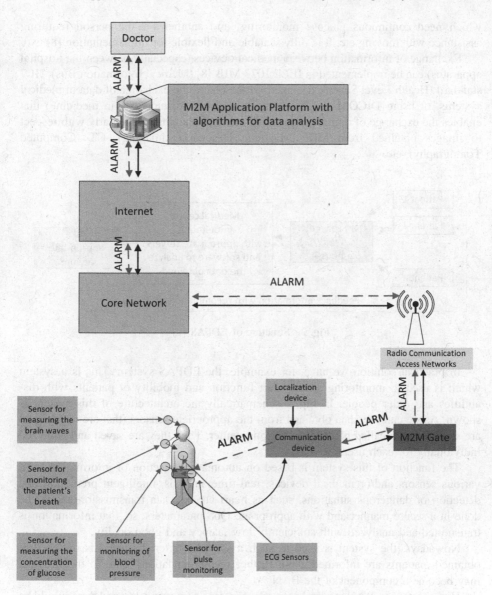

Fig. 1. BAN with M2M communication system [own study]

If any irregularities or problems are detected the M2M platform send alarm information to a doctor or rescuer and at the same time to this person which is monitored. Then the doctor may take appropriate action and thus save human life or health.

Note that the data correctly collected and analyzed through M2M platform, can be the basis to take action saving lives and health. A characteristic feature of the BAN network, with M2M communication, is that we can customize their structure and/or configuration for own the user's needs. Thus, different network structure we can plan for the people which need continuous glucose monitoring, and another for the person requiring assistance with moving etc. It is fully scalable and flexible for implementation [8].

Exchange of information between medical devices (especially between the hospital apparatus) can be implemented in IEEE 1073 MIB [8] (Medical Information Bus), HL7 standard (Health Level Seven), designed for the electronic exchange of data in medical systems, or using DICOM (Digital Imaging and Communications in medicine) that enables the exchange of digital imaging and medical imaging, particularly with respect to images obtained from MRI (Magnetic Resonance Imaging), CT (Computed Tomography), etc.

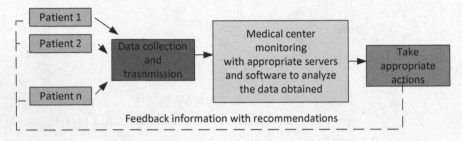

Fig. 2. Structure of EDFAS [9]

In practical solution we have, for example, the EDFAS system. This is a system which is used to monitoring assess heart function and mobility of patients with disabilities and older people. In Fig. 2 schematically the architecture of this system is shown. As we can see, data obtained from the appropriate sensors (that are in patients) are sent to the so-called medical monitoring center. Then they are saved and analyzed individually for each user.

The function of this system is based on automatic collection of information from various sensors and/or medical devices, real-time analysis, intelligent prediction and detection of dangerous situations, such as heart attack. Data transmission should be done in a secure manner and with appropriate QoS parameters, so, that information is transmitted and analyzed with sufficiently low latency and good quality.

Nowadays, the system is used in such a way that after the analysis of the data obtained, patients are informed about further recommendations. But, in the future it may become a component of the BAN.

However, using BAN medical networks can save someone's life and they should be designed in such a way as to become autonomous. Their action should be independent of the performance problems in a public network. The use for these purposes 5G networks is very promising, in which special attention is paid to their reliability, high quality of transmission, high transmission rate, low latency and high accuracy of localization [9, 10].

3 BAN Application Examples

The application of Body Area Networks is very important solution in transport systems because thanks to it we can prevent many accidents, which can be the result of sudden health deterioration.

We can find many examples of the use of BAN in M2M networks for help the people health. The first group is given below:

- remote heart and breathing monitoring for the people requiring no intensive care and send by wireless network automatic startup alarms to a doctor in an emergency situation,
- continuous monitoring people with heart disease and/or hypertension,
- automatic monitoring of hematological patients e.g. when they used chemotherapy or radiotherapy, if the patients are beyond medical institutions,
- remote transmission of results, e.g. ECG signals while driving,
- measurement photoplethysmographic signal (PPG),
- measurement galvanic skin response (GSR),
- stimulate the diaphragm of patients with congenital central hypoventilation or the so-called Ondine's Curse,
- automatic monitoring of glucose levels for people with diabetes and remote applying insulin in appropriate doses and, at the same time, begin the alarm in critical situations,
- remote continuous monitoring the results of electroencephalography for patients with epilepsy in order allowing determination of proper dosage of medicines, and in a way that would allow the elimination of attacks.

The second group of BAN users is very important for transport systems because it allows to provide help for physically or mentally disabled traffic participants. We can find here:

- supporting the movement and operation of blind or partially sighted people,
- automatic communication with disabled persons, for example, in railway stations, to get them necessary help,
- support to hearing impaired persons and deaf people [6].

However, the major BAN application for transport results from the possibility of equipping professional drivers with sensors which can support monitoring their vital functions in order to minimize the risk of an accident.

For example critical health problems increasing the risk of accident can result from:

- the loss of consciousness, due to a sharp decline in the level of glucose in the blood,
- heart attacks or stroke,
- falling asleep of a driver,
- other faints,
- test whether the driver is not under the influence of alcohol.

In addition, the BAN networks allow specific support to officers guarding public safety with specialized sensors, so that they will be able to remotely detect explosives and prevent tragedies. Also, BAN networks may also support the work of firefighters and emergency medical technicians. In the case of drivers, we can imagine the use sensors which monitor the level of alcohol in the breath of a driver, after entering the car. System can automatically lock the ability to run a car.

At this moment, the list of possible applications of the BAN network with M2M communication is very long. We can improve our safety, passenger comfort, increase the efficiency of health problems detection or improve our security. Additional communication of BAN networks with road infrastructure and some systems for prevent road accidents and road traffic violations [11] can further increase the safety on roads.

One can see that in transport systems telematics these networks will be of great importance, although at the moment there is little number of scientific publications on this research area. Probably after the implementation of 5G radio communication networks will be the rapid development of BAN systems in transport applications [4, 5].

4 Proposal of BAN Networks Implementation in Transport Systems as a Structural Part of V2X Systems

Implementation of BAN networks to transport systems as a structural part of V2X systems plays great role in development of transport systems for cars and other vehicles. Until 2024 will be implemented in practice many different types of this kind of wireless communication, under the common name V2X. V2X communication can be done between vehicles and other vehicles, pedestrians or elements (sensors) of road or cellular systems infrastructure [3, 12]. We can distinguish some subtypes of V2X communication as following:

- V2V – Vehicle-to-Vehicle,
- V2I – Vehicle-to-Infrastructure,
- V2P –Vehicle-to-Pedestrian,
- V2H – Vehicle to Home,
- In-V - In Vehicle Communications,
- Sometimes we are talking about Car-to-X communication as a part of previous mentioned.

The V2V communication means direct exchange of radio information between vehicles, V2I is to use the possibility of exchanging information between moving, stopped or parked cars and their surrounding infrastructure. V2P can be used to communication with pedestrians what is very important from the point of view of their safety. Interesting approach is the V2H communication in which is possible the communication between vehicles and buildings communication infrastructure e.g. in the case of smart-house concept (e.g. opening the gate). Additionally, in-V transmission allows transmission of information between different M2M sensors and control devices installed in a vehicle by which it is possible e.g. parking the car without a driver (the driver can not to be present inside a car) [3].

General concept of the proposed architecture of the system combining V2X technology with BAN is shown in Fig. 3. This concept is based on the integration of BAN functionality with V2X communication.

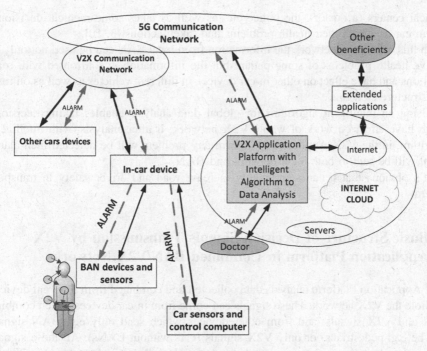

Fig. 3. Proposal of BAN networks implementation in transport systems based on V2X [own study]

In presented system architecture we can find a few important elements:

- BAN devices and sensors related to a person, especially car driver,
- V2V communication network for exchange the information between vehicles and some elements of road infrastructure,
- In-car device which receives and sends the information from/to V2V network and/or 5G radio communication network, the BAN network,
- V2X Application Platform with Intelligent Algorithms of Data Analysis for global analysis of data from different in-car devices,
- Car control computer,
- Servers for data analysis in a network cloud,
- Doctor (medical center) and other beneficiaries of the system, like police, fire brigades etc.

In-car device collects, combines, transmits and receives data from different sources: V2V network, local BAN network, Application Platform, doctor and others, as well as transmits information to car computer if necessary.

V2X Application Platform with Intelligent Algorithms of Data Analysis is located in internet cloud and enables global analysis of data from different in-car devices and other sources having effect on the V2X network. This platform takes decisions on road safety, road infrastructure control, emergency calls, and informing the fire brigades, medical centers (a doctor), the police etc. as well as takes some medical decisions, inform car driver on their health problems and taken actions [3, 10].

In this concept of network the information from local BAN can be used not only to resolve health problems of some patient but the information is confronted with road situations and have effect on other in-car devices in different vehicles as well as on road infrastructure elements.

Using an intelligent algorithm for global data analysis enables taking decisions which have effect on work of whole V2X network. If after analysis, with intelligent algorithm, in the V2X Application Platform any problem will be detected then alarm signal will be sent to both V2V network and BAN.

It's obvious that it can significantly increase road and driver safety in transport system.

5   Basic Structure of Decision Signals Transmission by V2X Application Platform in Combined BAN/V2X Networks

V2X Application Platform analysis data collected and combined from different devices in whole the V2X network. These signals can come from in-car devices which combine BAN and V2X signals, and from other devices which send only e.g. BAN signals (maybe e.g. pedestrians) or only V2X signals (cars without BANs). All these signals play some role in decision process made by artificial intelligence algorithms used in the platform [10].

V2X Application Platform can take different decisions having effect on a number of devices (cars, pedestrians) in V2X network. Of course taking into account signals from all devices working in V2X network is a nonsense. It is necessary to make some groups of devices placed in some list which is continuously actualized. In this list some devices are registered, and some devices are removed from the list. Thus, there is a need to create some rules for lists construction and it can be related to the rules of V2X network work.

In Fig. 4 the basic structure of signal transmission from V2X Application Platform is presented.

As we can see, sensors of local BAN placed e.g. in body of a driver inside a car will send information to In-car device and next to V2X communication systems. Finally this information is received by V2X Application Platform in which it is processed. Based on this information as well as another information from a network, the platform takes decisions and sent it using feedback decision information signals [10]. For instance, these decisions can be as follows:

- Safety stopping a car – information sent to expected in-car device, and next to the control computer of this car,
- Alarm information – sent to local in-car BAN networks (driver, passengers),
- Emergency information – sent to other cars connected to V2X network, which are placed on active list of devices,

- Emergency information – sent to road infrastructure elements connected to V2X network, e.g. change intelligent signs information, change of the lights on a crossroad, car-stop information for some parts of road etc.,
- Sending information to some pedestrians which are on the active list of devices,
- Speed limitation commands.

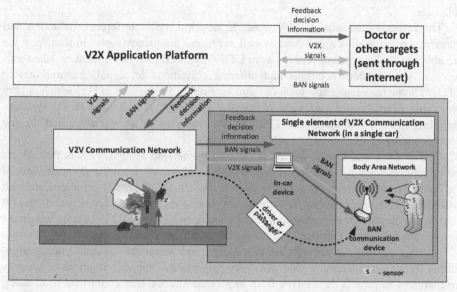

Fig. 4. Structure of signals transmission between, M2M Communication, V2X Systems and Body Area Networks [own study]

Thus, in an emergency, such as when the driver immediately stopped breathing, V2X Application Platform takes this information and activate set of actions in order to prevent a danger.

On the other hand the life of drivers and pedestrians is protected by acting other V2X functions because the car entering a crossroad has information about the upcoming cars, and in this privileged cars, and so he can avoid collision.

As shown, the concept of integration both BAN and V2X networks enables increasing of road safety. It seems that potentially increase the chance of drivers and passengers to experience accidents and collisions because even if the vehicle does not stop completely its speed is certainly reduced. Additionally, other vehicles takes actions of their speed limitation or stopping. Such cooperation will also reduce the number of accidents involving people with disabilities. Because, for example, if the V2P system get information from BAN, that a blind person approaching the pedestrian crossing, the system may force response in vehicles, limit the speed, launch beeps or light on the crossing [6, 7].

In December 2015, the 3GPP organization published technical report TR 25.885 (Release 14) [10] for use of LTE-V for the implementation of V2X communication services. On this document we have guidelines for using this solution to improve communication between vehicles, vehicles and parts of road infrastructure, as well as between vehicles and terminals carried by a pedestrian or cyclist.

As well known, in the LTE-V proposed system we have a problem with interferences as in LTE-Advanced. These interferences decrease transmission performance at a cell edge.

The implementation of LTE-V needs to develop new interference reduction and interference cancelation algorithms as well as rational frequency reuse. In some part we are able to use the same algorithms as in LTE-Advanced [14] but it must be taken into account some modern solutions and different conditions of signals transmission in LTE-V. Some interesting approach of frequency reuse we can find in [15] which is very promising from the point of view of spectral efficiency at cell edge in 5G systems, including LTE-V solutions for ITS. At the moment it is not specified in the documentation.

One of the requirements for a vehicle communication system V2X is that all car manufacturers must use a single standard. This is important from the point of view of the operation of the entire network. As all vehicles should be able to communicate with each other. It is unacceptable for vehicle manufacturers to use different standards. Of course, the manufacturer can expand the capabilities of the V2V system within its design, but it is important that it is compatible with the overall solution.

Another problem is appropriate procedures and select the channel coding algorithms for such transmission. But also very important is the development of security mechanisms to protect themselves from unwanted attacks, that could lead to disaster.

6   On-Body Sensors Example

In practical solutions the person using BAN (driver, pedestrian, and passengers) can wear a special smart vest, which is shown in Fig. 5 [13].

The vest has many additional sensors, for example, sensors for measuring temperature, blood pressure, heart rate. On this vest we have special belts for monitoring electrocardiography signals (ECG) and photoplethysmographic signal (PPG), galvanic skin response (GSR). These sensors give us change for monitoring physiological parameters for drivers. Additionally, the vest has the GPS module (Global Positioning Systems) [13]. So, this is the first step to implementation the BAN network for everyday use in a live of not only professional drivers [13].

At the moment, many companies are working on similar solutions. In general, it is intended to be consumer products. Although, it seems unlikely that these products will be used directly in transport systems but rather in medical monitoring systems. That is why it is so important to show manufacturers the possibility of using them in other systems, for example in BAN with M2M communications.

Particularly apply this type of clothes would be important at professional drivers and at the paramedics, firefighters and police officers. With high probability, this vest could protect drivers, for example, before falling asleep while driving.

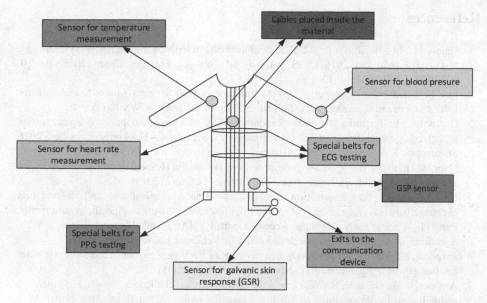

Fig. 5. Example of clothes for a driver adapted to the Body Area Network [13]

But also similar can carry pedestrians. This will allow the driver to be alerted to nearby drivers of the danger in case of an emergency situation. On the other hand, a person with Alzheimer's disease, equipped with such special clothes, can be monitored by the family with a GPS. However, police may also be informed about the patient's whereabouts. And pass this information through the V2X system with M2M communication to nearby drivers.

In the future, probably will be produced many different types of clothes with a variety of sensors which will give us new possibilities in Body Area Network [13]. But already existing ones at the moment can be used to protect the health and life of road users, both drivers, passengers and pedestrians.

7 Conclusion

Presented in the article, the concept of M2M communication systems, with proposed the implementation of Body Area Networks, as a structural part of V2X systems, for transport systems telematics, can significantly increase safety the all road traffic users (drivers, passengers and pedestrians). It allows the increasing of a functionality and working range of these systems.

The use of such proposed solutions allows, amongst others, permanent monitoring of the state of health of drivers and rapid response in case of danger. It undoubtedly can prevent road accidents and/or increase the safety of potential victims.

Enhanced applications of the proposed solution will appear when run on radio communications systems of 5G in which Ultra-Dense Networks will be implemented. Then we will have even more possibilities to protect users of transport system.

References

1. Gajewska, M.: Design of M2M communications interfaces in transport systems. In: Mikulski, J. (ed.) TST 2016. CCIS, vol. 640, pp. 149–162. Springer, Cham (2016). doi:10. 1007/978-3-319-49646-7_13
2. Liu, L., Gaedke, M., Koeppel, A.: M2M Interface: a Web Services—based framework for federated enterprise management. In: Proceedings of the IEEE ICWS (2005)
3. Gajewska, M.: Technika komunikacji radiowej V2X jako metoda poprawy bezpieczeństwa w ruchu drogowym. Przegląd Telekomunikacyjny i Wiadomości Telekomunikacyjne 6/2016
4. Statement of Nathaniel Beuse, Associate Administrator for Vehicle Safety Research, National Highway Traffic Safety Administration, Before the House Committee on Oversight and Government Reform Hearing on "The Internet of Cars" (2015)
5. IEEE Standard for Information Technology—Telecommunications and Information exchange between systems—Local and metropolitan area networks—Specific requirements, Part 11 Wireles LAN Medium Access Control (MAC) and Physical Layer (PHY), Specification Amendment 6: Wireless Access in Vehicular Enviroments (2010)
6. Jovanow, E., et al.: A wireless body area network of intelligent motion sensors for computer assisted physical rehabilitation. J. NeuroEng. Rehabili. (2005)
7. Antonescu, B., Basagni, S.: Wireless body area networks: challenges, trends and emerging technologies. In: Proceedings of 8th International Conference on Body Area Networks—BodyNets 2013, Boston, Massachusetts, pp. 1–7 (2013)
8. Kennelly, R.J., Gardner, R.M.: Perspectives on development of IEEE 1073: The Medical Information Bus (MIB) Standard. Int. J. Clin. Monit. Comput. Kluwer, Netherlands (1997)
9. Białoń, P., Klimasara, E.: Nowe paradygmaty przetwarzania danych w sieciach inteligencji otoczenia dużej skali. Monitorowanie osób starszych i projekt EDFAS, Praca nr 06300048, Warsaw (2008)
10. GPP TR 22.885 v 14.0 Study on LTE support for vehicle to everything (V2X) services (2015)
11. Gajewski, S., Gajewska, M., Sokol, M.: Architecture and basic assumptions of RSMAD. In: Mikulski, J. (ed.) TST 2011. CCIS, vol. 239, pp. 200–205. Springer, Heidelberg (2011). doi:10.1007/978-3-642-24660-9_23
12. Harding, J., et al.: Vehicle-to-Vehicle Communications, Readiness of V2V Technology for Application (2014)
13. Pandian, P.S., et al.: Smart vest: Wearable multi-parameter remote physiologital monitoring system. Med. Eng. Phys. **30** (2008)
14. Gajewski, S.: Soft-partial frequency reuse for LTE-A. Radioengineering **26**(1), 359–368 (2017)
15. Gajewski, S.: Throughput–coverage characteristics for soft and partial frequency reuse in the LTE Dow. In: Proceedings of 36th International Conference on Telecommunications and Signal Processing, TSP 2013, Rome, Italy, pp. 199–203 (2013)

Video System as a Psychological Aspect of Traffic Safety Increase

Ján Ondruš[1](✉) and Grzegorz Karoń[2]

[1] University of Zilina, Univerzitná 8251/1, 010 26 Zilina, Slovak Republic
jan.ondrus@fpedas.uniza.sk
[2] Silesian University of Technology, Akademicka 2A, 44-100 Gliwice, Poland
grzegorz.karon@polsl.pl

Abstract. The aim of the article is the introduction of the video system GEM CDU 2605 Zeus and the traffic flow analysis of the selected section, and specifically to show the extra value of the device. The respective video system is located close to the Zilina city centre (northern Slovakia). Based on the evaluated statistical data obtained from the video system it is possible to design the traffic precautions in the specific section, or to install the required technical equipment for traffic moderation. The use of these systems can increase the traffic safety and fluency including the traffic calming, traffic accidents number decrease, engine exhaust emissions decrease and an increase of the living standard in the towns and villages. Results of analysis and its evaluation and comparison are presented in the final part of this paper.

Keywords: Video system GEM CDU 2605 Zeus · Traffic flow · Traffic safety · Traffic intensity · Speed

1 Characteristics of the Town of Zilina

The town of Zilina is considered to be an administrative, economic and cultural centre of the northwest Slovakia. It is the fourth biggest town in Slovakia. The total area is 80.03 km². The number of citizens as of December 31, 2016 was 83 651, out of that 40 342 men and 43 309 women. Zilina is a regional capital of the Zilina region with the total area of 6 809 km², which represents almost 14% of the total area of the state. The northwest boarder of the region is at the same time a state boarder with the Czech Republic and in the north with Poland [1] (Fig. 1).

2 The Description of the Video System

The video system GEM CDU 2605 Zeus is a device for speed measuring, which shows the current speed of the vehicles and collects the individual statistics (Fig. 2). It also has installed an automatic wireless data transmission through a selected available GSM mobile operator network to the client. The software has interesting additional functions such as online automatic notifications of the sections in which the speed is being exceeded. The individual statistics can be used for defining the roads utilization with

© Springer International Publishing AG 2017
J. Mikulski (Ed.): TST 2017, CCIS 715, pp. 167–177, 2017.
DOI: 10.1007/978-3-319-66251-0_14

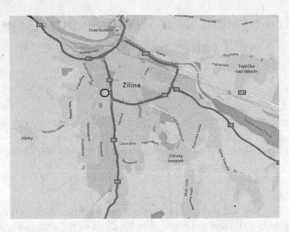

Fig. 1. Map of the town of Zilina with the solved area marked out [2]

a possibility of defining an hour and a day when the maximum permitted speed is most often exceeded. The operation of the device does not require maintenance for few years [3].

Fig. 2. Speed measuring device GEM CDU 2605 Zeus [own study]

The video system is an informing measuring device with an image record of the exceeded speed. It is designated for informational vehicles speed measuring. It contains a monitor with radar and shows the instantaneous speed of the coming vehicles, and a recording device for the predefined recording with an automatic wireless data transfer to the client. Measured speed data, time and place are showed in the image. The record contains a series of images with a detailed record of the vehicle exceeding the permitted speed. The video device is synchronized with the radar [4].

The main tasks of the device GEM CDU 2605 Zeus are the following:

- To show the driver the instantaneous speed of the vehicle,
- To record the traffic data,

- To provide the recorded images,
- To record the traffic violations,
- To share the data with the other devices [5].

The installation of the speed measuring system predicts that approximately 30% of drivers decelerate in front of the measuring device, and 60 to 90% vehicles decelerate after passing the device.

The main utilization of the device is on the spots with decreased or limited speed, or in places with higher rate of traffic accidents. It increases safety in places with high concentration of pedestrians, such as pedestrian crossings close to the schools and hospitals. The device shows the driver the instant speed, or informs the drivers about the speed of the vehicles driving in front of them, and in the distance of max. 80 m in front of the radar [3].

3 The Location of the Speed Measuring Device

The speed measuring device is located at the intersection of the streets Závodská cesta and Dolné Rudiny next to the city centre. The intersection defined as a three-armed, not controlled, regulated, joining intersection. It is situated between the two roundabouts - Rondel (1st roundabout) a Priemyselná/Závodská (2nd roundabout), (Fig. 3).

Fig. 3. Location of the speed measuring device [2]

This intersection is problematic mostly during the traffic peak hours due to the high traffic intensity not only of passenger vehicles, but also the freight vehicles and mass public transport. A high number of freight vehicles is caused by the industrial area located close to the respective intersection. A depot belonging to the Transport Undertaking of the town of Zilina is also located in the vicinity. The intersection also serves to trolley-busses providing the transport connection of the residential district Hájik with the city centre or with the other residential districts [6]. The device is fixed with a supporting construction on the street lamp, or on the contact trolley line pillar, at the height of three meters (Fig. 2). It is fixed in a standard way, with the non-corrosive constricting tapes Bandimex. The front of the board is turned towards the descending road from the Rondel. The pillar is situated approximately 1.5 m from the roadside. Should the pillar be too far from the roadside, the speed measurement accuracy and the display readability would worsen.

4   The Collection and Evaluation of the Statistical Data

The system GEM CDU 2605 Zeus is equipped with the software SYDO Traffic Tiny which collects and statistically evaluates the individual data. This application shows the speed indicators and the most important information. Due to a limited memory capacity the individual data must be imported onto a backup source where the software SYDO Traffic Tiny is installed. The application statistically shows the data as graphs or images, and the data can be exported also to other applications, such as Microsoft Excel [7]. The Fig. 4 shows an example of the graph "The vehicles speed", which shows the speed of the individual vehicles, as they passed the measuring spot. Each individual vehicle is depicted as one spot in the graph.

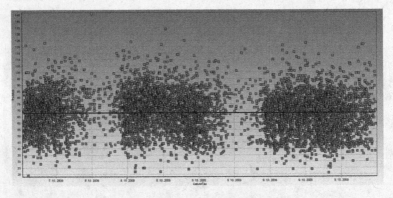

Fig. 4. The graph "The vehicles speed", SW SYDO Traffic Tiny [8]

The application also enables the set up or the change of the time axes and the depiction of the individual statistical data for the individual graphs. For printing, the software automatically generates a transparent data template.

Due to the accuracy of the collected data it is very important to correctly set the software. Mostly the vehicles classification requires a very precise set up through the default settings. The system automatically puts the vehicle into one of the categories (passenger vehicles, vans, freight vehicles, combination vehicles), and saves the respective image of the vehicle into the memory [7] (Fig. 5).

Vehicles classification is performed through the "virtual drive through gates", which, after the vehicle passes, evaluates the vehicle parameters of height, width and length. The gate which the vehicle passes, automatically evaluates the vehicle category based on the percentage occupancy of the dimensions, and sends the data file into the

Fig. 5. Online video images – vehicles classification [8]

control unit. It is necessary to point out, that should the virtual gates be incorrectly set, the data shall be misrepresented and not accurate. As a result the system can evaluate a passenger vehicle as a freight vehicle [9] (Fig. 6).

Fig. 6. Virtual drive through gates [8]

5 The Processing and Evaluation of the Measured Data

During the whole monitored period the system GEM CDU 2605 Zeus measured the individual data, which were consequently downloaded and collected in the application SYDO Traffic Tiny. This programme equipment enabled the access to data, which can be processed and evaluated through so called "reports".

The aim of this contribution was to evaluate the traffic intensity for the month of November 2016, and also the evaluation of the data such as the average speeds, exceeding of the individual speeds in the monitored section during the given time period, drivers' behaviour after spotting the speed measuring device, etc.

The vehicles are classed into three categories as follows "passenger vehicles PV" (passenger vehicles up to 3.5 t and vans), "freight vehicles FV" (vehicles over 3.5 t, trolley and busses) and "combination vehicles CV" (semitrailer trucks, trailer trucks and articulated busses or trolley). The traffic intensity is divided to the traffic intensity for the vehicles driving towards the measuring system, traffic intensity for the vehicles driving in the opposite direction and the total traffic intensity in both directions on the individual days of November 2016.

The presented Table 1 shows that the total traffic intensity in both directions for the month of November is at the level of 564 266 vehicles, out of which 288 973 are the

Table 1. Monthly traffic intensity [own study]

Date	The respective direction				The opposite direction				Total
	PV	FV	CV	Total	PV	FV	CV	Total	
1.11.2016	6 125	85	128	6 338	6 547	99	140	6 786	13 124
2.11.2016	10 052	566	239	10 857	9 154	603	302	10 059	20 916
3.11.2016	9 850	565	269	10 684	8 815	699	344	9 858	20 542
4.11.2016	10 028	589	284	10 901	9 245	701	348	10 294	21 195
5.11.2016	5 640	121	121	5 882	5 485	180	155	5 820	11 702
6.11.2016	4 725	99	82	4 906	4 668	89	105	4 862	9 768
7.11.2016	10 754	499	333	11 586	9 354	610	412	10 376	21 962
8.11.2016	10 255	547	315	11 117	9 100	599	485	10 184	21 301
9.11.2016	10 982	680	290	11 952	9 688	715	369	10 772	22 724
10.11.2016	10 822	580	340	11 742	9 840	633	488	10 961	22 703
11.11.2016	11 205	596	368	12 169	10 025	715	412	11 152	23 321
12.11.2016	5 549	157	95	5 801	5 147	188	159	5 494	11 295
13.11.2016	5 210	91	91	5 392	4 859	107	115	5 081	10 473
14.11.2016	10 856	601	315	11 772	10 405	728	381	11 514	23 286
15.11.2016	10 560	630	344	11 534	9 458	674	481	10 613	22 147
16.11.2016	10 999	680	377	12 056	10 455	755	504	11 714	23 770
17.11.2016	6 089	101	116	6 306	5 874	122	124	6 120	12 426
18.11.2016	11 258	605	324	12 187	10 225	680	451	11 356	23 543
19.11.2016	5 989	157	98	6 244	5 730	114	125	5 969	12 213
20.11.2016	4 890	88	111	5 089	4 778	129	144	5 051	10 140
21.11.2016	11 205	687	378	12 270	10 988	715	456	12 159	24 429
22.11.2016	10 588	655	408	11 651	9 785	766	509	11 060	22 711
23.11.2016	10 369	625	389	11 383	9 885	605	423	10 913	22 296
24.11.2016	10 887	572	371	11 830	10 450	680	411	11 541	23 371
25.11.2016	11 098	594	339	12 031	11 258	599	386	12 243	24 274
26.11.2016	4 599	105	113	4 817	4 512	125	133	4 770	9 587
27.11.2016	4 050	78	99	4 227	3 899	100	119	4 118	8 345
28.11.2016	10 908	606	358	11 872	10 466	588	399	11 453	23 325
29.11.2016	10 967	599	329	11 895	9 855	705	409	10 969	22 864
30.11.2016	11 456	638	388	12 482	10 888	688	455	12 031	24 513
Total	267 965	13 196	7 812	288 973	250 838	14 711	9 744	275 293	564 266

incoming vehicles (51.21%) and 275 293 are the outgoing vehicles (48.79%) of the total traffic intensity in the respective section.

The highest daily intensity in both directions was measured on 30.11.2016 (Wednesday) the number of vehicles was 24 513. The lowest daily intensity in both directions was measured on 27.11.2016 (Sunday) when the number of vehicles was 8 345.

The highest number of the freight vehicles in both directions was measured on 16.11.2016 (Wednesday) in the number of 1 435, the lowest number on 27.11.2016 (Sunday) in the number of 178.

The highest number of the combination vehicles in both directions was measured on 22.11.2016 (Tuesday) in the number of 917, the lowest number on 06.11.2016 (Sunday) in the number of 187.

The Table 1 also specifies the number of vehicles in individual categories. In the month of November 518 803 passenger vehicles, 27 907 freight vehicles and 17 556 combination vehicles passed the respective road section (Fig. 7).

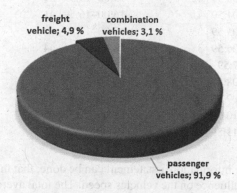

Fig. 7. The ratio of the individual vehicles categories [own study]

5.1 The Vehicles Speeds Evaluation

The primary function of the system GEM CDU 2605 Zeus is the vehicles speeds measurement, in the respective or the opposite direction of the radar head. Through the programme equipment of the SYDO Traffic Tiny the data of the average vehicles speed per hour, and of the speed exceeding, can be gained. The system is also able to measure or calculate, whether the respective vehicle increased or decreased its speed in front of the system and by how much the speed changed [9].

The most unprejudiced data are gained under the ideal conditions (ideal light conditions, low traffic intensity, good wind conditions, etc.). The equipment is highly precise, despite of this during the night or in the high traffic density the measured data can be deformed.

The vehicle speed, which is consequently processed and submitted to the equipment administrator, is defined by the video detection system, not the radar head. The radar head in this case serves primarily for the vehicle detection and displaying the actual vehicle speed on the radar board [7].

5.2 The Average Speed per Hour

When analysing the average speed per hour the software automatically sorts the individual measured speeds into the categories classified by the time frame in the 24-h format (Table 2).

The lowest average speed per hour, 31 km/h, is being reached during the afternoon peak hour, which is between 3 pm and 4 pm. On the other side, the highest average speed per hour is 46 km/h, measured during the night hours, when the traffic intensity is

Table 2. The average speed per hour [own study]

Time period	Average speed [km/h]	Time period	Average speed [km/h]
00:00–00:59	46	12:00–12:59	37
01:00–01:59	45	13:00–13:59	35
02:00–02:59	45	14:00–14:59	32
03:00–03:59	42	15:00–15:59	31
04:00–04:59	41	16:00–16:59	41
05:00–05:59	35	17:00–17:59	43
06:00–06:59	35	18:00–18:59	44
07:00–7:59	36	19:00–19:59	45
08:00–08:59	39	20:00–20:59	45
09:00–09:59	39	21:00–21:59	45
10:00–10:59	38	22:00–22:59	45
11:00–11:59	38	23:00–23:59	45

lowest. Based on these information a statement can be done, that the traffic intensity has a direct proportional influence on the vehicles speed. The total average speed per hour is on the level of 40.3 km/h, but during the day the values change in a certain interval.

5.3   The Speed Exceeding

The Software ŞYDO Traffic Tiny also classes the vehicles by the exceeded speed in the measured section. Based on the defined speed limits it creates the list of vehicles which exceeded the selected speed, in the direction of the system, in the opposite direction of the system, together in both directions, and subsequently creates the proportion of the individual categories in the respective month [9] (Table 3).

Table 3. Vehicles speed exceeding [own study]

Speed [km/h]	Vehicles no. in the direction	Vehicles no. in the opposite direction	Vehicles no. in both directions	Proportion [%]
>0	288 973	275 293	564 266	100.00
40–50	149 517	124 249	273 766	48.52
50–55	48 289	43 578	91 867	16.28
55–60	17 455	16 759	34 214	6.06
60–70	6 352	6 724	13 076	2.32
70–80	1 158	1 198	2 356	0.42
80–90	337	328	665	0.12
90–100	120	108	228	0.04
>100	41	39	80	0.01

Based on these information we can state, that the biggest part of vehicles, 48.52%, reached in the selected section the speed of 40–50 km/h. It is also interesting, that a quarter of the drivers, 25.25%, exceeded the maximum permitted speed, which is on the respective road set at 50 km/h. In the month of November 80 vehicles passed the section driving at 100 km/h, which means that they double exceeded the permitted speed. 16.28%, which represents 91 864 vehicles exceeded the permitted speed lightly, up to 5 km/h.

The table also shows that the vehicles driving towards the measuring equipment exceeded the speed more often than the vehicles driving in the opposite direction. There were 5 018 more vehicles exceeding the speed driving towards the equipment. This situation is very likely caused by the fact, that in the direction away from the Rondel (1st roundabout, Fig. 3) the road has a sharply decreasing profile.

5.4 The Speed Change of the Vehicles Exceeding the Maximum Permitted Speed

The system GEM CDU 2605 Zeus measures also the acceleration and the deceleration of the vehicles which exceeded the maximum permitted speed of 50 km/h in the direction towards the radar board. For this measurement the device compares two speeds of the vehicle, based on which it calculates the acceleration or the deceleration. The first speed of the vehicle is measured at the moment when the distance of the vehicle to the system is 70 m, which represents 5 s at the average speed of 50 km/h. The second speed is measured when the distance of the vehicle to the system is approximately 10 m [9].

The statistics shows the changes of the speed in front of the system, or, in other words, detects the value by which the vehicles accelerate or decelerate. The original speed, from which the vehicles decelerate or accelerate, is not shown, only the speed change value is specified.

The Table 4 shows the psychological effect induced by the measuring device. In the direction away from the Rondel (1st roundabout, Fig. 3) 72.2% of the vehicles decelerate by 0–5.9 km/h. In total 80.9% vehicles exceeding the speed of 50 km/h decelerate in front of the measuring device. This value clearly shows the positive influence of the measuring device on the vehicles speed and the road traffic safety.

Table 4. The vehicles speed change [own study]

Speed change [km/h]	Vehicles deceleration [%]	Vehicles acceleration [%]
0–1.9	41.7	4.2
2–5.9	30.5	5.8
6–9.9	4.2	3.9
10–14.9	2.3	3.2
15–19.9	0.7	0.8
20–24.9	0.6	0.4
25–29.9	0.4	0.3
30–34.9	0.4	0.3
35 and more	0.1	0.2
Total	80.9	19.1

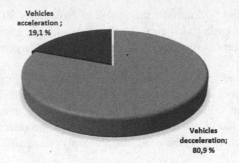

Fig. 8. Graphical illustration of changing the speed in front of the measuring device [own study]

The drivers driving slower than the maximum permitted speed, after finding out this fact, accelerate to the permitted speed limit. In many cases they even exceed the permitted speed. This is the main reason that 10.0% of vehicles accelerate in front of the device by 0–5.9 km/h.

The extreme cases present the vehicles which accelerate by more than 30 km/h, these represent 0.5%. This most likely happens during the night hours, when some vehicles exceed the speed of 100 km/h in the town area. 19.1% out of all vehicles, which exceeded the maximum permitted speed, accelerated even more in front of the device (Fig. 8).

6   Conclusion

The aim of this contribution was the analysis of the traffic flow on the selected intersection in the town of Zilina. The measuring system GEM CDU 2605 Zeus was used to collect the individual traffic data, such as the number of the vehicles, traffic intensity, the speed, etc. The part of the system is created by the statistical module for the traffic data and the evaluation software SYDO Traffic Tiny. The system is equipped with the automatic wireless data transfer to the client, through the modem or similar equipment. The record contains a series of images (a sequence) with the detailed record of the vehicle exceeding the speed.

The system is aimed to improve the state of the traffic in the towns and villages, increase the traffic safety in the certain areas, improve the traffic fluency, decrease the number of traffic accidents, decrease the number of fatalities on the roads, decrease the amount of the emissions and to improve the living standard in the towns and villages [10]. This video system serves as a light decelerator in the preventive precautions in the road traffic. Placing these systems into the most disposed locations shall enable the creation of the traffic flow maps, including their sharing on the internet and navigation systems.

Acknowledgement. Centre of excellence for systems and services of intelligent transport II, ITMS 26220120050 supported by the Research & Development Operational Programme funded by the ERDF.

References

1. Gnap, J., Konečný, V., Poliak, M.: Demand elasticity of public transport. In: Ekonomický časopis, Roč. 54, č. 7, pp. 668–684 (2006)
2. www.maps.google.sk. Accessed 14 Feb 2017
3. http://www.elektronika.sk/gem-cdu-2605-zeus-lpr. Accessed 14 Feb 2017
4. Kalašová, A., Mikulski, J., Kubíková, S.: The impact of intelligent transport systems on an accident rate of the chosen part of road communication network in the Slovak Republic. In: Mikulski, J. (ed.) TST 2016. CCIS, vol. 640, pp. 47–58. Springer, Cham (2016). doi:10.1007/978-3-319-49646-7_5
5. http://www.vecturatrade.cz/pristroje-na-mereni-rychlosti.html. Accessed 14 Feb 2017
6. Černický, Ľ., Kalašová, A., Kapusta, J.: Signal controlled junctions calculations in traffic-capacity assessment - Aimsun, OmniTrans, Webster and TP 10/2010 results comparison. Transp. Probl. **11**(1), 121–130 (2016)
7. Internal materials, ZTS Elektronika SKS s.r.o
8. Software SYDO Traffic Tiny. Accessed 14 Feb 2017
9. GEMOS CZ, SYDO Traffic (instruction). Accessed 14 Feb 2017
10. http://www.regionservis.cz/document/filename/2659/Prezentace_spol._GEMOS_CZ_s.r.o..pdf. Accessed 14 Feb 2017

The Impact of Telematics on the Functioning of TSL Sector Entities

Ryszard Janecki[✉]

University of Economics in Katowice, 1 Maja 50, 40-287 Katowice, Poland
ryszard.janecki@ue.katowice.pl

Abstract. Modern technologies of ITS have a varying impact on the TSL sector activity. Mutual relations in the system modern telematics systems - sector entities transportation, spedition and logistics can be considered taking into account, among others, functioning of TSL entities. Essential for their activities are the key challenges of the present. The article considers the impact of modern telematics systems for business activity of transport, spedition and logistics enterprises in accordance with the principles of sustainable mobility and corporate social responsibility. In content of the article were pointed out the essence of both concepts possible to implement in the TSL, were determined their objectives and benefits they generate for the entities sphere of the sector, including its customers. It also presents a proposal to assess the impact of telematics on TSL sector entities of using a specific set of criteria.

Keywords: TSL sector · Social responsibility · Sustainability mobility · Modern telematics systems

1 Introduction

The process of functioning of the TSL sector entities is a complex and multi-faceted problem. Nowadays activity of the TSL entities sphere should be developed as a series of economic processes in the context of a comprehensive logistics chain. It includes not only transportation and storage, but also within the required range formal and legal and customs service. This implies the need to harmonize the many activities carried out often by various TSL entities. With this approach it is possible to achieve the optimal flow of goods, fulfilling the criteria related to time, reliability, and efficiency offered by the services sector and to provide their consumers with the widely understood convenience.

Great importance in the economic activity of TSL sector is the key challenges of the present. These should include, among other things concepts of corporate social responsibility (Corporate Social Responsibility, CSR), wider social responsibility (Social Responsibility, SR) and sustainable mobility (Sustainability Mobility, SM). The implementation of each of them needs to become application solutions using modern telematics systems, as illustrated in Fig. 1.

From this point of view, it is advisable to present the essence possible to implement in the TSL concept. It should be emphasized the need to use modern telematics systems that enable implementation of the objectives of both concepts. We should also pay attention to

© Springer International Publishing AG 2017
J. Mikulski (Ed.): TST 2017, CCIS 715, pp. 178–190, 2017.
DOI: 10.1007/978-3-319-66251-0_15

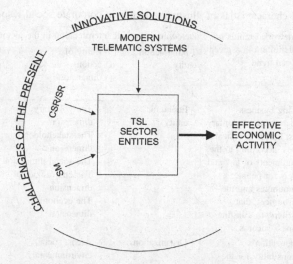

Fig. 1. The operation of the TSL sector in the conditions of implementation of the concept of corporate social responsibility and sustainability mobility [own study]

the benefits and synergy of effects from the implementation of telematics solutions that support both concepts of TSL enterprises operation. Economic activity in accordance with the principles of social responsibility and taking account the process of sustainability mobility is increasingly important attribute of the TSL sector subjective sphere.

2   The Concepts of Corporate Social Responsibility (CSR/SR) and Sustainability Mobility (SM) as the Contemporary Challenges to the TSL Sector

In the development of corporate social responsibility can be distinguished two trends:

- first - relating social responsibility only for commercial entities - market enterprises, so the sphere of business, known as corporate social responsibility (Corporate Social Responsibility, CSR)
- second, developing in the last decade, expanding social responsibility beyond the realm of business, including the whole social space (Social Responsibility, SR).

General characteristics of both the concept of social responsibility are presented in Table 1.

The use of the concept of CSR and SR in business strategies requires the determination of objective dimensions, which are recognized in seven areas, as shown in Fig. 2.

Economic activity of TSL enterprises taking into account that highlighted subject frames of both concepts must be as systemic actions. It is possible, in turn, when are formulated objectives to be achieved in each of the subject areas. Concretised objectives include [5–12]:

Table 1. General characteristics of illustrative trends in corporate social responsibility [2–4]

Name of pictorial trend	Framework formula of SR definition in a given pictorial trend	Determinants of SR interpretation in the pictorial trend		
		Taking action entity	Entity of actions - the key dimensions	Beneficiaries of the undertaken activities
1	2	3	4	5
1. Corporate social responsibility (CSR)	Making business decisions on a voluntary basis, linked to ethical values, according to the requirements of law and respect for people, communities and the environment, that contribute to a lasting business success	– Economic entity	– The voluntary dimension – The stakeholders dimension – The social dimension – The environmental dimension – The economic dimension	– Internal and external companies stakeholders – Local communities – Society
2. Social responsibility (SR)	Organization's responsibility for the consequences of their decisions and activities on society and the environment through transparent and ethical behaviour which: – Contributes to sustainable development, including health and the welfare of society, – Takes into account the expectations of stakeholders, – Is in accordance with the law and international norms of behaviour, – Is integrated throughout the organization and practiced in its external relations	– Organizations	– Areas: social, environmental, economic and of stakeholders	– Stakeholders – Society

- creating opportunities to employees of training facilities and acquire the powers related to transport, forwarding and logistics,
- maintaining the required number of employees in companies in the TSL sector,
- providing priority approach to the health and safety of workers,
- reducing emission and reducing waste in the TSL in passenger transport,
- reducing energy consumption by TSL sector supporting passenger transport,
- implementing organizational solutions in urban transport of people, contributing to reduce the negative impact on the environment,
- implementing innovative ecological solutions in the transport of people,
- reducing the negative impact of the TSL sector on the environment through proper management of supply chains,

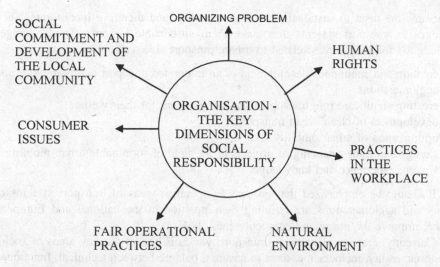

Fig. 2. Key dimensions of the subject in the concept of social responsibility [1]

- developing environmentally conscious logistics potential,
- providing economic, environmental and social operational efficiency,
- adapting service offer of TSL companies to customer requirements, taking into account the life cycle of their products with reduced emissivity,
- developing a reliable, professional and with the use of innovative supply chain management, allowing customers of TSL sector companies to build a competitive advantage in their markets,
- promoting public transport and offering its services as an alternative to the passenger car because of the quality of life of the cities and agglomerations residents,
- adjusting the offer of public transport to the needs of social groups at risk of exclusion,
- sharing knowledge about transportation, freight forwarding and logistics with residents,
- conducting and supporting social action regarding TSL issues.

From placed specifications it can be concluded that the important positions for the purposes of social responsibility in the TSL occupy these ones that relate to the environment, quality of life, operational efficiency and innovation of TSL. This gives rise to a wide spectrum of activities in the field of intelligent transport systems and intelligent logistics, and therefore the implementation of modern telematics systems. Protection of the environment, the level of renewable energy consumption, functioning process modernity, innovative vehicles, the modal split in passenger transport, the priorities of the movement of vehicles and sustainable mobility are just some of the possible areas of activity of modern TSL enterprises in the implementation principles conditions of corporate social responsibility.

The way to a better quality of life in cities and urban areas and their functional environment is satisfactory meeting the mobility needs, which is the main task of the second concept, which has serious influence on the functioning of the TSL sector. This generates the need to sustainability urban mobility, and therefore incorporates such changes in transport systems that make them sustainable systems. These changes include the following areas related to urban transport [13]:

- creation and continuous development of an integrated transport policy in cities and agglomerations
- creating significant role transport users and concern for their welfare,
- development of clean urban transport,
- optimization of urban mobility,
- strengthening the financing of projects in the field of sustainable urban mobility,
- sharing experience and knowledge.

It should be emphasized that projects in specified areas of transport systems of cities and agglomerations are building their position in the national and European space, improve the image and competitiveness [14].

Currently, creating sustainable transport, which is one of the key areas of socio-economic reality, includes measures to ensure a balance between technical, functional, spatial, economic, social and environmental factors, characterizing the transport systems subject to sustainability. Sustainable intelligent transport must therefore be planned, managed, secure, low-carbon, serviced on line and intermodal [15–17].

Transport system in which are created continuously and permanently mentioned attributes, provides a constant increase in the sustainability of mobility. It is understood as the ability to meet the needs of society in terms of freedom of movement, access to key activities and transport services, trade and establish relationships without sacrificing other values of human and environmental resources needed today and in the future [18–20].

Component connecting both concepts in complex solutions implemented in the TSL sector are modern telematics systems. In further considerations have been synthetically introduced innovations with the use of telematics in the TSL sector.

3   Innovations Using Telematics in the TSL Sector Entities

Practical solutions in telematics are strongly determined by the areas of their applications. So it is in the TSL sector, hence the talk of sector telematics, the development of which, thanks to many advantages, is used increasingly modern telematics systems. Because in transport and logistics it is difficult to extract a unified technology group, the division of TSL telematics on transport telematics and logistics telematics is justified [21].

An example specification of telematics service systems in transport contains Table 2. Using the technical - subjective criterion can be distinguished three groups of innovative solutions:

- Intelligent Transport Systems (ITS) in the modes of transport and intermodal systems,
- ITS in urban transport,
- intelligent systems in logistics.

Their extensive list is given in [21].

Table 2. Type of services provided by telematics systems of transport [22–24]

Areas of application	Services
1	2
Traffic and travel management	Information before traveling
	Information for drivers while driving
	Information about the journey by public mean of transport and booking
	Passenger service information
	Traffic control
	Accident actions management
	Demand management on traveling service Control of emissions and their reduction
	Checking the intersections of roads and railways
	Supporting enforcement regulations
	Infrastructure maintenance management
	Driving and navigation
Public transport management	Management of transport operation and fleet
	Electronic Ticket Service
	Travel information about transit
	Personalized public transit
	Public travel safety
	"On demand" transport management (also multimodal)
Transport payments	Electronic payment services, including the collection of fees for the use of transport infrastructure
Cargo transport vehicles actions (commercial)	Electronic commercial vehicles briefing
	Automatic inspection of road safety
	On-board safety measurement
	Administrative processes for commercial vehicles
	Control of hazardous materials transport
	Fleet management of commercial vehicles
	Transport planning support (obtaining orders, consignments completing etc.)
	Consignments monitoring
Accidents management	Accident notification (official notification to the competent institutions about the accident) and safety of people
	Management of emergency services vehicles
	Notification of hazardous materials transport
Advanced vehicle safety systems	Preventing longitudinal and lateral collisions
	Preventing collisions at intersections
	Against collision vision systems
	Emergency security
	Against the crash immune systems
	Automatic vehicles handling
	Against crushing protection
Security systems	Safety of public traveling (including pedestrians)
	Safety of disabled road users
	Smart intersections
Information control	The use of archived data
Management of construction and road infrastructure maintenance	Constructional renovations actions
	Maintenance operations, especially in winter

The specification indicates the following phenomena that currently accompany the development of the telematics TSL sector. Firstly, transport telematics (including shipping) integrates telecommunications and information technology to manage transportation systems. Thus providing support for both operational activities (technical) and organization associated with the movement of people and goods. The primary purpose of these solutions is also increasing efficiency and improving the level of transport safety. Among the proposed measures are also formulated directly demand the use of ITS solutions aimed at sustainability mobility and the creation of intelligent urban transport.

On the other hand, logistics telematics is characterized by versatility, providing not only smooth operation of TSL enterprises in the areas of organizational, economic and social, but above all the opportunity to achieve a competitive advantage in the market [25]. Logistics telematics promotes proactive approach to logistics services. It is expressed as a need for virtual planning of supply chains carried out through the use of e-logistics. These solutions also provide coordination of all ongoing logistics processes in logistic systems [21, 26].

As a summary of this part of the synthetic considerations (due to the volume of publications), arise the following assertions:

- it should be emphasized occurring in the TSL sector very strong relationships between the innovation and the use of transport telematics solutions and logistics telematics in the sector,
- modern telematics systems are necessary to meet specific goals and objectives of CSR/SR and SM in the TSL sector; they allow the integration of both concepts and thus achieve the beneficial effects of actions synergies.

4   The Impact of Telematics Technologies on TSL Entities Function

From the previous considerations appears that telematics is a tool to realize many of the objectives of social responsibility and sustainability mobility TSL sector entities. Thus, the growth of the sector innovation:

- creates better possibilities for implementing the principles of each considered concept by TSL entities in their business activity,
- is a factor integrating the two concepts, allowing for actions synergies and increase the benefits available to TSL entities and beneficiaries of social responsibility and sustainability mobility.

Considering the context of the instrumental role of innovative solutions using modern telematics systems in the TSL sector can be formulated answers to questions related to their influence on the functioning of the subjective sphere of the sector according to the principles of implemented concept of corporate social responsibility and sustainability mobility. Under these conditions, the impact of telematics has allocation in the three areas of economic activities relating to:

– introduction of the principles of corporate social responsibility and sustainable mobility to practice of TSL sector entities, depending on the level of advancement of the implementation of both concepts,
– improvement the use of telematics projects in the process of implementation and operation of corporate social responsibility and sustainability mobility in the TSL sector,
– adaptation of telematics projects to the rules of implemented concepts.

In the first of the highlighted areas of impact of telematics, the use of modern and comprehensive telematics projects ensure realization of the objectives in five of the seven subject areas of social responsibility [27].

There are changes in the economic activity of TSL entities. They are focused on the implementation and then realization of the principles of social responsibility and/or sustainable mobility. On the market is being shaped range of logistics services satisfying the requirements of the selected or both operating concepts.

In the case of complex ITS solutions for urban transport, taking into account the link between the two concepts under consideration (Fig. 3), are being obtained better use of projects and increase the benefits of introduced innovations. Can be mentioned in this field inter alia [27]:

– generating synergy effects,
– increasing importance of solutions for public transport, enabling the simultaneous achievement of the SR and SM objectives at an environmental nature and related to local communities,
– changing currently expressed opinions of minor importance of ITS projects for local governments, other entities making investment decisions in this area and for transport users,

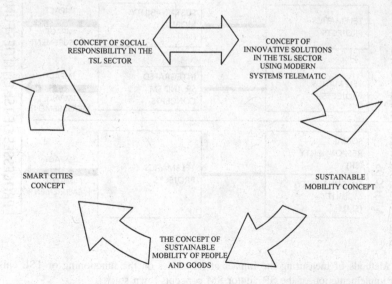

Fig. 3. Relationship between the applicable concepts in the TSL sector [27]

– growing importance of ITS projects in cities, agglomerations and regions regarding the issues of increasing traffic and the increasing movement congestion,
– obtaining benefits for transport users (residents, economy entities, other transport users) in the cities, agglomerations and regions in which have been deployed ITS systems.

Effective action of the TSL sector entities to improve the use of telematics solutions are determined by adequacy of design solutions, the quality of the investment process implementing the project, and the course of operation processes. Catalogue of requirements dedicated within this range of telematics projects is extensive [27].

Speaking about the impact of telematics innovation, must also be indicated the need for implementation by TSL entities actions to ensure the compatibility of the proposed telematics solutions with the principles of corporate social responsibility and sustainable mobility [27–29].

Under these conditions, a necessity in the process of managing the operation of TSL entities is the need to use instruments to measure the impact of telematics solutions. Implementation of telematics projects are integrated with the development of the TSL sector entities managing socially responsible (CSR/SR), and measures aimed at sustainability mobility (SM concept).

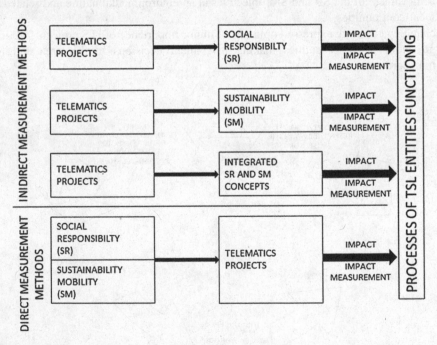

Fig. 4. Methods of measuring the impact of telematics on the functioning of TSL entities in terms of implementation of the SR and/or SM concepts [own study]

Considering these phenomena, the proposed instruments measure the impact of telematics on the functioning of the TSL sector entities in variant terms, using existing methods, is as follows:

- an intermediate variant, in which is carried out a separate measurement of the impact of actions, consistent with the principles of corporate social responsibility or sustainability mobility in the implementation of which are being used modern telematics systems [30–32],

Table 3. Examples of indicators of the impact of telematics solutions on TSL entities functioning [own study]

Group of impact indicators	Examples of impact indicators
1	2
1. Telematics projects in urban transport	– The scale of the increase in capacity of linear and spot elements in the street-road system in [%]
	– Level of reduction of time losses in the road-street network in [%]
	– The scale of the air emissions reduction in [%]
	– Accident rate in [the number of people killed (injured)/100 k registered vehicles]
	– Density of accidents [number of accidents/km]
	– Scale of fuel consumption reduction [%]
	– Scale of changes in the size of passenger transport by public transport, thanks to applications of modern telematics systems in terms of corporate social responsibility and sustainability mobility in [%]
	– Scale of changes in the size of cargo traffic through the use of advanced telematics systems in terms of social responsibility and sustainability mobility by TSL entities in [%]
	– Share of long transport routes on which are installed intelligent transport systems in [%]
	– Growth rate of the number of people sharing the information required to smooth prepare and comfortable travel in [%]
	– The share of the drivers number violating traffic rules identified by monitoring events in selected traffic parameters or driver behaviour in [%]
2. Telematics projects in logistics	– Scale of changes in the production volume of logistics services in terms of applications of telematics systems and implementing the principles of SR and/or SM in [%]
	– Scale of changes in stock volume in the supply chain in terms of applications of telematics systems and implementing the principles of SR and/or SM in [%]
	– Share of shutdowns caused by insufficient supply in [%]
	– Rate of on-time deliveries in [%]
	– Scale of the value of orders rejected by the customers in [%]
	– Changes in the size of the actual duration of cargo operations in [%]
	– The average fuel consumption per unit of exploitation labour in [l/vehicle-km] or [l/vehicle-h]
	– The average mileage of the fleet on the hour crossing in the network [km/h]
	– Share of mistakes in picking the total number of this type of operation in [%]
	– The average flow values through the warehouse in [zł/work-h]
	– Indicator of the supply without failures in [%]

- an intermediate variant, in which is carried out an integrated measurement of the impact actions in accordance with the principles of both concepts, in the implementation of which are being used modern telematics systems,
- a direct variant; the measure of the impact of telematics applications on the functioning of the TSL sector entities in the conditions of concept implementation of corporate social responsibility and/or sustainable mobility.

Figure 4 shows practicable, approach to measuring the impact of telematics on the functioning of the TSL sector entities. However, Table 3 shows examples of indicators of influence determined for direct measurement.

The method for calculating the measure values for the considered methods of measurement has one thing in common. It is the regularity pointing to the fact that all the parameters occurring in the calculation formulas counters relate to economic activities of TSL entities under implementation of the principles of corporate social responsibility and sustainability mobility. These processes are supported by modern applications of telematics systems.

5   Conclusion

It can be noticed that the current space of, positive functioning of TSL entities, actions in the implementation of advanced telematics systems are becoming increasingly important. At the same time economic decisions bringing TSL companies economic success must be socially responsible and lead to sustainability mobility.

On the basis of considerations regarding the role of telematics in the functioning of the entities TSL in the conditions of socially responsible business, focused on the sustainability of mobility, can be formulated following conclusions:

- there is a large variety of possible areas of use and multiplicity of solutions for innovative projects within the TSL sectors, using modern telematics systems,
- innovative projects in the TSL sector are an important instrument for achieving the sectoral objectives of corporate social responsibility and sustainability mobility,
- integration of TSL sector innovation concepts, with the concepts of corporate social responsibility and sustainability mobility creates the conditions for doing business according to the rules of both concepts; currently this synergistic system is not functioning, which requires rapid changes this state of affairs,
- there are effective methods of measuring the impact of telematics on the functioning of TSL sector entities; this tool should be used, because it allows the selection of economically, socially and environmentally beneficial solutions that integrate modern telematics systems with the principles of corporate social responsibility and sustainability mobility.

References

1. Polish Standard (PN) – ISO 26000, pp. 31–32
2. Dahlsrud, A.: How corporate social responsibility is defined: an analysis of 37 definitions. Corp. Soc. Responsib. Environ. Manag. 15(1), 7–11 (2008)

3. What is Social Responsibility (SR). http://asq.org/learn-about-quality/social-responsibility. Accessed 10 Jan 2017
4. Duckworth, H.A., Moore, R.A.: Social Responsibility: Failure Mode, Effects and Analysis, pp. 2–3, 25–26. CRC Press, Taylor & Francis Group, Boca Raton (2010)
5. Gableta, M.: Zakres przedmiotowy gospodarowania potencjałem pracy. W: Gabelta, M.W. (ed.) Potencjał pracy przedsiębiorstwa. Wydawnictwo AE we Wrocławiu, Wrocław, pp. 22–25 (2006)
6. Tarantino, A. (ed.): Governance, Risk and Compliance Handbook, Technology, Finance, Environmental and International Guidance and Best Practices. Wiley, Hoboken (2008)
7. Zieliński, M.: Korzyści z wdrożenia koncepcji CSR w zarządzaniu zasobami ludzkimi. Zeszyty Naukowe Politechniki Śląskiej Seria: Organizacja i Zarządzanie 74, 656–657 (2014)
8. Matuszczak, A.E., Myszak, M.: Nowa kultura mobilności jako kierunek rozwoju transportu miejskiego i regionalnego w województwie śląskim. W: Michałowska M. (red.): Współczesne uwarunkowania rozwoju transportu w regionie. Zeszyty Naukowe Wydziałowe Uniwersytetu Ekonomicznego w Katowicach Studia Ekonomiczne 143, pp. 142–143 (2013)
9. Jasińska–Bilczak, A.: Odpowiedzialność społeczna przedsiębiorstw sektora TSL. Logistyka 5, 1860–1866 (2014)
10. Ciechorzewska, M.: Kształtowanie kultury bezpieczeństwa jako element działań CSR w firmach transportowych. Logistyka 3, 713–720c (2015)
11. Stawiarska, E.: Wykorzystanie współczesnych technologii informatycznych w kształtowaniu odpowiedzialnych społecznie spedytorów i transportowców. Zeszyty Naukowe Politechniki Śląskiej seria organizacja i Zarządzania 80, 239–251 (2015)
12. Wyzwania i szanse w sektorze transportu i logistyki. www.csrprofit.com/id.100022. Accessed 10 Jan 2017
13. Plan działania na rzecz mobilności w miastach. Komunikat Komisji do Parlamentu Europejskiego, rady Europejskiej Komitetu Ekonomiczno-Społecznego i Komitetu Regionów COM (2009) 490. Bruksela, pp. 5–12 (2009)
14. Janecki, R.: Nowa kultura mobilności jako kierunek rozwoju transportu miejskiego i regionalnego w województwie śląskim. W: Michałowska M. (red.): Współczesne uwarunkowania rozwoju transportu w regionie. Zeszyty Naukowe Wydziałowe Uniwersytetu Ekonomicznego w Katowicach Studia Ekonomiczne 143, pp. 142–143 (2013)
15. Gudmudson, H.: Sustainable transport and performance indicators. W: Hester R.E., Harrison R.M. (eds.): Transport and the Environment. Issues in Environmental Science and Technology 20, pp. 35–52 (2004)
16. Rudnicki, A.: Uwarunkowania przestrzenne polityki transportowej. W: Więcławowicz – Bilska E., Zusiak Z. (eds.): Planowanie przestrzenne a wyrównanie szans w obszarach rozszerzonej Unii Europejskiej. Czasopismo Techniczne 2, pp. 117–125 (2005)
17. Black, W.R.: Sustainable Transportation: Problems and Solutions. The Guilford Press, New York (2010)
18. Mobility for development, facts and trends. The World Business Council for Sustainable Development, Geneva (2007)
19. Mobility 20130: Meeting the challenges to sustainability. The Sustainable Mobility Project, Full Report 2004. WBC SD, Geneva (2004)
20. Williams R.: A Definition of Sustainable Mobility 2007. http://www.carbonsmart.com/files/definition_of_sustainable_mobility_2007nr15.pdf. Accessed 10 Jan 2017
21. Mikulski, J.: Innowacje z zastosowaniem telematyki w sektorze TSL. W: Michałowska M. (ed.): Aktywność podmiotów sektora TSL w tworzeniu i realizacji strategii społecznej odpowiedzialności biznesu. Wydawnictwo Uniwersytetu Ekonomicznego w Katowicach, p. 83 (2016)

22. Wydro, K.B.: Usługi i systemy telematyczne w transporcie. Telekomunikacja i Techniki Informacyjne **3–4**, 24 (2008)
23. ROSETTA – Real Opportunities for Explotation of Transport Telematics Applications. http://www.org.trg.sot.acx.uk. Accessed 10 Jan 2017
24. Wojewódzka–Król, K., Rolbiecki, R.: Inteligentne systemy transportowe w świetle europejskiej polityki transportowej. W: E-gospodarka w Polsce. Stan obecny i perspektywy rozwoju. Zeszyty Naukowe Uniwersytetu Szczecińskiego Ekonomiczne Problemy Usług 57, p. 70 (2010)
25. Bujak, A., Cieśliński, W.: Prototypowanie optymalizacji zintegrowanych procesów logistycznych z wykorzystaniem paradygmatu SOA (Sernice Orientem Architecture) i Business Process Inteligence (BPI). LOGITRANS – VII Konferencja Naukowo-techniczna LOGISTYKA, SYSTEMY TRABNSPORTOWE, BEZPEICZEŃSTWO W TRANSPROCIE. Logistyka 2, p. 14 (2014)
26. Kaźmierska–Grębosz, M., Grębosz, M.: "Wspomaganie komputerowe w kontekście rozwoju inteligentnej logistyki". Mechanik 67, p. 404 (2015)
27. Janecki, R.: Zastosowania nowoczesnych systemów telepatycznych w sektora TSL na tle zasad CSR. W: Michałowska M. (ed.): Aktywność podmiotów sektora TSL w tworzeniu i realizacji strategii społecznej odpowiedzialności biznesu. Wydawnictwo Uniwersytetu Ekonomicznego w Katowicach, p. 117 (2016)
28. Zieliński, M.: Korzyści z wdrożenia koncepcji CSR w zarządzaniu zasobami ludzkimi. Zeszyty Naukowe Politechniki Śląskiej seria: Organizacja i Zarządzanie **74**, 655 (2014)
29. Aluchna, M.: Ramy teoretyczne i metodologiczne badania. W: Płoszajski P. (ed.): Strategie społecznej odpowiedzialności polskich spółek giełdowych. Oficyna Wydawnicza SGH, Warszawa, pp. 16–17 (2013)
30. Rok, B.: Odpowiedzialny biznes w nieodpowiedzialnym świecie. Forum Odpowiedzialnego Biznesu, Warszawa, pp. 44–45 (2004)
31. Żemigała, M.: Społeczna odpowiedzialność przedsiębiorstwa. Wydawnictwo Wolters Kluwer Polska, Kraków, pp. 198–220 (2007)
32. Bartniczak, B.: Zrównoważony transport na poziomie regionalnym jako przedmiot pomiaru wskaźnikowego. Studia Ekonomiczne Zeszyty Naukowe Wydziałowe Uniwersytetu Ekonomicznego w Katowicach 143, p. 16 (2013)

Autonomous Vehicles and Road Safety

Maria Michałowska[1] and Mariusz Ogłoziński[2(✉)]

[1] University of Economics in Katowice, 1 Maja 50, 40-287 Katowice, Poland
mariamic@ue.katowice.pl
[2] Institute for Transport Safety, Górnicza 16 lok. 6, Chorzów, Poland
mariusz.oglozinski@gmail.com

Abstract. The article is about participation of autonomous vehicles in traffic and their impact on road safety. The aim of the article is to present the opportunities of using autonomous vehicles to meet the needs related to the movement of persons and goods, in the context of improving road safety. Road transport is the most popular branch of the transport, which is reflected in its share in generating fatalities at all transport processes. Entry into service of the autonomous vehicles for transport of people and goods, can contribute to improving road safety indicators and, consequently, to reduce the social, economic and environmental costs incurred in connection with road traffic. Reducing or eliminating the impact of the human factor from the decision-making process regarding the quality and way of participating in road traffic may prove to be a landmark step in reducing road deaths.

Keywords: Road safety · Autonomous driving

1 Introduction

The availability of road infrastructure, relatively low purchase costs of means of transport and ease of gaining permission to drive road vehicles cause, that road transport is one of the most frequently used modes of transport. Having regard to all the advantages of this branch, one cannot forget about the costs incurred in connection with road traffic, i.e. social, economic, environmental. Research indicate the dominant share of road transport in generating fatalities at all transport processes related to the movement of people and goods. To ensure the safety of transport processes, including road safety, with increasing number of journeys, it seems currently the most urgent challenge facing the road users, manufacturers of means of transport and operators managing road infrastructure. The concept of safety is equated with a lack of risk, confidence, serenity and certainty. Abraham Maslow puts the need for safety as one of the fundamental just after physiological needs. On the other hand, the contemporary society have strongly established need for mobility that is associated with the risk of safety threats.

2 Losses Related to Road Traffic

The level of road safety in highly developed countries is seen as one of the elements of the quality and comfort of life [2]. According the World Bank data, in connection with traffic accidents during the year in the world nearly 50 million people are injured, and

© Springer International Publishing AG 2017
J. Mikulski (Ed.): TST 2017, CCIS 715, pp. 191–202, 2017.
DOI: 10.1007/978-3-319-66251-0_16

about 1.2 million people die, of which 70% in developing countries. According to WHO, the lack of determined actions to improve road safety in the next 10 years in developing countries will bring about to die more than 6 million people, and 60 million will be injured. Traffic accidents in 2004 accounted for the ninth cause of premature death, and in 2030 will be the third (Fig. 1).

The leading causes of burden of disease, world, 2004 and 2030[1]

2004	% DALYs	Rank		Rank	% DALYs	2030
Lower respiratory infections	6.2	1.		1.	6.2	Unipolar depressive
Diarrhoeal diseases	4.8	2.		2.	5.5	Ischaemic heart disease
Unipolar depressive	4.3	3.		3.	4.9	Road traffic accidents
Ischaemic heart disease	4.1	4.		4.	4.3	Cerebrovascular disease
HIV/AIDS	3.8	5.		5.	3.8	COPD
Cerebrovascular disease	3.1	6.		6.	3.2	Lower respiratory infections
Prematurity and low birth weight	2.9	7.		7.	2.9	Hearing loss, adult onset
Birth asphyxia and birth trauma	2.7	8.		8.	2.7	Refractive errors
Road traffic accidents	2.7	9.		9.	2.5	HIV/AIDS
Neonatal infections and other	2.7	10.		10.	2.3	Diabetes mellitus
COPD	2.0	13.		11.	1.9	Neonatal infections and
Refractive errors	1.8	14.		12.	1.9	Prematurity and low birth weight
Hearing loss, adult onset	1.8	15.		15.	1.9	Birth asphyxia and birth trauma
Diabetes mellitus	1.3	19.		18.	1.6	Diarrhoeal diseases

Fig. 1. The leading causes of premature death in the world [1] (Calculations based on the rate of DALYs (Disability Adjusted Life Years), which is the sum of years of potential life lost due to premature death and the years of productive life lost due to disability.)

Road traffic accidents are also economic losses, which globally give approximately 500 billion US dollar a year. In the European Union, road accidents are the first cause of external reasons death of people up to 45 years of age, generating a yearly loss of over 200 million EUR. In Poland, road accidents cost more than 30 billion PLN a year, that is about 2% of GDP [3]. All this makes road transport as the most dangerous branch of transport, posing the greatest risk of loss of life while during move, as well as generating significant losses in the economic dimension.

3   The Safe System

In the IV European road safety programme, announced in 2010: "Towards a European road safety area: policy orientations on road safety 2011–2020", it is assumed that the road users are the primary part of road safety system, and such a system should take into account the human errors and inappropriate behaviour and correct it as far as possible [4]. The idea of a safe system of man-vehicle-road is also the main goal of the Global Plan For The Decade Of Action For Road Safety 2011–2020, adopted by the General Assembly of the United Nations in 2010. A safe system means the creation and development of the road transport system, which is better able to adapt to the man, his errors and weaknesses. The starting point is to accept the fact that a man, as a road user, makes mistakes, and accidents cannot be completely eliminated. Presented approach assumes that human limitations should provide the basis for creating a system of road transport, and road infrastructure and vehicles should cooperate, taking into account these limitations. Safe system of man-vehicle-road requires integration through speed management systems, vehicles, and the design of road infrastructure. The safety system approach assumes a significant shift of responsibility for road safety from road users on those who create the road transport system, for example road operators or vehicle manufacturers.

The results of the research carried out by Volvo shows that the main causes of road accidents are combinations of factors: in 90% driver related, in 30% associated with the road and its surroundings, in 10% associated with the vehicle. This is confirmed by the results of research into the factors affecting the formation of road accidents, carried out at the Institute of Transport Economics in Oslo (TØI). It shows that by a combination of factors, the cause of traffic accidents is a man in 91.5%, road in 26.3%, and a vehicle in 6.7% (Fig. 2).

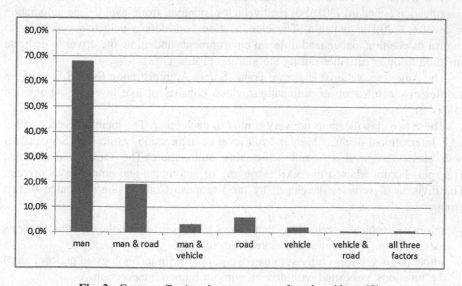

Fig. 2. Factors affecting the occurrence of road accidents [5]

Fully autonomous driving means transferring the driving task to a computer system and thus eliminating the human factor which is at the root of many road accidents. According to the European Commission, automated driving will increase road safety significantly, as human error is involved in more than 90% of all traffic accidents on Europe's roads; in which more than 40,000 people are killed and 1.5 million injured every year [6]. The data clearly show that the man is the main factor affecting the formation of road accidents. Reduce or eliminate its impact from the decision-making process regarding the quality and way of participating in road traffic may prove to be a landmark step in efforts to improve road safety.

4   Automated Vehicle Classifications

In the literature many different definitions are used: *automated, autonomous, self-driving, driverless* vehicles. *Automated vehicles* are those that use on-board equipment to perform one or more driving tasks automatically. *Self-driving* vehicles are designed to drive autonomously, without the control of a human driver. That means, that *self-driving* vehicles are a wider family of automated vehicles. Another distinction is the degree to which the automated vehicle is *autonomous*, relying solely on its on-board equipment to collect information, take decisions and inform tasks, or *connected*, i.e. in communication with other vehicles, personal devices (e.g. smart phones) or the surrounding traffic infrastructure to collect information and perform driving tasks [7]. Although some differences, all these concepts are strictly linked with each other. In common sense, the term *autonomous vehicle* applies to the vehicle, which certain functions associated with its move are carried out automatically, without human intervention, or with his limited participation. Having this on mind there are several classifications of automated vehicles, proposed by different sources.

The Society of Automotive Engineers (SAE International) created a six level classification (standard J3016) of road vehicles spanning from level 0 – no automation to level 6 – full automation. The classification considers a vehicle's capability to control its position, understand different environments and allow the driver to dedicate attention to other activities during the journey (Table 1) [8].

The American National Highway Traffic Safety Administration (NHTSA) provides a different classification of automation, which consists of five levels of automation (Table 2).

These two classifications are very similar to each other. The main difference is that SAE International distincts high and full level of automation, while NHTSA consider both classes as level 4 – full self-driving automation. The OECD International Transport Forum adapted the SAE taxonomy of automated and autonomous driving. The division of vehicles developed by the European Commission is limited to two groups:

- automated vehicle – technology that allows the driver to pass the on-board systems part of the responsibilities associated with driving,
- autonomous vehicle – fully automated, equipped with technology to perform all of the functions associated with driving without human intervention.

Table 1. Levels of driving automation for on-road vehicles by SAE's standard J3016 [7]

Level	Description
Level 0: no automation	The full time performance by the human driver of all aspects of the dynamic driving task, even when enhanced by warning or intervention systems
Level 1: driver assistance	The driving mode-specific execution by a driver assistance system of either steering or acceleration/deceleration using information about the driving environment and with the expectation that the human driver perform all remaining aspects of the dynamic driving task
Level 2: partial automation	The driving mode-specific execution by one or more driver assistance systems of both steering and acceleration/deceleration using information about the driving environment and with the expectation that the human driver perform all remaining aspects of the dynamic driving task
Level 3: conditional automation	The driving mode-specific performance by an automated driving system of all aspects of the dynamic driving task with the expectation that the human driver will respond appropriately to a request of intervene
Level 4: high automation	The driving mode-specific performance by an automated driving system of all aspects of the dynamic driving task, even if a human driver does not not respond appropriately to a request of intervene
Level 5: full automation	The full time performance by an automated driving system of all aspects of the dynamic driving task under all roadway and environmental conditions that can be managed by a human driver

Table 2. Levels of autonomous driving by NHTSA [9]

Level	Description
Level 0: no automation	The driver is in complete and sole control of the primary vehicle controls (brake, steering, throttle, and motive power) at all times, and is solely responsible for monitoring the roadway and safe operation of all vehicle controls
Level 1: function-specific automation	Automation at this level involves one or more specific control functions; if multiple functions are automated, they operate independently of each other. The driver has overall control, and is solely responsible for safe operation, but can choose to cede limited authority over a primary control (as in adaptive cruise control), the vehicle can automatically assume limited authority over a primary control (as in electronic stability control), or the automated system can provide added control to aid the driver in certain normal driving or crash-imminent situations (e.g., dynamic brake support in emergencies
Level 2: combined-function automation	This level involves automation of at least two primary control functions designed to work in unison to relieve the driver of controlling those functions. The driver is still responsible for monitoring the roadway and safe operation, and is expected to be available to take control at all times and on short notice (e.g. adaptive cruise control and automated steering working together to guide the car's movements)
Level 3: limited self-driving automation	Vehicles at this level of automation enable the driver to cede full control of all safety-critical functions under certain traffic or environmental conditions, and in those conditions to rely heavily on the vehicle to monitor for changes in those conditions requiring transition back to driver control. The driver is expected to be available for occasional control, but with sufficiently comfortable transition time
Level 4: full self-driving automation	The vehicle is designed to perform all safety-critical driving functions and monitor road-way conditions for an entire trip

The European Road Transport Research Advisory Council (ERTRAC) has drafted an Automated Driving Roadmap providing definitions of the different automation systems and the expected date of their possible deployment. According to the roadmap, fully autonomous vehicles may be deployed in 2026–2030 (Table 3).

Table 3. Automated driving roadmap by ERTRAC [10]

Level number	Automation performance	Date of possible deployment	Specification
Level 0	Park distance control	Already deployed	The system assists the driver to manoeuvre into tight spaces by communicating distance from obstacles by means of acoustic or optical signals
Level 1	Park assist	Already deployed	The system automatically steers the car into parallel and bay parking spaces, and also out of parallel parking spaces. The driver retains control of the car at all times
Level 2	Traffic jam assist	2015/2016	The function controls the vehicle longitudinally to follow the traffic flow in low speeds (lower than 30 km/h). The system can be seen as an extension of the Adaptive Cruise Control with Stop & Go functionality, i.e. no lane change support
Level 3	Traffic jam chauffeur	2017–2018	Conditional automated ariving up to 60 km/h on motorways or similar roads. The system can be activated in a traffic jam scenario. It detects a slow-driving vehicle in front and then handles the vehicle both longitudinally and laterally. Might include lane change functionality
Level 4	Highway pilot	2020–2024	Automated driving up to 130 km/h on motorways or motorway-like roads from entrance to exit, on all lanes, including overtaking movements. The driver must deliberately activate the system, but does not have to monitor it constantly. Vehicle-to-vehicle communication, cooperative systems, ad-hoc convoys can be created
Level 5	Fully automated vehicle	2026–2030	Able to handle all driving without any input from the passenger

All provided ratings indicate that fully autonomous vehicles currently do not exist, because each of them requires specific human support. However, in literature the term of *autonomous vehicle* is being used for determining vehicle movement process, there is somewhat automated.

5   Possibilities and Limitations Associated with the Use of On-Road Autonomous Vehicles

The use of on-road autonomous vehicles to meet the everyday needs of mobility raises many emotions, questions and discussions on ethical, legal, financial, economic and technical dimensions. The first death involving the autonomous vehicle took place on the 7th of May 2016 in the city of Williston, Florida, United States. Tesla Model S vehicle with autopilot function enabled was on the dual carriageway, while a truck with a semi-trailer was traveling through a junction across its direction. Tesla hit the trailer, as the on-board devices of the vehicle did not detect the white trailer against the bright sky. An additional adverse factor in this situation was high location of the trailer in conjunction with its position relative to the road. A speeding vehicle entered under the trailer, and the driver, Joshua Brown, who at the time watched movie on the onboard DVD player, was killed on the spot. It should be noted, that by the time of this event the Tesla autonomous vehicles (level two by NHTSA) overcame a total route length of 130 million miles. According to data from the NHTSA, casualty on American roads happens statistically every 97 million miles.

An event which happened in Williston is important for at least two reasons. The first, it shows the weaknesses of used technology indicating, that it is still the early stage of its development. The second, it shows the interdisciplinarity of road safety issues, which in combination of man-vehicle-ITS-road takes on a new dimension. It is therefore possible to determine that the proliferation of autonomous vehicles will be the solution to the problem of road traffic accidents?

Admission to traffic the autonomous vehicles is associated with a number of restrictions.

The first is the issue of the adjustment of traffic rules. The United Nations Economic Commission for Europe (UNECE) has modified the record of Article 8 of the Vienna Convection on Road Traffic, regulating the issues of vehicle roadworthiness. According to the amendment, the autonomous vehicles can be permitted to road traffic, provided that they meet the construction requirements stated in the UNECE regulations, and the driver will be able to take control of the vehicle and turn off the autopilot device. Also the regulation was changed, which instructs to turn off the autonomous driving mode when speed of 10 km/h is exceeded [12, 13]. So far in the United States it is permitted by law for autonomous vehicles to participate in road traffic in California, Nevada, Tennessee, Michigan and Florida (Fig. 3).

In Europe, the legal provisions allowing autonomous vehicles to participate in road traffic are introduced in Spain, Italy, Greece, Sweden, the United Kingdom and Finland. Still unregulated issue is additional marking of autonomous vehicles, for example in Nevada (US) it is the red color of the number plate. The challenge seems to be also determining the responsibility in case of a road traffic accident. This will also require changes in driving education and licensing.

In the behavioural context, unknown remains the possible human reaction to depriving him the possibility of autonomous decision, as regards style and way of driving, and use of road infrastructure. The authors of a report prepared by European Transport Safety Council (ETSC), Brussels based non-governmental organization,

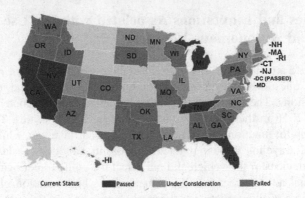

Fig. 3. Legal permission of autonomous vehicles to road traffic in the USA [9, 11]

indicate that in order to the deployed ITS systems could work effectively in improving road safety, there is a need for an in-depth analysis of the mechanisms of behavioural compensation by the driver of this state of affairs [14]. There is a concern that ridding the driver the decision making capabilities while participating in road traffic will affect negatively his psyche and can induce him to extreme behaviour, which could result in potentially dangerous situations.

The limitation of the development of autonomous mobility can also be different behavioural factor, which is the sense of safety of the driver and passengers while driving. The results of tests carried out in 2016, by the editors of one of the oldest British car magazines WhatCar? on a group of 4000 drivers indicate that 27% of them felt that dangerous, and 24% was felt that very dangerous if they had traveled using an autonomous vehicle. Safe or very safe was felt to be less than 25% of those polled. 19.5% of respondents have expressed an interest in the possibility of driving in an autonomous vehicle, while 45% considered this possibility as very little attractive and 23% as not enough attractive [15].

Another restriction of participation of autonomous vehicles in road traffic, can be the choice of "lesser evil" in case of an emergency situation: to protect the driver or the other person, including vulnerable road users, i.e. pedestrians, cyclists, motorcyclists. The results of the tests carried out by a team of researchers from the Center for Research in Management in the Toulouse School of Economics, the Department of Psychology at the University of Oregon and the Media Laboratory of the Massachusetts Institute of Technology, show that 75% of the interviewees considered that the autonomous vehicle should take into account the limitation of the number of victims of a possible accident, or the vehicle should turn and crash in a way, that only the driver was killed, saving 10 people. 50% of respondents felt that the driver should die even if he saves only one human being. However, the results of the study were entirely different when they had to decide from the position of the driver and a passerby. In this case, regardless of the position taken, all the test persons put their safety first. Among the interviewed people there were also ones, that stated that if an autonomous vehicle would have to behave so, that in this case the driver was to be killed instead of the other road users, they would not like to use such a vehicle [16].

An important road safety related restriction with participation of autonomous vehicles indicated researchers with the Transport Research Institute at the University of Michigan. The results of their research indicate that experienced drivers use eye contact and register other subtle signs in order to assess the intentions of the other road users. The lack of such suggestions may cause misunderstanding of the intent of the autonomous vehicle, and consequently dangerous situations [17].

As a serious threat that is accompanying autonomous vehicles in road traffic, it is indicated to be the safety of pedestrians, cyclists and other vulnerable road users. Their behaviour is often unpredictable and sudden, that autonomous vehicles may not cope with so many variables in dense city traffic.

Another serious threat may be a matter of data protection and cybersecurity. In research conducted by the American National Highway Traffic Safety Administration, 50% of respondents expressed their concern before the acquisition of control over the autonomous vehicle by unauthorized people and causing a threat to themselves and other road users [17]. Protection of personal data and privacy is also one of the determining factors for the successful deployment of autonomous driving. Users must have assurance that personal data are not a commodity, and know they can effectively control how and for what purposes their data are being used [6].

Dissemination of autonomous vehicles in road traffic may result in improving of road safety. However achieving a positive effect in this area, will depend on smooth dealing with many obstacles with using autonomous vehicles in traffic and understanding restrictions of human cooperation with machines. All of this poses a serious challenge for man, as a participant of road traffic, but also for vehicle manufacturers, designers and managers of road infrastructure. It will be necessary to create a technical standardisation for international compatibility and interoperability.

In terms of road safety, the most difficult may be the initial phase of dissemination of autonomous vehicles and transitional period, when, on the one hand, technology will require further refinement, the participants of road traffic will have to get used to the new conditions, and, on the other hand, both autonomous and non-autonomous vehicles will participate in traffic. However, researchers from the British Transport Research Laboratory (TRL) indicate that in the longer term refined technology of autonomous vehicles will bring a revolution in the field of road safety. It will reduce the number of fatalities among pedestrians by about 20%, and a significant reduction in the number of road collisions. They also assume, that the number of injuries due to traffic accidents will be reduced to a similar extent as after introduction of obligatory use of safety belts [18].

The possibility of improving safety and fluency of road traffic in the context of fully autonomous vehicles may be the change of the current concept of intersections. Traffic light can be eliminated, and the movement of vehicles in all directions will be able to take place continuously [19]. This solution also reduces the problem of congestion. The potential weakness of this concept is the movement of pedestrians, cyclists, motorcyclists and other vulnerable participants, who can disturb the traffic order posing a threat to themselves and other road users. The success of this concept will also require them to adapt to the prevailing traffic rules. Participation of autonomous vehicles in road traffic may also extort a change in road markings, what may increase the transparency and will be beneficial also for the other road users. Potentially beneficial in

Table 4. Possibilities and limitations associated with the use of on-road autonomous vehicles [own study]

Possibilities	Limitations
Improving of road safety records	Behavioural and ethical issues: human reactions and compensation, eye contact and other subtle signs while assessing the intentions of the other road users
Improving safety and fluency of road traffic	Safety of pedestrians, cyclists and other vulnerable road users
Eliminating of traffic lights – the continuous movement of vehicles possible in all directions	Legal framework: traffic rules, vehicle roadworthiness driving education and licensing responsibility on case of an accident
Reducing the problem of congestion	Software and data processing: the choice of "lesser evil" in case of emergency data protection technical standardisation for international compatibility and interoperability
Change of the current concept of intersections: change of road markings, what will make it more transparent and thus beneficial for the other road users	Cybersecurity and its threats
Increased safety: the ability of communication vehicle-to-vehicle (V2V), vehicle-to-infrastructure (V2I), and vehicle-to-any other object (V2X)	Infrastructure – needs to be improved

this area, in terms of road safety related to autonomous vehicles, is the ability of communication vehicle-to-vehicle (V2V) and vehicle-to-infrastructure (V2I) (Table 4).

The analysis of capabilities and limitations related to the use of autonomous vehicles show a long way yet to overcome, to benefit from the positive effects of such vehicles on the improvement of road safety. 90% of what takes place in road traffic is related to a man, who is the most important, but at the same time, the most sensitive element of the safe system of man-vehicle-road. Even fully autonomous traffic will not be able to eliminate all dangerous traffic incidents and avoid victims of traffic accidents. In road traffic there are not only drivers involved, but also pedestrians, cyclists and other vulnerable road users, which cannot be replaced or eliminated. This would be contrary to the idea of personal freedom, and also to the general trend of promoting foot and bicycle mobility.

6 Conclusion

The current state of development of the technology used by the autonomous vehicles does not allow total to stave off the problem of fatalities and persons injured in road accidents. An intermediate state, which is a period of testing, and deployment of technology does not bring a radical improvement of road safety. A consistent and safe transport system requires the harmonisation of rules related to the use of autonomous vehicles: standardisation of on-board equipment, software, possibility human intervention, as well as rules of liability: driver-user, the vehicle manufacturer, the manufacturer of the vehicle software, road infrastructure administration.

Radical improvement in road safety would require substantial reduction or even elimination of human factor from the decision-making process. Only the introduction and dissemination of vehicles of the fourth level of autonomous driving can largely solve this problem and, consequently, reduce the social, economic and environmental costs connected with road traffic. Fully programmable road traffic is now unacceptable, and the obstacle is not the lack of technology, but behavioural factors underlying human behaviour. Human emotions do not exist in machines. An attempt of revolution in this area by giving ways to autonomous vehicles, and dehumanisation of road traffic can have far reaching consequences not only for the human mental health, but also for the automotive industry, and eventually for the whole global economy. The matter is whether people are ready for ħt.

References

1. WHO: The global burden of disease: 2004 update (2004). Accessed 10 Feb 2017
2. Ogłoziński, M.: Efektywność narzędzi poprawy bezpieczeństwa ruchu drogowego w Polsce. UE Katowice (2015)
3. Jaździk-Osmólska, A. (red.): Metoda oraz wycena kosztów wypadków i kolizji drogowych na sieci dróg w Polsce na koniec roku 2012. Instytut Badawczy Dróg i Mostów, Warszawa (2013)
4. EC: On the implementation of objective 6 of the European Commission's policy orientations on road safety 2011–2020, First milestone towards an injury strategy, European Commission, Brussels, 19 Mar 2013. SWD (2013) 94 final (2013)
5. Ulleberg, P.: Human factors in traffic safety promotion. Institute of Transport Economics, Oslo (1997)
6. Pillath, S.: Automated vehicles in the EU, Briefing, European Parliamentary Research Service (EPRS) (2016)
7. EP: Research for TRAN Committee – Self-piloted cars: The future of road transport? Study, Directorate-General for Internal Policies Policy Department B: Structural and Cohesion Policies, European Parliament, Committee on Transport and Tourism, Brussels (2016)
8. SAE, SAE International: On-Road Automated Vehicle Standards Committee, Taxonomy and Definitions for Terms Related to On-Road Motor Vehicle Automated Driving Systems, Information Report (2014)
9. www.nhtsa.gov/technology-innovation/automated-vehicles. Accessed 02 Mar 2017
10. Automated Driving Roadmap, European Road Transport Research Advisory Council, ERTRAC Task Force 'Connectivity and Automated Driving', Version 5.0 (2015)
11. The Center for Internet Society. www.cyberlaw.stanford.edu/wiki/index.php/Automated_ Driving:_Legislative_and_Regulatory_Action. Accessed 02 Mar 2017
12. UN, United Nations: Report of the sixty-eighth session of the Working Party on road Traffic Safety, Economic and Social Council, Geneva, 24–26 March 2014. www.unece.org/info/ media/presscurrent-press-h/transport/2016/unece-paves-the-way-for-automated-driving-by-updating-un-international-convention/doc.html. Accessed 28 Feb 2017

13. ECE, Economic Commission for Europe: Inland Transport Committee, Automated driving, Geneva, 19–22 Sept 2016
14. ETSC: Intelligent transportation systems and road safety, Brussels (1999)
15. www.whatcar.com/news/are-autonomous-cars-safe-2/. Accessed 02 Mar 2017
16. Bonnefon, J.F., Shari, A., Rahwan, I.: Autonomous vehicles need experimental ethics: are we ready for utilitarian cars? 13 Oct 2015. www.researchgate.net/publication/282843902_Autonomous_Vehicles_Need_Experimental_Ethics_Are_We_Ready_for_Utilitarian_Cars. Accessed 29 Feb 2017
17. Autonomous Vehicles. The legal landscape in the US and Germany. A Norton Rose Fulbright whitepaper. www.nortonrosefulbright.com/knowledge/publications/141954/autonomous-vehicles-the-legal-landscape-in-the-us. Accessed 02 Mar 2017
18. www.trl.co.uk. Accessed 01 Mar 2017
19. Humanes, P.: Autonomous vehicles: increasing road safety? www.compass.ptvgroup.com/2016/08/autonomous-vehicles-increasing-road-safety/?lang=en. Accessed 02 Mar 2017

Perspectives of Transport Systems Development in the Light of Radio Communication Systems Evolution Towards 5G

Slawomir Gajewski[✉]

Faculty of Electronics, Telecommunications and Informatics,
Gdansk University of Technology, Narutowicza Str. 11/12,
80-233 Gdańsk, Poland
slagaj@eti.pg.gda.pl

Abstract. In the paper conditions of development and implementation of transport systems with reference to the development of radio communication networks towards 5G are presented. First, general properties of next generation systems are mentioned and their architecture. Moreover, planned characteristics of B4G and 5G systems are depicted which can significantly contribute to the promotion and development of transport systems. In particular the paper covers new transmission mechanisms resulting from the development of the Internet of Things and M2M technology as well as related, for example: V2V or V2X. On the other hand, the issue has been analyzed against the background of the development of the so-called heterogeneous networks and used physical resource management mechanisms, reduction and coordination of inter-cell interference and handover problems. In particular attention was also drawn to the conditions of radio signals transmission in the context of high mobility of mobile terminals in transport systems.

Keywords: B4G · 5G · HetNet · ITS · UDN · V2V · M2M · Systems evolution

1 Introduction

The major aspect of the evolution of radio communication systems towards 5G is topological conversion of cellular networks to development of global so-called heterogeneous (HetNets) Ultra-Dense Networks (UDN). In this type of network there will be implemented hierarchical cell structure in which the macro-cells will cover large areas and small cells will be implemented on the areas of macro-cells. In the group of small cells we can use pico-cells as dominant groups used to improve the general network performance measured by e.g. throughput and latency of signals transmission. Additionally, femto-cells can be introduced for communication with very small ranges. The major difference between homogeneous and heterogeneous networks is the growing diversity of the various functional cell types. It consists of e.g. the separation of the control plane from the data plane performance, and the introduction of aspects of the transmission of a distributed more efficient character, in contrast to the centralized models.

Nowadays, the first step here is the change of 4G network architecture in relation to 3G, in which abandoned the use of radio network controllers (RNCs), for direct

© Springer International Publishing AG 2017
J. Mikulski (Ed.): TST 2017, CCIS 715, pp. 203–215, 2017.
DOI: 10.1007/978-3-319-66251-0_17

communication of base stations (eNodeBs) with nodes located in a core network. The second step is the introduction in a standard of 4G system (LTE-Advanced) so-called phantom cells in which the transmission of user data relates, in general, only a small area of cells (e.g. pico-cells or femto-cells). While control data transmission, and, in general, the management of physical resources, as well as possible additional transmission of user data, are performed by the macro-cell eNodeBs (or micro-cell). These base stations also cover the areas of phantom cells of small size. This approach results in a greater spectrum efficiency of data transmission, due to small ranges and reduced levels of inter-cell interference and due to decreasing the signaling traffic associated with handover in closely spaced, small cells [1].

In 5G networks the concept of small cells is one of the major directions of network development. In this case, so-called Ultra-Dense Networks will be introduced which are the heterogeneous networks. In UDNs advanced methods of resource management and very large number of small cells located with high density will be used.

The novel model of radio access network architecture is of great importance to transport systems and other types of systems in which the communication between devices is necessary, in relation to the problem of Machine-to-Machine (M2M) communications in Internet-of-Things concept. All projections indicate that in the coming years the communication between machines (M2M and related techniques) will become dominant in the radio communication networks. The traffic generated in radio communication networks by different machines, devices and others, becomes larger than the traffic generated by humans. In telematics it means a revolution in the availability of new services, in the scope of systems work and technical capabilities for the development of various sensor and communication infrastructure.

2   Evolution of Radio Communications Towards B4G/5G

The directions of network evolution are connected with exponential growth of the number of mobile and fixed users and the number of devices connected to the network. It causes additional requirements and challenges providing higher network capacity (throughput), lower latency, greater reliability and high mobility support compared to 4G. Great problems of communication over 2G, 3G and 4G networks results from:

- Non-uniform and often low transmission rates available for users.
- Subjective user experience of non-unified, poor quality of service.
- Non-satisfied end-to-end transmission performance due to high latency and non-satisfied transmission rate.
- Poor coverage and quality of indoor connections due to high additional loss of signals compared to outdoor connections.
- Insufficient high mobility connections performance.
- Too large battery consumption and, in general, too large energy consumption in a network.

Thus, some requirements are defined for 5G network which are the answer for these problems indicated in previous networks.

2.1   Requirements and Challenges for 5G Networks

Nowadays, underway works aimed at identifying and systematizing the main objectives that should be achieved ultimately by the 5G radio communication networks. The most important of these, defined by ITU (International Telecommunication Union), are as follows [2]:

- Very low latency and great reliability for services realized by peoples what is very important in relation to temporary, subjective felt quality of services, including planned realization of these services in internet cloud, and taking into account the importance growth of multimedia services and the services using virtual reality.
- Very low latency and great reliability for services oriented to the transmission between machines. In this case, particular importance can be real time transmission in transport telematics, e.g. real time transmission to and from vehicles in motion (cars), the autonomous vehicles (without drivers) control and communication, unmanned vehicles communication and control, safety services and response in distress as well as real time communication between devices of road infrastructure, etc.
- The service of large number of users/machines including very high density of their location, which will be possible provided that a satisfactory transmission quality for end-users under specific conditions, e.g. during fairs, concerts, etc.
- High quality of connections at high mobility of terminals and large speed of motion, including invested in very high-speed trains and cars. This problem was analyzed in 3G and 4G networks but it was not resolved. In 5G, there is a problem with high quality communication, especially for V2V (Vehicle-to-Vehicle) real-time communication services. The success of the concept of global communication between vehicles (V2V) or between vehicles and road infrastructure (Vehicle-to-Infrastructure – V2I) is conditioned by reliability of transmission of high quality at high speed of vehicles.
- Enhanced multimedia services, including those related to medical assistance, safety and security, which also forms part of the network of WBAN type (Wireless Body Area Networks) dedicated just for such purposes, especially in transport applications.
- Internet of Things in wireless networks, where difficultly predictable and continually growing number of devices will be connected to the network and will often continuously send information to other devices and people with different and varying requirements for energy consumption, power of signals transmitted and introduced delays etc.
- Highly accurate location-based services, which will be able to be used for precise localization and navigation of devices, such as: drones and other unmanned vehicles, autonomous cars, etc.

Thus, one can see, the challenges for 5G compared to actually implemented networks are as follows:

- Very high transmission rate for connections, large cell throughput and larger stability of transmission rate at whole area of cell.
- Very low latency for services achieved, among others, by distribution of network elements, realization of large number of services in internet cloud, cell virtualization

(it means that e.g. communication nodes or access points can realize only front-end functions but the reception and decoding of signals can be done in some servers in a cloud.
– Ubiquitous things communication.
– Scalable and flexible infrastructure and network architecture.
– Excellent quality subjective perceived by users.
– Larger number of types of real-time services of high stability and reliability.
– Advanced physical resource methods implementation based on frequency reuse [4], distributed scheduling and coordinated multipoint transmission [2].

2.2   General Concept of 5G Architecture

Analysis of the needs of 5G network at current stage of mobile network development indicates that important aspects of research are the techniques of multiple user access to radio communication channel. In fact it is not only the multiple access method, but additionally, a number of techniques of physical system resources management. These resource management techniques are now (e.g. in LTE and LTE-Advanced), and will be in the future, allocated both in time and frequency domains, and perhaps, also in signal power domain, as proposed for example in the case of multiple access methods called NOMA (Non-Orthogonal Multiple Access). It should be expected that OFDMA (Orthogonal Frequency Division Multiple Access) will be extensively utilized for many years but may be used in some modified versions.

However, the fulfillment of the 5G networks requirements needs significant changes in the approach to network architecture [3]. It should be the fundamental changes. It also follows from the expected changes of cellular network functions. In these networks voice services were initially dominated, currently dominate user data transmission services (e.g. internet services), but in the near future most likely radio communication traffic probably will be dominated by M2M data transmission between machines or D2D (Device-to-Device) transmission between devices.

First, one can see that the 5G network is heterogeneous UDN network of high density of small cells what is presented in Fig. 1. In UDNs we can allocate a number of small cells of different type, and it can be fixed location cells and mobile cells. For instance we have:

– Macro-cells which will cover a number of small cells.
– Fixed or ad hoc small cells outside buildings – outdoor cells.
– Fixed or ad hoc small cells inside buildings – indoor cells.
– Mobile small cells in-cars.
– Mobile small cells in-trains, etc.

From this point of view the global concept of small cells is very important in transport applications and it can be modern view on development of transport systems for communication and telematics as well as for sensor applications.

The use of small cells will be possible and really efficient if the antenna techniques of Massive MIMO (Massive Multiple Input Multiple Output) will be implemented in 5G system.

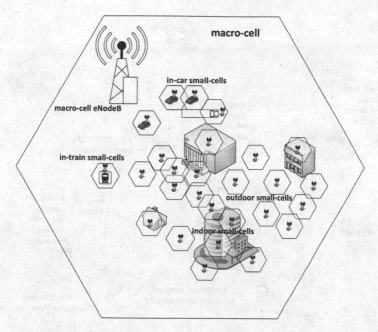

Fig. 1. General cell structure for 5G ultra-dense networks [own study]

It is assumed that the architecture of 5G network will consist of major two logical layers:

- Radio Access Network for realization of wireless communication between base stations, access points, relays, nodes and user terminals as well as different network elements, vehicles and machines, consists various cellular structures and network functions.
- Backbone network (core network) which can be located in "an internet cloud" (network cloud) where the target is distributed architecture of the core unlike the cores of 2G, 3G and 4G networks (but note that it is not well known when it will be possible to practical implementation).

Example functional architecture of 5G network is presented in Fig. 2. As we can see, the 5G network combines different solutions and subnets for different purposes. In a heterogeneous 5G network the concept of small cells will be crucial. Also, mobile users will benefit from the small cells in cars and trains, through the use of massive MIMO techniques in which we can use large number of antennas (tens or maybe hundreds of antennas).

In general, one can say that the 5G network architecture includes:

- Cells in which the massive MIMO technique is used to network performance improvement dedicated to various practical applications and local networks.
- Home networks providing access to telecommunications services and control devices (smart-house devices).

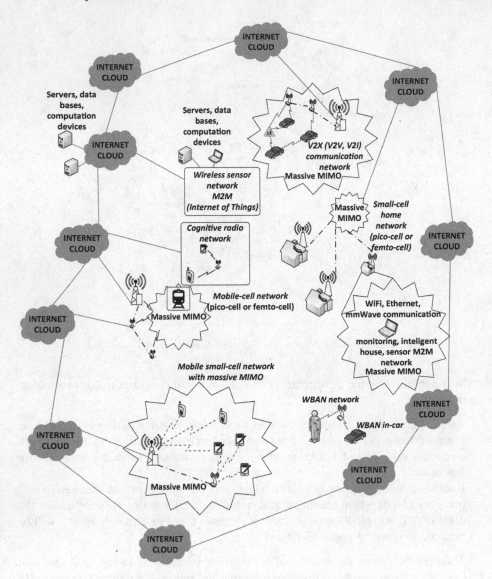

Fig. 2. General architecture of 5G network [own study]

- Mobile-cells networks using pico-cell or femto-cell networks in trains, buses, cars etc.
- Wireless sensor networks for communication between machines (M2M) devices (D2D), advanced V2X networks etc.
- WBANs for health monitoring, together with pico-cell and femto-cell networks, created ad hoc for the purpose e.g. rescue, and also in-car WBANs (or in-train WBANs), which can be used to health monitoring applications of TIR or train drivers.

- Intelligent transport networks and multi-modal systems for communication between vehicles and the road infrastructure, for improving road safety, rail, etc.
- Cognitive networks.

There is no doubt that the construction of networks will be changed primarily, pursuing these objectives, which will manifest in the dominant importance of small cells operated using advanced techniques of massive MIMO. Moreover, data processing backbone network (core network) or some functions of this network will probably be held in an internet cloud.

In the future, it is possible that also signal processing will be transferred to the cloud from eNodeBs, access points, relays etc. Thus, these nodes will made only functions related to the transmission and reception of signals in RF circuits (front-end). While the signal processing and the receiving process will be implemented in a cloud. This can facilitate the coordination of interference between cells and improve mechanisms for managing physical resources in the network.

2.3 New Concept of In-building and In-vehicle Communications

In present mobile networks, each mobile station, regardless of whether it is inside or outside buildings or vehicles, it must communicate directly with a base station located in the center of cell area. For indoor users, walls are extra obstacles for signals due to the need to their transmission through the wall, resulting in a sharp increase in signal loss and quality. From personal experience, everyone knows what it means in practice and how big entails consequences. It causes decreasing of spectral efficiency, data rate, both base and mobile station coverage, objective and subjective quality of service, additional energy consumption, and also in very frequent cases, the drastic increase in the transmission delay and connections dropping.

Therefore in 5G networks there are proposed the separation of transmission mechanisms inside and outside the buildings and vehicles [2, 3] through the creation of local networks using small-cells (pico-cells or femto-cells). In this case, attenuation of walls of buildings and vehicles will no longer have such a large impact on the transmitted signals loss. It is proposed here to use a variety of techniques, such as: transmissions via relay nodes, using the so-called Radio Resource Equipment (RRE), in the form of additional antennas placed closer to the mobile stations (not close to access point or eNodeB location) or implementation of massive MIMO antennas at different locations.

It means, despite primary eNodeB antennas, the implementation of additional antennas spaced from it at greater distances, but associated with the same base station. These antennas can be connected via e.g. optical fiber-link or radio-link. Its ultimate aim is the use of tens or hundreds antennas connected via massive MIMO. In principle, in the case of 5G networks, it is assumed that the users inside buildings or vehicles will not directly communicate with outdoor eNodeB, but only with in-building or in-vehicle access points, and the antenna arrays of massive MIMO will be installed inside and outside buildings. For indoor communication it can be used additional signal transmission techniques, such as WiFi etc.

2.4   Resource Management and Interference Coordination in B4G/5G Networks

New structures of cells in heterogeneous networks make it necessary to work on the mechanisms of radio resource management. A major problem with common using of frequency band in cells is management of frequency band in the way which allows reduction of received power of inter-cell interfering signals (so-called inter-cell interference – ICI). For this purpose, the mechanisms of frequency reuse can be implemented, but they may not be sufficient to achieve high performance of both B4G and 5G networks. Therefore, the key to achieving high performance is to use techniques enabling inter-cell interference coordination.

In heterogeneous networks very important are methods of so-called coordinated multipoint transmission (CoMP) and advanced data scheduling. These methods allow intelligent resource coordination and reception of signals in hierarchical cell structures which are characteristic for heterogeneous networks as well as in each radio access network using the same frequency channels in different, neighboring cells. The concept of small cells gives the chance for improving the efficiency of signals transmission, especially at cell edges (cell boundaries). It is possible because cells are small, and these cells do not always have to come into contact with each other, and, additionally, the powers of signals transmitted are very small. It significantly reduces inter-cell interference and facilitates the interference coordination.

These problems are very important in high-rate transport systems in which dominated is telecommunication traffic based on M2M or V2X communication. It means that transport systems based on M2M will have much better conditions of their work in next generation radio communication networks compared to previous generation of networks. Big hopes are also related to transport communication in phantom cells because one of major target of phantom cells implementation is reduction of problems with channel handover and signalization traffic during handover process in small cells, especially in the case of high mobility of terminals and their movement with high speed what is significant in transport telematics and transport communications systems.

3   Novelties of B4G/5G Network for Transport Applications

The novelties in the concept of 5G radio communication networks have great impact on transport systems. We can take into account different transport applications:

- V2X (Vehicle-to-Everything; it means V2V, V2I and others) communication of cars in a city related to smart city concepts [5].
- V2X communication of vehicles at highways and express roads.
- Autonomous vehicle control (cars, unmanned drones, and other vehicles).
- Advanced sensor technologies for smart-cities.
- In-vehicle communication for driver and passengers (in cars, in buses, in trains).
- BAN networks for road safety.
- Some safety applications, like e-call etc.
- Intelligent transport systems in roads, intelligent road signs, sensors etc.

A large part of these solutions is already practical application. But the most important are the solutions based on V2X communication and local BANs, which can be used to improvement of road safety and rescue operations support.

3.1 Transport Safety Challenges

In this case the true technological challenge is the implementation of V2X networks [5]. In principle, the functioning of such a network is based on the automatic collection of information from various sensors and/or medical devices and their analysis. By using this analysis it will be possible ongoing exchange of information between vehicles (V2V) and between vehicles and road or rail infrastructure (V2I).

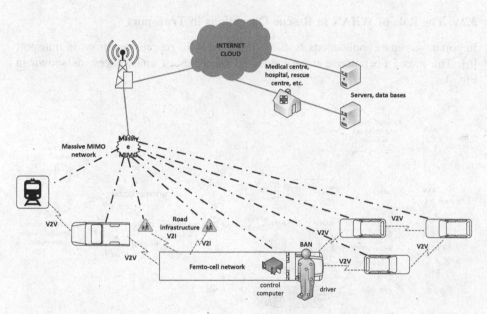

Fig. 3. Example of V2X network with BAN as application of 5G network [own study]

The real-time exchange of information between the vehicles will be a breakthrough in terms of road safety. Moreover, the use of additional WBAN network for drivers, especially large vehicles, will enable continuous health parameters monitoring of their bodies. The importance of WBAN networks in a transport can be either completely crucial for the road and rail safety. Figure 3 shows an example in which the WBAN network can detect that a driver fainted or fallen asleep, which starts the whole cycle of automatic actions.

First, the vehicle control computer also obtains this information, which allows the emergency stop of a vehicle. In the future, it can be possible to switch the control computer to autonomous/emergency mode of car driving (without driver). As a result, the vehicle can then be safely stopped or even parked by autonomous unit of the vehicle. In addition, the information on the need for an emergency stop of the vehicle may also reach out to other vehicles following behind the analyzed vehicle or oncoming from opposite direction. In addition, the elements of road infrastructure may obtain such information and such e.g. switch traffic lights on a crossroad or stop vehicle traffic near the place of an incident etc.

To support the service of described V2X network is necessary ultra-dense cell structure with numerous and dense located antennas placed in small distances (massive MIMO) as well as intelligent road sensor infrastructure enabling the control of road traffic, and high accuracy positioning services available in 5G networks. Only then it will be possible to achieve a sufficient level of vehicle network reliability. Therefore, for these applications, it is necessary to use the network 5G with massive MIMO and small-cell concept.

3.2 The Role of WBAN in Rescue Operations in Transport

Important group of applications is using of WBANs for rescue operations in transport [6]. The rescue operation can be conducted by teams of ambulances, as shown in Fig. 4.

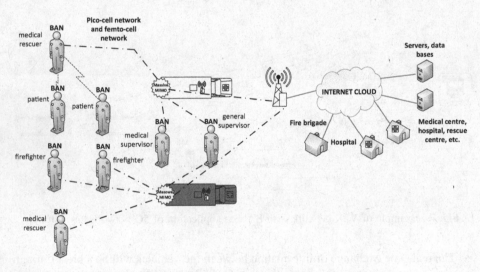

Fig. 4. Example VBAN for rescue operation [own study]

In this case, for example, patients who are the victims of accidents are constantly monitored by a doctor or medical rescuer. Both patients and rescuers are equipped with their own sensor WBAN networks, which, however, perform different functions. The WBANs of patients are in general for monitoring medical parameters of their organisms [7]. While the WBAN networks of rescuers enable them to monitor the status of patients, but also to communicate between themselves and communication with the supervisors of the rescue operation. In this case, the access network based on massive MIMO and access points are located in an ambulance. It allows immediate formation of an ad hoc local pico-cell or femto-cell network at place of location of an ambulance. The data we are talking about is also sent to the emergency center, hospital emergency department, etc., which allow remote support of rescuers work and notification of a hospital, which can advance to prepare for the admission of critically injured patients.

In the case of fire brigade, it can be changed the functional range of sensors fitted to each firefighter. These sensors can be used for communication, location, monitoring the parameters of a body, but also it can be used to measure environmental parameters surrounding a firefighter. Therefore functions here are twofold. Firstly, related to health and safety of the firefighter, but also to support his work. For example, there may be used sensors for smoke, carbon monoxide, temperature, etc., and also the cameras transmitting image, if needed. As we can see, as for the ambulance service, the access point is located in the fire engine and creates an ad hoc cell of a small size. Such a cell may be created in every car and create the associated self-organized network elements. Firefighters can communicate with each other, and all the data goes to the general supervisor and the management center, which can be located in one of vehicles. In addition, monitoring data can also reach the emergency centers, department of fire brigade, a hospital etc.

3.3 Intelligent Transport Networks and Autonomous Vehicle Driving

The Ultra-Dense Networks with small cells can significantly support different systems of intelligent transport in which the M2M or V2X communication is dominated. We can great hope for a major breakthrough in the development of such systems inter alia thanks to fuller availability of wireless networks of high transmission rate and general performance along roads, particularly in urban areas, but also in sub-urban areas. Of course the same breakthrough is possible for rail communication and unmanned aerial vehicles control.

For instance, in analysed type of road systems we can find [8]:

- Vehicle systems for road safety improvement (emergency systems like e-call, advanced anti-crash systems, advanced cruise systems, etc.).
- Systems for communication between cars (vehicles) and road infrastructure for safety and others.
- Systems for limitation of road violations, cameras, sensors and traffic enforcement cameras with data transfer.
- Automatic speed limiting systems.
- Automatic car driving systems for autonomous traffic.

Vehicle systems for road safety improvement are still implemented in cars but their use is limited to a single car, and they cannot communicate with other cars and road infrastructure. Development of 5G networks is very good starting point for their development and dissemination.

Systems for communication between cars (vehicles) and road infrastructure for safety can revolutionize transport systems. Intelligent road signs and sensors for adaptation of traffic characteristics to environment conditions, road traffic, weather conditions etc., and possible communication and sending information to cars, can be excellent means to improving traffic flow and road safety.

Systems for limitation of road violations, cameras, sensors and traffic enforcement cameras with data transfer as well as systems for automatic speed limitation are not good information for drivers which like high speed of their cars. But from the point of view of road safety they are invaluable.

Very interesting from the point of view of 5G networks development are automatic car driving systems for autonomous traffic. They can be related to driver safety using WBANs but the major sense of its development is car-without-driver communication. Dense networks with small cells and expanded sensor systems gives the chance for their disseminating in advanced forms unlike simple systems actually deployed in some countries like USA. Additionally, very high accuracy localization services which are planned to introduce in 5G networks are necessary.

4   Conclusion

Over the next 10 years 5G networks will become widespread in the world and in Poland. This will open a new chapter in the history of telecommunications and transport, which will probably be crucial for the services of mass data transmission between devices, machines, vehicles and people.

Modern concept of 5G network architectures gives new possibility to the implementation of various transport-oriented systems, safety systems, emergency communication, systems supporting autonomous vehicle driving, and many others.

Interesting perspective for transport systems is the use of WBANs for safety improvement. WBAN networks represent only a small part of the gigantic whole, but they can also be of great importance, especially in the field of discussed in the article applications for rescue and transport safety. Planned for widespread implementation, vehicle safety systems and communication systems between vehicles may be supplemented by locally established WBAN networks for drivers and rescuers. They can monitor the state of health of a driver, but also e.g. the state of a driver sobriety. Doubtless this will substantially improve traffic safety and significantly reduce the number of road accidents.

References

1. Lokhandwala, H., Sathya, V.B., Tamma, R.: Phantom cell realization in LTE and its performance analysis. In: Proceedings of IEEE International Conference on Advanced Networks and Telecommunications Systems (ANTS) (2014)
2. Gupta, A., Kumar Rakesh, J.: A survey of 5G network: architecture and emerging technologies. IEEE Access **3**, 1206–1232 (2015). Special Section in IEEE Access: Recent Advances in Software Defined Networking for 5G Networks
3. GPP PPP. 5G architecture. White paper (2016). https://5g-ppp.eu/white-papers/
4. Gajewski, S.: Throughput-coverage characteristics for soft and partial frequency reuse in the LTE downlink. In: Proceedings of 36th International Conference on Telecommunications and Signal Processing, TSP 2013, Rome, Italy, pp. 199–203 (2013)

5. Gajewska, M.: Design of M2M communications interfaces in transport systems. In: Mikulski, J. (ed.) TST 2016. CCIS, vol. 640, pp. 149–162. Springer, Cham (2016). doi:10.1007/978-3-319-49646-7_13

6. Antonescu, B., Basagni, S.: Wireless body area networks: challenges, trends and emerging technologies. In: Proceedings of 8th International Conference on Body Area Networks—BodyNets 2013, Boston, Massachusetts, pp. 1–7 (2013)

7. Otto, C., et al.: System architecture of a wireless body area sensor network for ubiquitous health monitoring. J. Mob. Multimed. **1**(4), 307–326 (2006)

8. Gajewski, S., Gajewska, M., Sokol, M.: Architecture and basic assumptions of RSMAD. In: Mikulski, J. (ed.) TST 2011. CCIS, vol. 239, pp. 200–205. Springer, Heidelberg (2011). doi:10.1007/978-3-642-24660-9_23

Guidelines for Multi-system Shipborne Radionavigation Receivers Dealing with the Harmonized Provision of PNT Data

Adam Weintrit[1] and Paweł Zalewski[2(✉)]

[1] Gdynia Maritime University, ul. Morska 81 - 87, 81-225 Gdynia, Poland
weintrit@am.gdynia.pl
[2] Maritime University of Szczecin, Wały Chrobrego 1, 70-500 Szczecin, Poland
p.zalewski@am.szczecin.pl

Abstract. In the paper the Authors present draft Guidelines for Shipborne Position, Navigation and Timing Data Processing elaborated by the International Maritime Organization (IMO) [1]. The purpose of these Guidelines is to enhance the safety and efficiency of navigation by improved provision of position, navigation and timing (PNT) data to bridge teams (including pilots) and shipboard applications (e.g. AIS, ECDIS, etc.). The shipborne provision of resilient PNT data is realized through the combined use of on-board hardware and software components. The shipborne PNT Data Processing (PNP-DP) is the core repository for principles and functions used for the provision of reliable and resilient PNT data. These Guidelines define principles and functions for on-board PNT data processing taking into account the scalability of PNT-DP. Within the e-Navigation strategy the IMO has identified the user need on improved reliability, resilience and integrity of bridge equipment and navigation information as one of the five prioritized e-Navigation solutions, whereby the resilient provision of PNT data acts as Risk Control Option.

Keywords: Marine navigation · PNT · IMO · Data processing · e-Navigation

1 Introduction

The International Maritime Organization (IMO) has defined mariners' needs with respect to safety of navigation. Within the e-Navigation strategy the IMO has identified the user need on improved reliability, resilience and integrity of bridge equipment and navigation information as one of the five prioritized e-Navigation solutions [2], whereby the resilient provision of PNT data acts as Risk Control Option [3]. In the end of 2016 the IMO Correspondence Group developed draft Guidelines for Shipborne Position, Navigation and Timing Data Processing [1], which will be presented and discussed on 4th NCSR Sub-Committee session in March 2017.

Although these draft Guidelines are directly associated with the performance standards for multi-system ship-borne radio-navigation receivers [4], the scope of application covers all shipborne navigation equipment and systems applying or providing Position, Navigation and Timing data and associated integrity and status data. In principle,

© Springer International Publishing AG 2017
J. Mikulski (Ed.): TST 2017, CCIS 715, pp. 216–233, 2017.
DOI: 10.1007/978-3-319-66251-0_18

the PNT data processing is considered as integral part of on-board systems like the Multi-system Shipborne Radionavigation Receiver (MSR), RADAR/ARPA, ECDIS or INS.

The document [1] is structured according to the rules on the modular concept of Performance Standards published by the Organization as IMO Guidelines for Application of the Modular Concept to Performance Standards (SN.1/Circ. 274) [5]. It comprises the following modules: Module A – data input: sensors, services, and sources; Module B – functional aspects; Module C – operational aspects; Module D – interfaces; and Module E – documentation. Methods and functions are designed in a hierarchical composition to meet different levels of data processing and performance requirements. Application grades as well as accuracy and integrity level are introduced.

2 Objectives of the Guidelines

The purpose of the presented Guidelines [1] is to enhance the safety and efficiency of navigation by improved provision of position, navigation and timing (PNT) data to bridge teams (including pilots) and shipboard applications (e.g. AIS, ECDIS, etc.). The shipborne provision of resilient PNT data and associated integrity (I) and status data (S) is realized through the combined use of on-board hardware (HW) and software (SW) components. The shipborne PNT Data Processing (PNT-DP) is the core repository for principles and functions used for the provision of reliable and resilient PNT data. The PNT-DP specified within these Guidelines is defined as a set of functions facilitating:

– multiple, sources of data provided by PNT-relevant sensors and services (e.g. GNSS receiver, DGNSS corrections) and further on-board sensors and systems (e.g. radar, gyro, SDME, echo-sounder providing real-time data) to exploit existing redundancy in the PNT-relevant input data; and
– multi-system- and multi-sensor-based techniques for enhanced provision of PNT data.

The Guidelines aim to establish a modular framework for further enhancement of shipborne PNT data provision, by supporting:

– consolidation and standardization of requirements on shipborne PNT data provision considering the diversity of ship types, nautical tasks, nautical applications, and the changing complexity of situations up to customized levels of support – scalability of PNT-DP;
– the identification of dependencies between PNT relevant data sources (sensors and services), applicable PNT data processing techniques (methods and thresholds) and achievable performance levels of provided PNT data (accuracy, integrity, continuity and availability);
– harmonization and improvement of on-board PNT data processing based on a modular approach to facilitate changing performance requirements in relation to nautical tasks, variety of ship types, nautical applications, and under consideration of user needs;

- the consequent and coordinated introduction of data and system integrity as a smart means to protect PNT data generation against disturbances, errors, and malfunctions (safety) as well as intrusions by malicious actors; and
- standardization of PNT output data including integrity and status data.

They are recommended for equipment manufacturers, shipyards, ship owners and managers responsible for on-board equipment and systems used for PNT data provision.

2.1   Architecture of Shipborne PNT-DP

Generally, a shipborne PNT-DP is made up of three functional blocks:

- Pre-processing;
- Main processing; and
- Post-processing.

The pre-processing function extracts, evaluates, selects and synchronizes input (sensor and service) data (including the associated integrity data) to preselect the applicable techniques to determine PNT and integrity output data. The proposed architecture of the PNT-DP is shown in Fig. 1.

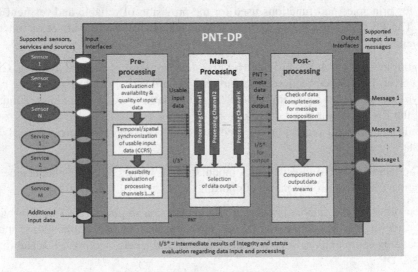

Fig. 1. Architecture of PNT-DP [1]

The main processing function generates the PNT output data and associated integrity and status data. The post-processing function generates the output messages by coding the PNT output data (PNT, integrity, and status data) into specified data protocols.

2.2 Integration of PNT-DP

The shipborne Position, Navigation and Timing Data Processing (PNT-DP) can be integrated as software into ship's navigation systems such as INS (Integrated Navigation System), ECDIS (Electronic Chart Display and Information System) or RADAR as pictured in Fig. 2.

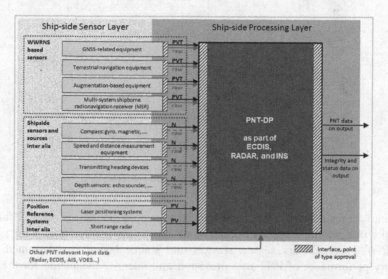

Fig. 2. PNT-DP unit integrated as software into INS, ECDIS, or RADAR [1, 6]

The Multi-system Shipborne Radionavigation Receiver (MSR) is appropriate to facilitate the combined use of WWRNS to improve the provision of PVT (Position, Velocity, Timing) data and related integrity data. The application of enhanced processing techniques can be realized by the MSR itself or by the INS (Fig. 2).

3 Modular Structure of Guidelines

The Guidelines [1] are structured according to the rules on the modular concept.

3.1 Module A – Data Input: Sensors, Services and Sources

Different PNT data processing functions need comprehensive input data to keep the PNT-DP running as specified in the Guidelines. The Guidelines define how the shipborne PNT-DP should provide output data by processing input data (from sensors and/or services and/or sources) while availability and performance of input data may vary temporally and spatially (see Fig. 3).

The desired level of PNT data output depends on currently available inputs that may independently vary over a short or long period of time. In a minimum configuration, PNT uses the minimum number and type of sensors as defined by SOLAS.

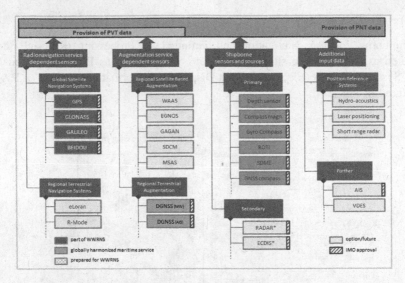

Fig. 3. Sensors, services, and sources [1]

The manufacturer may add inputs and outputs to support additional nautical functions and tasks that require better performance or more information (e.g. with integrity indication). The necessary sensor, service, and source layout is determined by the necessary performance of PNT data provision and integrity evaluation for the subsequent nautical functions and tasks. The proposed classification is as follows:

A.1 Type of Services for Positioning
Services are classified by grade/type as follows:

- **Radionavigation services** provide navigation signals and data, which enable the determination of ships position, velocity and time.
- **Augmentation services** provide additional correction and/or integrity data to enable improvement of radionavigation based determination of ships position, velocity and time.

Services are classified regarding its geographical coverage:

- **Global services** are characterized by their world-wide coverage. They may have limitations regarding usability for different phases of navigation due to signal disturbances reducing the availability or performance of transmitted signals and/or provided data.
- **Regional services** (and may be local services) are only available in dedicated service areas. They may be used to improve the performance of ships' navigational data in terms of accuracy, integrity, continuity and availability even in demanding operations.

A.2 Types of Sensors and Sources

The type approved sensors and data sources are distinguished into the following categories:

- **Service dependent sensors** rely on any service from outside the ship provided by human effort. They cannot be used on board without at least a satellite-based or terrestrial communication link to the service provider (mainly used to provide data of ships position, velocity and time).
- Shipborne sensors and sources:
  - **Primary sensors** use a physical principle, e.g. earth rotation or water characteristics and are independent of any human applied service provision (mainly used to provide data of ships attitude and movement);
  - **Secondary sensors and sources** may be used to provide additional data for the verification of PNT data, e.g. water depth at known position from an ENC, line of position, or directions and distances provided by on-board RADAR.

The above described sensors are considered to be usable world-wide and free of any rebilling user charge.

A.3 Additional Input Data

In addition to sensors, services and sources listed in A.1 and A.2 further PNT-relevant data may be used for shipborne PNT data provision to increase redundancy or to evaluate plausibility and consistency of data input (ship sensed position e.g. by position reference systems). Such data may be provided via AIS or VHF Data Exchange System (VDES), see Fig. 4.

A.4 Requirements on Input Data

All sensors, services and data sources used as input for the shipborne PNT-DP, should comply with the relevant IMO Performance Standards.

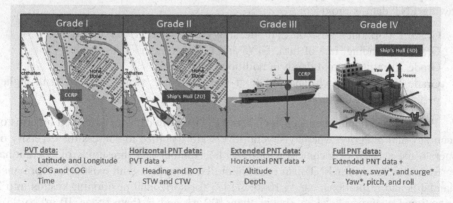

Fig. 4. Application grades of PNT-DP (*provided with improved accuracy) [1, 6]

3.2   Module B – Functional Aspects

B.1 General (objective, functional architecture and requirements)

The overarching objective of the shipborne PNT-DP is the resilient provision of PNT data including associated integrity and status data.

In this context resilience is:

- the immunity against relevant failures and malfunctions in data acquisition and processing to meet the specified performance requirements on PNT data for accuracy and integrity with respect to continuity and availability under nominal conditions; and
- the ability to detect, mitigate and compensate malfunctions and failures based on supported redundancy in data acquisition and processing to avoid loss or degradation in functionality of PNT-DP.

The requirements on data output of PNT-DP are specified by the application grade of PNT-DP defining the amount and types of output data, and the supported performance level of provided PNT data regarding accuracy and integrity.

The following application grades of a PNT-DP are used to define different requirements on the amount and types of PNT data output (Fig. 4):

- Grade I supports the description of position and movement of a single on-board point (e.g. antenna location of a single GNSS receiver);
- Grade II ensures that horizontal attitude and movement of ship's hull are unambiguously described;
- Grade III provides additional information for vertical position of a single on-board point and depth;
- Grade IV is prepared for the extended need on PNT data e.g. to monitor or control ship's position and movement in three-dimensional space.

Depending on the supported application grade of an on-board PNT-DP, the following PNT data is provided:

- Grade I: horizontal position (latitude, longitude), SOG, COG, and time;
- Grade II: heading, rate of turn, STW and CTW in addition to Grade IV;
- Grade III: altitude, and depth in addition to Grade II;
- Grade IV: heave, pitch, and roll (and may be surge, sway, and yaw with higher performance) in addition to Grade III.

Performance requirements on each set of PNT output data are described in terms of accuracy and integrity, whereby several levels are specified to address the diversity of operational as well as technical requirements. Number and thresholds of operational performance levels per PNT data type should be compliant with existing performance standards and resolutions, e.g. A.1046(27) [7] for horizontal positioning results into 2 operational accuracy level: A (better than 100 m) and B (better than 10 m) to 95% confidence; A.915(22) [8] specifies the future need on two additional operational accuracy level: C (better than 1 m) and D (better than 0.1 m). In addition, the introduction of technical performance levels (A.1, A.2, B.1, B.2, …) enables a graduated specification of task- and application-related requirements on PNT data. Furthermore it prepares a need-driven evaluation and indication of accuracy.

Integrity data per each individual PNT output data should be provided to indicate the further usability of data. The value of included integrity information depends on applied principles of integrity evaluation in relation to a dedicated accuracy level:

- None: Unavailable integrity evaluation;
- Low: Integrity evaluation based on plausibility and consistency checks of data provided by single sensors, systems, services, or sources;
- Medium: Integrity evaluation based on consistency checks of data provided by different sensors, systems, services, and sources with uncorrelated error parts6 as far as possible;
- High: Integrity evaluation based on estimated accuracy (protection level).

As a result of preceding paragraphs the performance of an individual PNT output data (requirement as well as result of evaluation) should be defined by specified accuracy and integrity levels. Accuracy and integrity levels should be defined for all PNT data of the supported application grade or a combination of them to ensure that the requirements on data output of a PNT-DP are comprehensively specified (Fig. 5).

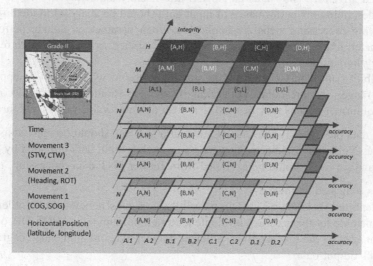

Fig. 5. Composition of requirements on PNT/I output data (application grade II as example) [1]

B.2 Pre-processing (functional and methodical aspects, results of pre-processing)

The pre-processing prepares the input data for main processing and pre-evaluates the feasibility of data processing methods supported by main processing under current conditions.

Data streams received from input data providing entities should be time-stamped with the time of reception using system time of the PNT-DP. The system time should be synchronised with a common time base by using the input data of an appropriate source, preferably UTC. Incoming data provided by sensors, systems and services should be evaluated with respect to:

– completeness and correctness of transmission; and
– plausibility and consistency of data content.

The evaluation of a data stream received from an input data providing entity should comprise the following methods:

– The correctness of transmitted input data should be checked with respect to the rules of the protocol in use (completeness, parity, etc.). Incorrect data should be excluded from further processing.
– It should be checked if the expected data update rate, as needed for main processing, is met. If the determined update rate implies a latency violation the data should be marked accordingly.

The evaluation of data content should comprise the following methods:

– Parameters describing the characteristics of the input data providing entity should be analysed to identify which following processing steps are applicable. Such parameters include performance parameter, like number and type of measurements (e.g. GPS/DGPS); and status parameter, like healthy/unhealthy.
– Data describing the performance of input data should be analysed to identify which following processing steps are applicable. Such parameters include performance parameter, like UERE, HPL; and time of data validity, as available, with respect to latency limitations.
– Plausibility and consistency of data should be tested with respect to appropriate value ranges and thresholds. Data failing those tests should be marked accordingly. Data of former epochs may be used to realise dynamic value ranges and thresholds.

Input data provided by sensors, systems, and services should be marked as invalid if the data sources have indicated that they are invalid. Input data provided by sensors, systems and services should be excluded from further PNT data processing, if:

– data is indicated as invalid;
– the identified violation of latency, plausibility, or consistency is in an order, which is intolerable for the accuracy level intended in minimum by the PNT-DP, or cannot be managed by the PNT-DP in a sufficient manner to avoid unintended degradations of PNT output data.

Input data which have passed the evaluation tests should be adjusted spatially and temporally within a Consistent Common Reference System (CCRS) where required to meet the specified accuracy level. The method for the time synchronization should provide a common time scale referenced to the system time of the PNT-DP, preferably given in UTC. The resolution of time synchronization shall not degrade that of input data. The time scale used for time synchronization should also be used to trigger the complete data processing: pre-processing, main processing, and post-processing. All spatially related information should use a CCRP. If CCRP transformation fails, this should be indicated by corresponding status data.

B.3 Main Processing (functional and methodical aspects of PNT data generation, and PNT data output selection, results of main processing)

The main processing serves to improve PNT data provision by applying appropriate methods for completion, refinement and/or integrity evaluation. The results of main processing are:

- the selected PNT data for output;
- associated integrity data;
- metadata to describe the characteristics of selected output data (e.g. source and processing identifier);
- status data describing the current status of main processing;
- internal status data for controlling of post-processing; and
- internal integrity data contributing to integrity data at output of PNT-DP.

PNT data currently determined by the main processing may be fed back into pre-processing to support the evaluation of the subsequent sensor, system and service data.

Within main processing the pre-evaluated input data (from sensors, system, services, and sources) should be used to feed at least one data processing channel. A single processing channel should provide some or all intended PNT data and associated integrity data. The number of processing channels operated in parallel should ensure at least the provision of all PNT output data in the designated application grade and the supported accuracy and integrity levels. The methods provided by an individual processing channel should at least ensure that the intended PNT output data are provided with the intended accuracy and integrity when the requirements on data input are met (nominal conditions).

More than one processing channel should be supported for the provision of one type of PNT data and associated integrity data:

- if different accuracy and integrity levels are supported by application of different methods for data processing, or
- if an increase of reliability and resilience is aimed by parallel processing of largely independent input data with the same methods.

Parallel processing channels should differ in used input data, or applied methods, or both. These differences may result into measurable differences in PNT data output:

- The additional use of augmentation data should improve the accuracy of PNT output data by application of corrections, or should enhance the integrity evaluation with independent evaluation results, or should serve both.
- If parallel processing channels are equipped with the same methods and are fed with largely independent input data, the results of those channels should cover the same types/set of PNT data. The PNT data can be used alternatively for data output due to its independence and should be used internally for integrity evaluation.
- Enhanced processing channels should combine multiple types of input data to enable the application of effective methods during data processing such as:
  - self-controlling (e.g. detection and exclusion of outliers), self-evaluation (e.g. consistency tests or estimation of protection level as overestimate of expected inaccuracies),

- self-correction (e.g. dual-frequency GNSS signal processing to correct iono-spheric path delays; noise reduction by filtering), and/or,
- self-management (e.g. failure compensation by interpolation or extrapolation in a common model of movement).
- The capability of enhanced processing channels can be increased, if redundancy in data input enables the simultaneous and coordinated use of effective methods such as self-correction, self-controlling, self-evaluation, and self-management.

The need for the provision of reliable and resilient PNT data requires that at least a parallel processing channel should be implemented as a fall-back solution for an enhanced processing channel, which is more sensitive to availability of data input (Fall-back may not be available after loss of sensible input data).

Ultimately, the number and types of parallel processing channels is determined by the supported application grade as well as supported accuracy and integrity levels of aimed PNT data output, arranging of data processing methods to single channels, the aimed level of reliability and resilience of PNT data specifying the residual need for fall-back solutions per application grade and assigned accuracy and integrity levels.

An improvement to accuracy for several or all PNT data types by a processing channel is achieved if one, or a combination of the following methods is applied:

1. Methods applying augmentation data provided by recognized services and external sources (if available and indicated as usable):

   - to improve the accuracy of data by error correction (e.g. GNSS range and range rate corrections);
   - to exclude faulty or disturbed data taking into account integrity evaluation results (e.g. health indicator of GNSS signals provided by Beacon or SBAS);
   - to apply performance indicators provided for individual data to control its influence on potential PNT data output (e.g. weighting within data processing);

2. Methods utilizing redundancy in the database:

   - for self-determination of corrections and application (e.g. dual-frequency signal processing to correct ionospheric path delays);
   - for self-reliant detection and exclusion of faulty data (e.g. FDE by RAIM); and
   - for self-determination of performance indicators for used/derived data to weight its influence on potential PNT data output;

3. Methods utilizing redundancy in database for application of enhanced algorithm such as:

   - equalization calculus based on an overdetermined set of input data, and
   - filtering with adaptive and/or assisted measurement and transition models (e.g. deeply coupled GNSS/INS positioning).

Fall-back solutions should be provided by simultaneously operated processing channel(s) providing the same PNT data with a lower accuracy level by application of methods using less input data (to reduce the sensitivity to completeness of data input), and methods using other input data (to reduce the sensitivity to availability of specific input data).

A redundant solution for a single processing channel should be supported by at least one simultaneously operated processing channel providing independent PNT data types with the same accuracy levels by applying of methods operating with different input data to ensure independency in relation to data input providing systems, services or sensors, and/or methods differing in error influences in relation to data input and processing.

Both, fall-back and redundant solutions should provide an improved resilience of PNT data provision by using fall-back solutions with an acceptable limit of loss of data accuracy, and using redundant solutions with respect to continuity and reliability of PNT data provision in relation to each supported accuracy level.

Integrity evaluation should be based on methods that test the plausibility or consistency of potential PNT output data or methods to estimate the current size and behaviour of its individual errors (e.g. noise), error budgets (e.g. ranging error), or resulting errors (e.g. inaccuracy of SOG). An integrity evaluation should be assigned to each processing channel in relation to the nominally designated PNT data output (taking into account currently used data input). Generally, the applied method of integrity evaluation determines the achieved integrity level:

1. Level None: Failed, unavailable or incomplete integrity evaluation by the processing channel methods and should be regarded as having no integrity.
2. Level Low: The integrity evaluation of the processing channels, dealing with the refinement or completion of data provided by single sensors or measuring systems, should only be based on plausibility and consistency tests in relation to models of the individual sensor and system:

   – Plausibility tests should prove if data types are within an expected value range (e.g. ship's speed). The expected value range should ultimately determine the detectability of errors (e.g. indicated speed over ground is much higher than ship's maximum speed).
   – Simple consistency tests should prove either, if successive data follows an expected time behaviour (e.g. range and range rate) or if multiple outputs of data are compliant within a common measurement model (e.g. position and speed determined by different methods). Consistency should be assumed, if the difference between compared values is smaller than a specified threshold describing the tolerable relative error between both.
   – Enhanced consistency tests should evaluate the expected consistency between used input data and achieved processing result, whereby thresholds used (e.g. in statistical hypothesis tests) should be conditioned in relation to accuracy requirements on output data.
   – Enhanced consistency tests should be applied iteratively with methods detecting and excluding most likely faulty input data or intermediate processing results, if supported redundancy of input data enables the application of such tests. This is an appropriate method to improve accuracy and integrity of output data (e.g. RAIM).

3. Level Medium: If the capability of simple as well as enhanced consistency tests should be increased, the tests should be performed with data provided from different sensors and measuring systems with largely uncorrelated error influences:

   - If the degree of correlation in the error margin as well as in the data itself is not taken into consideration, the difference of compared values should not be considered as an estimate of absolute accuracy.
   - If the error margin of compared values is completely uncorrelated, the difference between both values has to be smaller than the sum of tolerable inaccuracies per considered value. In this case the consistency test serves the evaluation, if pre-specified accuracy levels are met.

   Largely uncorrelated data may inherit partially correlated errors. These errors remain undetected by consistency checks. If the thresholds used during evaluation take the existing uncertainties into account the consistency tests should continue as method to evaluate the fulfilment of certain accuracy levels.
4. Level High: The highest performance of integrity evaluation should provide a reliable estimate of the inaccuracy of a single PNT data type. This implicates the necessity to determine the absolute magnitude of significant errors and resulting consequences for the accuracy limits of single PNT output data.

Hardware redundancy in sensors, systems, and services enables the application of further methods dealing with alternative generation of standard PNT output data (e.g. heading determination with data from 2 or 3 GNSS receivers) and/or the provision of further data types for PNT output (e.g. torsion monitoring of ship's hull). Methods for alternative generation of standard PNT output data should only be applied, if the resilience of PNT data provision is significantly increased. Aspects of error correlation and propagation should be considered carefully, if methods are being operated on the same database. Any further methods may be applied to generate additional PNT output data, as long as performance degradation of required PNT data provision is avoided. It is recommended to facilitate those methods by implementing additional processing channels.

Status data should be considered as part of the potential PNT data output; to report current usability of available sensors, systems, and services as well as the feasibility and performance of supported data processing channels and methods. Each processing channel should support the generation of status data at PNT data output by application of own methods to describe or update the status based on:

1. Checking if status data provided by pre-processing is available. In case of:

   - the unfeasibility of intended data processing the incoming status data should be forwarded; and
   - degradation of intended data processing the status data should be amended by additional information from performed processing;

2. Checking of tolerated changes in nominal input data in relation to changes in data output; and the reporting of

- faults in the augmentation input data resulting in the seamless switching to lower accuracy and/or integrity level (e.g. methods of absolute error estimation are no longer applicable);
- loss in redundancy on input data resulting in, the seamless switching to lower accuracy and/or integrity level (e.g. methods for consistency checks and/or plausibility checks are no longer applicable); and
- loss in over-determination of input data (e.g. full GNSS processing is reduced to GNSS processing of four satellites, RAIM FDE is replaced by no RAIM) – Status indications should be raised accordingly;

3. Checking if processing is started or operated by the processing channels as expected (e.g. watchdog on certain steps during processing to ensure detection of system faults); and
4. Checking if designated output data is supplied in the corresponding time intervals (nominal update rate is fully available). Test and reporting should include detection of timely incoherent data rates on the input into main processing as well as real-time losses during main processing caused by system failures.

The selection of a PNT data output should be based on data provided by active processing channels that are operated in parallel. The supported combination of processing channels defines the specific method to be applied for selecting the PNT output data including associated integrity and status data.

The selection process should comprise an evaluation of the results of each individual processing channel regarding its intended performance level of PNT/I data provision, consistency checks between results of individual processing channels on the basis of a common PNT data model, and the selection of a single set of PNT/I output data based on predefined assessment rules (redundancy & degradation).

The method for performing the selection process requires an unambiguous classification and ranking system of intended results of each processing channel under normal operating conditions, and degraded results of each processing channel in the case of disturbed operating conditions (as results of degradations and/or breakdowns of data input and processing), in relation to its potential utilization for PNT data output. The method should analyse associated integrity and status data as real-time indicator for the current functionality and performance of each processing channel.

The classification of data performance should be based on accuracy and integrity levels used for the specification of operational and technical requirements per single type of PNT data.

For each type of PNT data the ranking system defines the relationship between certain accuracy and integrity levels and "best"/"worst" PNT data output:

1. If a certain accuracy and integrity level is only supported by a single processing channel, the achieved integrity level should dominate the selection like illustrated in Fig. 6.
2. If a certain accuracy and integrity level is supported by more than one channel, under nominal operation conditions the selection of data should follow the configured prioritisation and in case of performance degradations the selection is done in compliance with the prioritization illustrated in Fig. 5.

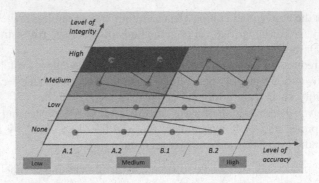

Fig. 6. Ranking list for safety-relevant [1]

3. If the same accuracy/integrity level is met by two or more processing channels, the priority should be given to the results of the processing channels operated under nominal conditions.

The selection process should ensure that PNT data and related integrity data are associated by selecting data provided by the same or assigned processing channel. The selection process should be considered as failed:

- if the pre-processing detects the unfeasibility of data processing for all supported processing channels; or
- if none of the processing channels provide any type of PNT data with an increase of accuracy and/or integrity.

A failed selection process should be indicated by status data marking the current output data as unusable.

B.4 Post-processing (functional and methodical aspects, results of post-processing)

The post-processing checks completeness of selected PNT output data (PNT data, integrity data, and status data) from main processing and generates output data streams.

The PNT integrity and status data, which has been selected by main processing for output, should be checked using the following methods:

- check of completeness and timeliness of selected output data in accordance with the nominal configuration of the PNT-DP (application grade, accuracy and integrity level, update intervals, intended status reporting);
- check if the required update interval is achieved per output data of PNT-DP; and
- check of availability of output data in relation to supported message formats.

The results of applied checks should be used to update/complete the status data for output. Each of the composed messages should contain PNT system time and, preferably UTC. A source indication for provided PNT data should be included. An important benefit of PNT-DP is the provision of integrity data associated with the PNT data at output. Therefore the messages at output should support the provision of additional integrity data, whereby:

– the integrity data per provided PNT data type should include a reference to the supported accuracy and integrity level;
– additional metadata may flag the used integrity method; and
– the provided integrity data should include the result of the integrity evaluation process performed. Such data should contain at least parameters of error distribution.

Results of post-processing should comprise:

– Messages carrying the selected PNT data together with associated integrity data in a specified message format. Both enable the subsequent connected equipment to identify whether the provided data is usable for its dedicated nautical application (e.g. automated track-control); and
– Status messages reflecting the health status of the entire PNT data processing unit.

3.3 Module C – Operational Aspects

C.1 Configuration
The intended application grade including the required accuracy and integrity level determines the minimum requirements on the data input and configuration of PNT-DP. The configuration should include the specification of thresholds and value ranges used for integrity evaluation and system controlling (e.g. in relation to operational and technical accuracy levels as well as applied integrity evaluation techniques).

C.2 Operation Management (Automatic Operation, User Interaction)
The concept of the PNT-DP is based on automated processing (pre-processing, main processing, and post-processing) to adapt the functionality on current data availability and usability. The PNT-DP is embedded software contributing to the Bridge Alert Management (BAM) of the mothering system by provision of status and integrity data. It does not generate alerts by itself. Since the shipborne PNT-DP has a residual risk regarding total loss of all functionalities, the operational environment e.g. the mothering system, should ensure, by a bypass, that available sensor and service data are available for applications.

The knowledge of users regarding the usability and integrity of input devices (sensors and services) may result in the user decision to exclude data of these sensors and services from PNT data processing. However, the manual exclusion of input data is only possible on the mothering system by controlling, opening, and closing of data interfaces. Due to automatic operation, there is no difference between a user exclusion of data input or a failed data input for the PNT-DP.

3.4 Module D – Data Communication Interfacing

Where possible, standardized and approved communication protocols for interfacing should be used, refer to publication IEC 61162 [9].

D.1 Input Data

The communication protocol for input data should allow the implementation of the supported functions for the intended application grade and performance level. In particular, this includes reception of all PNT relevant data (raw or processed); the data received should be marked either by the source itself or with a unique source identifier within the PNT-DP.

D.2 Output Data

The communication protocol for output interfacing should allow the transmission of selected PNT data including integrity and status data.

D.3 Configuration Interfacing

The manufacturer should provide data interfacing with the mothering system for configuration.

3.5   Module E – Documentation

The documentation of a PNT-DP should cover at least:

- operating manual;
- installation manual;
- configuration manual;
- failure analysis, and
- onboard familiarization material.

The documentation should be provided, preferably in an electronic format.

4   Conclusion

The IMO Correspondence Group elaborated draft guidelines for shipborne Position, Navigation and Timing (PNT) data processing to be discussed on the 4th session of NCSR Sub-Committee in March 2017. The document defines principles and functions for on-board PNT data processing taking into account the scalability of PNT-DP. The Guidelines provide rules to handle differences regarding installed equipment, current system in use, feasibility of tasks and related functions, performance of data sources as well as usability in specific regions and situations. A structured approach for the stepwise introduction of integrity is developed to achieve resilient PNT data provision in relation to the application grades and supported performance levels. These Guidelines aim to achieve standardized and integrity tested PNT output data to enhance user awareness regarding achieved performance level. It is worth noting that the research on EGNOS implementation in the maritime domain, in line with the proposed guidelines, has been performed by Maritime University of Szczecin, DLR and ESA [10].

References

1. IMO NCSR 4/6: Guidelines Associated with Multi-system Shipborne Radionavigation Receivers Dealing with the Harmonized Provision of PNT Data and Integrity Information. Report of the Correspondence Group on the Development of the Guidelines for Shipborne Position, Navigation and Timing Data Processing, submitted by Germany. Sub-committee on Navigation, Communications and Search and Rescue, International Maritime Organization, London, 15 November 2016
2. Weintrit, A.: Prioritized main potential solutions for the e-Navigation concept. TransNav Int. J. Mar. Navig. Saf. Sea Transp. **7**(1), 27–38 (2013)
3. IMO NCSR 1/9: Development of an e-Navigation Strategy Implementation Plan. Report of the Correspondence Group on e-Navigation, submitted by Norway. Sub-committee on Navigation, Communications and Search and Rescue, International Maritime Organization, London, 28 March 2014
4. IMO Resolution MSC.401(95): Performance Standards For Multi-system Shipborne Radionavigation Receivers. International Maritime Organization, London, 8 June 2015
5. IMO SN.1/Circ. 274: Guidelines for Application of the Modular Concept to Performance Standards. International Maritime Organization, London, 10 December 2008
6. Engler, E., et al.: Guidelines for the coordinated enhancement of the maritime position, navigation and time data system. Sci. J. Marit. Univ. Szczec. **45**(117), 44–53 (2016)
7. IMO Resolution A.1046(27): Worldwide Radionavigation System. International Maritime Organization, London, 20 December 2011
8. IMO Resolution A.915(22): Revised Maritime Policy and Requirements for a Future Global Navigation Satellite System (GNSS). International Maritime Organization, London, 22 January 2002
9. IEC 61162: Maritime navigation and radiocommunication equipment and systems. International Electrotechnical Commission, Geneva 2016
10. Zalewski, P., et al.: Concept of EGNOS implementation in the maritime domain. In: ESA Navitec 2016 Final Programme and Proceedings. IEEE, December 2016

Development of a Weight-in-Motion Measurement System with an Optical Sensor

Aleš Janota$^{(\boxtimes)}$, Vojtech Šimák, and Jozef Hrbček

Faculty of Electrical Engineering, University of Žilina,
Univerzitná 8215/1, 010 26 Žilina, Slovakia
{ales.janota,vojtech.simak,jozef.hrbcek}@fel.uniza.sk

Abstract. The paper presents outputs of an introductory part of the applied research project called OPTIWIM aimed at development of a new weight-in-motion measurements system based on the use of an optical sensor. The brief introduction to the framework and objectives of the project is followed by analysis of state-of-art of a selected segment of sensors applied to dynamic measurements of road vehicles. This makes possible to define expected properties of the new measurement system to be developed. Since new solution requires an approach based on an optical sensor usage, the final part of the paper summarizes different optical-based principles being under considerations of the development team.

Keywords: Weight · Motion · Measurement · Optical · Sensor · Development

1 Introduction

Our world is confronted with steep increase in road vehicles over recent years. That makes it necessary to manage traffic flows more efficiently to keep transport sustainable, i.e. to optimize traffic flows, avoid traffic jams, reduce negative consequence to our environment, etc. Intelligent transport systems that can be seen as a part of the smart city agenda play a significant role in this process. Using sensors to monitor roads, bridges, and buildings provides awareness that enables a more efficient use of resources [1]. High quality of road networks is a basic prerequisite for safe and fluent traffic operation. Therefore, detection of overloaded vehicles is needed to avoid or minimize damaging of pavement surface. Advanced sensing is a key to smart infrastructures. Various measurement approaches are available to estimate a vehicle's weight. A survey of measuring methods, detection principles and commercial products existing at the market can be found e.g. in [2]. To ensure fluency of the road traffic flows, dynamic measurements are preferred, realized with the help of so-called Weigh-in-Motion (WiM) systems that are able to capture vehicle gross and axle weights. These systems are unmanned, non-intrusive, and capable of collecting the data in real time automatically and sending them to the central system. Collected information is crucial for safety and enforcement, asset management, freight management and/or traffic operations. The primary function of the WiM system is usually preselection of offending vehicles in front of vehicle inspection checkpoints, with check by the police on approved static scale, up to direct enforcement. In addition, collected statistic data give

© Springer International Publishing AG 2017
J. Mikulski (Ed.): TST 2017, CCIS 715, pp. 234–247, 2017.
DOI: 10.1007/978-3-319-66251-0_19

us insight into capability and performance through monitoring traffic aggressiveness and is important for pavement management, i.e. pavement preservation.

This paper has been written with motivation to present the first outputs of the new research project solved at the authors' workplace, presented in the context of their several years' intention to develop a new dynamic measuring system based on usage of optical sensors.

1.1 Research Project Framework

The project belongs to the category of applied research, focused on design, optimization and creation of a device for weight measurement of a vehicle (or its axle) moving according to the currently valid traffic regulations on the road or highway. The development team consists of researchers from 3 UNIZA departments – *Physics*; *Control & Information Systems*; and *Multimedia and ICT* who cooperate with partner research division of the Slovak *Betamont, Ltd.* company. Their ambition is to bring an innovative solution fitting to the existing Measure-in-Motion® system [3], previously designed by the project partner, ensuring compatibility of the used optical sensor output with an interface of the existing processing unit. List of planned activities together with their % distribution is depicted in Fig. 1.

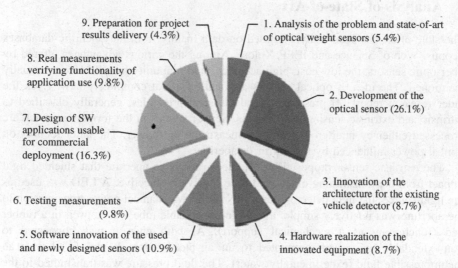

Fig. 1. Distribution of research activities during project duration [own study]

Activities follow each other respectively; however, some of them are overlapping in time. The first one was completed in Nov 2016; development of the optical sensors has actually been under way and will finish in 2018. Total length of the project is 4 years. Development costs are to be partially covered by the grant obtained from the Slovak Research and Development Agency (Slovak abbrev. APVV).

1.2   Project Objectives

The primary goal of the project is to develop and construct a high-tech and unique measurement device competitive within the region of the European Union which will allow weighting of the moving vehicles. The device will consist of an optoelectronic pressure sensor, hardware and software parts allowing its interconnection with the existing installations. The system will provide recording of the passing vehicles with pre-defined classification of the overloaded vehicles or oversized cargo in all road lanes. If needed, the system can be extended for the speed measurement. Information about the critical vehicle passing will be presented in the form of an image with identification data (event date and time, registration number in the form of both a text and an image detail) and information about the exceeded parameter (weight, dimension or speed). In order to achieve correct processing of the measured data from the traffic flow it is necessary to obtain correct and precise readings from all technological subsystems. Therefore such new more-suitable constellation of the existing solutions has to be designed which will provide relevant data to the main processing system. In order to achieve competitiveness of the final system it is needed to remove any drawbacks from each technological subsystem of the measurement device.

2   Analysis of State-of-Art

The state-of-art has been analyzed based on data mainly from the scientific databases Scopus, Web of Science and IEEE Xplore. Among the various advantages offered by fiber-optic sensors, the low cost, high accuracy, and immunity from EMI were usually mentioned. The idea of optical sensors application is not completely new. Even the older research report [4] presents several physical principles, generally classified to intrinsic and extrinsic sensing mechanisms. Sensor signals in the former approach are processed either by interferometry or by measuring the power. The latter focuses on optical power influenced by the mirror or aperture.

The extrinsic sensor proposed by [4] used a movable aperture that shielded light impact from the transmitting objective to the receiving objective. A LED was used as the light source; the light detector was a PIN diode. Mechanical power transmission to the aperture was relatively simple: a flexible expandable tube was housed in a rubber pad (functioning as a mechanical support). At both ends it was connected to non-extendable hoses and terminated to an air-bleed valve. It was filled with an incompressible fluid (experimentally water). The fluid pressure was transmitted to the diaphragm within the optical sensor assembly. Results of the experimental tests were presented in the form of frequency characteristics, linearity measurements and comparison to piezoelectric sensing. In addition to advantages mentioned above, the proposed WiM sensor could be installed on the surface of the pavement and did not require its disruption.

An intrinsic optical sensor based on the microbend principle of a multi-mode optical fiber was proposed in [5]. The fiber was placed in a mechanical deformer which modified (in dependency on external load) the optical fiber geometry by the periodic changes. The microbend takes the form of a very small sharp bend (a kink) in the optical fiber. Microbending losses cause the propagating light intensity to be coupled out of the core. Thus the vehicle weight could be obtained through measuring the variation of light intensity in optical fiber. The paper provides math analysis and equations needed to calculate period of the toothed jaw, its thickness and damping of the bent fiber. The source of IR light (1550 nm) had the optical power 10 mW. Received power was measured by a photodiode. The proposed sensor was able to measure static forces up to 30 kN with acceptable linearity and maximum error 3.82%. For dynamic effects (vehicles in the move) time responses were given together with wavelet methods for filtering of undesired effects such as noise, transitions effects caused by wheel running to/from the sensor. Dynamical tests of the sensor proved acceptable properties for speeds up to 15 km/h with error lower than 5%. The WiM accuracy is usually represented as follows:

$$A_{WiM} = (W_d - W_s)/W_s \cdot 100\% \qquad (1)$$

where A_{WiM} is a WiM measurement accuracy, W_d is axle weight or gross weight measured by the WiM, and W_s is axle weight or gross weight measured by the static scale.

The review paper [6] discussed possible applications of optical fibers in WiM systems. Various factors were taken into considerations – the way of sensor installation, nonlinearity, hysteresis, inertia, temperature effects, inertia forces of a vibrating vehicle, extremely short time of wheel's effect on the sensor. Examples of signals recorded by a commercial sensor (by Sensor Line GmbH) were used to document errors from measurements repeated at constant and variable vehicle's speeds. Potential causes of measurement errors were analyzed together with methods leading to accuracy increase. Then two approaches to calculations of the total vehicle's axle weight were considered – one based on knowledge of the impact (rubber-pavement contact) area and pressure in the tire. The following formula was used to calculate the total weight of the axis:

$$W_{a/2} = A \cdot P \qquad (2)$$

where $W_{a/2}$ is a weight on half axle, A is an area of the tire footprint, and P is an air pressure inside the tire. The other approach used time integration of the signal from fiber optics.

Combination of both methods was proposed in the study [7]. Basic parameters (impact area, tire pressure) were calculated from time course of the signal from the commercially available optical sensor (Sensor Line PUR by Sensor Line GmbH). The output optical power of the sensor decreased with increasing force affecting the fiber due to external force causing fiber bending. To enable comparison, the power was re-calculated to the dimensionless parameter (visibility loss). A series of static measurements performed under various temperatures made possible to get temperature dependency and approximate it by the 2nd order polynomial. The authors also approximated the static transfer sensor characteristics for various widths of the testing load platform by the exponential function. They tried to eliminate nonzero reaction time and sensor inertia by series of measurements for various impact speeds by a power approximation curve. To make calculation of the impact area accurate, two sensors were used – the 1st one placed transverse and the 2nd one diagonal to the pavement axis. Additional calculation improvement resulted from elimination of the force cause by the rolling friction of the tire. It helped to work only with the gravitation force of the vehicle axle. Thus the impact area could be determined by length of output (filtered) pulses from the sensor and the tire pressure was determined by the pulse amplitude. Experimental measurements of ten-wheeler proved applicability of the proposed methods for the speed range from 50 to 90 km/h with maximum error 10%.

An optic fiber sensor based on Fiber Bragg Grating (FBG) was applied in [8]. The sensor, packaged by fiber reinforced polymer (FRP), was designed as a 3-axis sensor capable of sensing vertical, axial and lateral forces in regard to the pavement. The sensor was built-in the concrete pavement. Force impact was expected in the vertical direction. However, experimental results demonstrated that for this kind of pavement the component forces were non-negligible in lateral and axial directions. The cause of this effect was identified as bending stress applied in the pavement. Quality of measurements was declined by a measuring instrument with too low sampling rate (only 10 Hz). As consequence the error range was between 0.06% and 81%.

Another solution based on Fiber Loop Ring-Down Spectroscopy was described in [9]. The sensor itself was principally identical with the microbend sensor [5]. In this case there were two sensors installed in the known distance. They were interconnected by optical fibers into the loop creating an optical resonator. Through the coupler it was linked to the source of optical pulses 10 ns wide with the power 10 mW. Another coupler was used to connect a photodetector. The optical resonator became excited by the pulse while the resonator ring-down depended on damping of the microbend fiber sensor. Design was complemented with math analysis and relevant equations. Field tests demonstrated high accuracy – better than 2% for speeds up to 100 km/h and acceptable 10% for speeds up to 55 km/h.

The sources analyzed above represent typical known examples of how fiber optics can be applied in the WiM systems. Their authors provide relatively detailed description of measurement principles, mathematical background, results of laboratory and/or field tests. For more general and comprehensive review of the literature for fiber optic sensors, performance criteria (precision, accuracy and durability), their applications in WiM systems or recommended criteria for testing and evaluation one could recommend e.g. [10]. Typical problems of optic fiber in WiM applications are covered in [11]. In addition, road transportation is not the only suitable application domain – [12] presents fiber optic sensing system for WiM and wheel flat detection (WFD) in railway assets.

3   Design Concept

This chapter gives some introductory notes on the current initial state, requirements specification for a new solution and a few remarks on considered physical approaches.

3.1   Current Situation

The project partner deploys and operates the WiM systems Measure in Motion®. Actually installed sensors are based on the piezoelectric principle, either for dynamic or static measurements. A typical layout of particular elements is shown in Fig. 2.

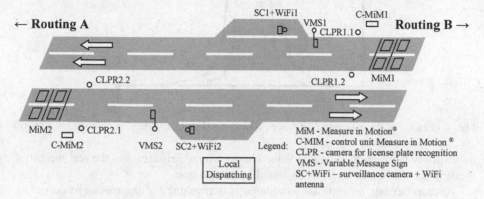

Fig. 2. Graphical representation of the sensors layout – present state [own study]

Figure 3 shows typical curves representing measured forces caused by a moving vehicle and acting on the active area of the sensor. These curves depend not only on the vehicle weight but also on length of the wheel track, tire pressure and a width of the sensor itself. However, we can see that that the area under the curve for a vehicle with the same weight is still the same for various curves. This is a key idea: to find

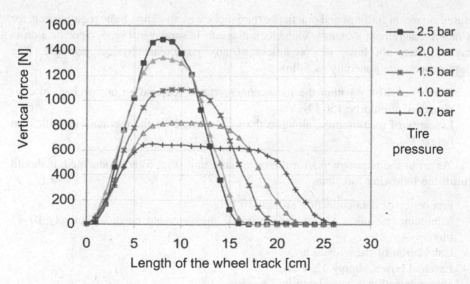

Fig. 3. Dependency of the force (measured by the sensor) on tire pressure [own study]

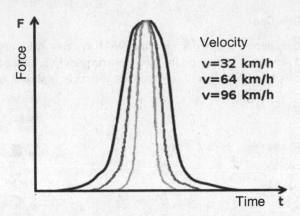

Fig. 4. Change of the area under the force curve $F = f(t)$ in dependency on velocity [own study]

dependency between the numeric value characterizing this area and the real weight of the vehicle. The exact area can be found by integration.

Another serious factor to be considered is dependency of the measured curves on vehicle velocity depicted in Fig. 4. For correct interpretation of measured and calculated values the key knowledge is knowledge of vehicle's velocity. This value is obtained using loop detectors or could be calculated from the time interval when capturing the same axle by two successive WiM sensors, or even by using any other suitable method.

Designing the algorithm for calculation of the total weight all the mentioned factors must be taken into account and a proper way of interpretation must be chosen.

3.2 Requirements Specification

Based on recent findings and practical experiences we can define basic requirements for a new measurement system whose block diagram is shown in Fig. 5. Specification is saying what, not how, i.e. regardless of any particular physical principle. The requirements are generally as follows:

- Sensor sensitivity within the measurement range (depending on mechanical construction) should be 150 kN.
- Linearity of measurement along to the whole length of the measurement platform below 1%.

As far as the requirements for the processing unit (Fig. 5) are concerned, it should fulfill the following functions:

- Frequency of data collection at least 10 kHz
- Minimum resolution 12 bits for the whole measurement range (effectively 10–11 bits)
- Calculation of the integral area
- External power supply 12 V DC
- Communication – considered in 2 modes:

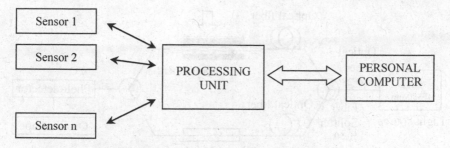

Fig. 5. The basic block diagram [own study]

Mode 1 (standard): after signal processing the unit will send secured information about integral area, initial – maximum amplitude, number of samples and time stamp, or other agreed factors. The total width of the transmitted word is assumed 20B per one sensor (for complex information about the integral).

Mode 2: the unit will send each measured (secured) sample. That will require communication with the personal computer (PC) via bus that will enable real-time transmission of information from 16 sensors concurrently each 100 µs. Time synchronization will be needed.

– Number of channels per the processing unit is not specially defined.

3.3 Selected Sensor Principles Under Considerations

In the initial phase two main physical principles have been studied, both based on detection of the modulated optical signal:

– Detection of the output of fiber interferometer deformed by a moving vehicle;
– Detection of optical birefringence of a proper polymer caused by a moving vehicle;
– Detection of light passing through the optical fiber that is reflected or transmitted by the Fiber Bragg Grating (FBG).

The first approach has been under investigation during several recent years and made platform for multiple qualification theses solved at the Dept. of Physics (UNIZA).

The principle is clear from the scheme (Fig. 6). As the light source, LED working at the wavelength 1550 nm is used. After splitting 50 to 50, one optical fiber branch (the upper) undergoes pressure perturbations while the other not. Photodetector receives signals from both branches and we can observe the result e.g. at the oscilloscope. The fiber elongates due to the applied pressure (caused by the vehicle weight), thus the optical path elongates as well.

Performed experiments have proved that elongation of the deformed fiber significantly depends on mechanical rigidity of the bending fiber. What is more, for low spaces used in deformer constructions there is minimal elongation of the optical path since it is compensated by shortening of fibers in spaces where the fiber is not stressed. There are several approaches how to solve this problem, e.g. to use fibers with lower rigidity – either a plastic fiber or a stripped glass fiber; and/or to increase the size of spaces; and/or to bend the fiber in the spaces.

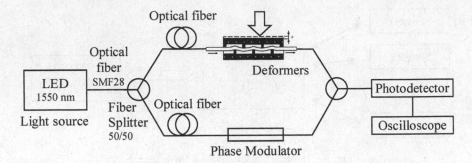

Fig. 6. The principle of fiber interferometer deformation [own study]

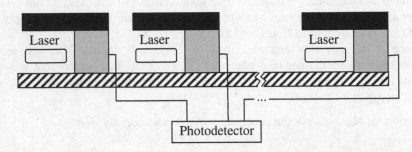

Fig. 7. The principle of optical birefringence of a proper polymer [own study]

Another important and proved finding is that it is important to have linear dependency of output quantity on input pressure. Measured pressure depends on width of the tire; however, that value is (usually) unknown. For vehicles of the same weight wider tire means lower pressure. If we have linear dependency (thanks to the flexible element - a layer between the sensor and the tire itself) it is compensated by the length of the actual area where fiber (and optical path) is elongated.

The second approach based on detection of optical birefringence rate observed in a proper polymer (Fig. 7) also followed the previous pre-research performed by the members of the project team from the Dept. of Physics (UNIZA). Particularly they studied the polydimethylsiloxane (PDMS) as a proper polymer.

Detailed results of investigations of the effect of mechanical stress on the coefficient of absorption and the refractive index of PDMS in the NIR region are available in [13], results of previous experimental studies of photoelasticity in its deformation can be found in [14], both references published by members of our research team. As in the previous case, this solution uses linear dependency of the output signal on input deformation. Experimental laboratory tests performed with PDMS and PMMA (Poly methyl methacrylate or Plexiglas) showed that one sample is not usable for performing measurement on long distance (e.g. within the road width). The reason is a high attenuation of light (dB/cm).

The potential solution of this problem seems to be usage of a higher number of samples applied on short distance each. A proper arrangement of samples may bring

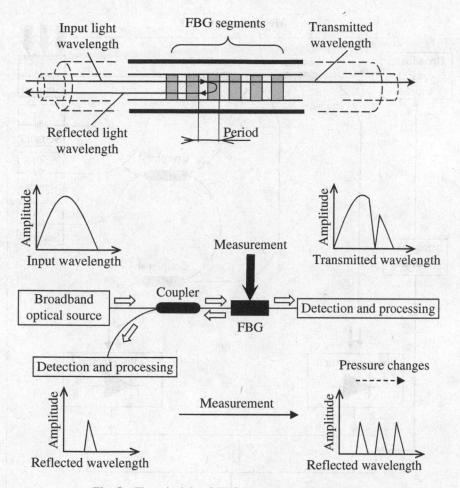

Fig. 8. The principle of FBG measurement [own study]

another advantage: if we install them skew across the road, there is possibility to measure velocity of the measured vehicle as a benefit to original measurement of vehicle weight.

The last considered approach is based on detection of the light that passes through the optical fiber and then is reflected or transmitted by the FBG (Fig. 8). The sensor consists of a row of segments where each segment has a certain refraction index that changes periodically. The grating period is typically half of the input light wavelength. The incident light propagated through the optical fiber is either reflected by the gratings or transmitted through them. Signals of all wavelengths reflected at the segments are composed into the only reflected signal. A part of the light radiation is not reflected and propagates further through the gratings. Measurement of load dependency is possible since the gratings change with their deformations. Gratings under the load modify their average wavelength of the reflected light signal. The light overlaps with the light reflected by the other gratings. The final bandwidth of the whole reflected light signal is wider under the load than without the load. Therefore, a simple photodetector is able to

Measurement Unit

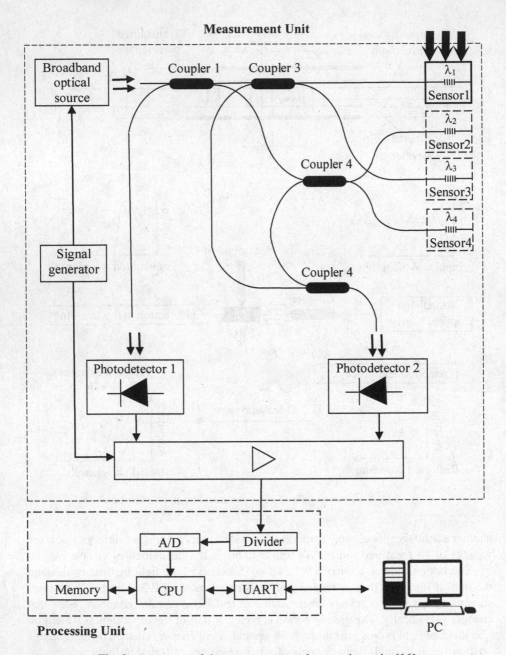

Fig. 9. A concept of the measurement and processing units [16]

detect changes of reflected light intensity. More details about the sensing principles are discussed e.g. in [15].

Utilizing properties of the FBG sensors, a potential concept of the measurement unit is depicted in Fig. 9. The block diagram consists of two basic units – the measurement unit and the processing unit [16]. The former one is responsible for weighing the vehicles and processes signals coming from the optical fiber sensors installed in the road body. The unit includes four sensing FBG sensors (indicated as Sensor 1 up to Sensor 4). The signal from the wideband optical source propagates to the sensors via couplers (connecting devices). It reflects from sensing gratings towards individual photodetectors (PIN diodes) detecting changed intensity of the reflected light.

The processing unit separates the detected signal and transmits it through signal processing circuits to the CPU processor. It uses a specially designed algorithm calculating an applied load and sends the relevant signal to the personal computer. Thus, the processing unit is used for processing, archiving and visualizing measured data. The computer saves and displays data based on operator's requirements.

The presented concept contains four sensors; however, their number is variable and may be changed according to an actual and needed configuration, e.g. for measurement of vehicle in either one or two traffic lanes. If needed, the block diagram may be extended for other devices, for example induction loops that may bring an added value – measurement of vehicles speed. In addition, the system may also be equipped with enforcement and/or surveillance cameras. Their resolution and response times should be sufficient to ensure capturing images of weighed vehicles and specially images with their readable license plate numbers. If the system is intended to provide enforcement functionality, it must contain some extra hardware and software support, e.g. a database server for running evaluation software with a user interface, automatic reading of license plate numbers, database of measured data and communication interface used to transfer data into the supervisor computer. The server could also include other software modules such as a module for automatic vehicles counting, a module for vehicles classification (using different classification schemes), and a module for detection of vehicle with dangerous goods, etc. Obtained results are then directly visualized (through a defined HMI), stored in the memory, sent via a defined bus to a superior level (e.g. a computer server). They may be used to monitor traffic operation and informative collecting of data on vehicles, or to control a traffic flow depending up its components.

4   Conclusion

The paper summarizes and presents reflections from the first project phase realized in 2016. That is why analysis of the state-of-art is included together with brief characteristic of the project in question and requirements specification for a newly developed WiM sensor. The most details have been given to three considered approaches based on different physical sensing principles. One of them will potentially be adopted for new sensor construction. The final decision about the winning concept has not been selected definitely yet. From time point of view the new sensor solution will be under development till 2018.

In 2017 the project activities will continue with advanced studies and further development of an optical fiber sensor. At the end of the year a new activity focused on upgrade of Measure-in-Motion® architecture will become open and an effort will be made to optimize actual WiM sensors layout (placement, shape, construction parameters, etc.). Communication capabilities will include Internet connections, remote control/monitoring etc.

A special attention will have to be paid to challenging selection of the material creating a contact layer between the sensor itself and the vehicle tire. A special problem seems to be elimination of waves propagating through the road together with the rolling tire and their negative effects on the sensing process and its accuracy. These waves are generated by the moving vehicle, but apart from its weigh they also depend on the structure and material of the road surface (asphalt, concrete, stone pavement, clay, etc.), installation arrangement of the sensor and some other dynamical factors as well.

Acknowledgement. The paper has been written with the support of the Slovak Research and Development Agency within the applied research project APVV-15-0441 "OPTIWIM – Measurement system with optical sensor for WiM systems".

References

1. Hancke, G.P., Silva, B.C., Hancke Jr., G.P.: The role of advanced sensing in smart cities. Sensors **13**(1), 393–425. www.mdpi.com/1424-8220/13/1/393. Accessed 12 Jan 2003
2. Janota, A., Nemec, D., Hruboš, M., Pirník, R.: Knowledge-based approach to selection of weight-in-motion equipment. In: Mikulski, J. (ed.) TST 2016. CCIS, vol. 640, pp. 1–12. Springer, Cham (2016). doi:10.1007/978-3-319-49646-7_1
3. Betamont: Traffic Network Technologies, Vehicle Detector Measure-in-Motion®. www.betamont.sk/road_systems/measure_in_motion.htm. Accessed 4 Feb 2017
4. Safaai-Jazi, A., Ardekani, S.A., Mehdikhani, M.: A low-cost fiber optic weigh-in-motion sensor, strategic highway research program. Research report SHRP-ID/UFR-90-002, National Academy of Sciences, Washington, D.C. (1990)
5. Bin, M., Xinguo. Z.: Study of vehicle weight-in-motion system based on fiber-optic microbend sensor. In: Proceedings of the International Conference on Intelligent Computation Technology and Automation, Changsha, China, 11th–12th May 2010, pp. 458–461 (2010)
6. Batenko, A., et al.: Weight-in-motion (WiM) measurements by fiber optic sensor: problems and solutions. Transp. Telecommun. **12**(4), 27–33 (2011)
7. Grakovski, A., et al.: Weight-in-motion estimation based on reconstruction of tyre footprint's geometry by group of fibre optic sensors. Transp. Telecommun. **15**(4), 97–110 (2014)
8. Huang, Y., et al.: Real-time weigh-in-motion measurement using fiber bragg grating sensors. In: Proceedings of SPIE 9061, Sensors and Smart Structures Technologies for Civil, Mechanical, and Aerospace Systems, p. 906109, 8 March 2014
9. Songlin, Y., et al.: Pressure sensor for weight-in-motion measurement based on fiber loop ring-down spectroscopy. In: Proceedings of SPIE 10158, Optical Communication, Optical Fiber Sensors, and Optical Memories for Big Data Storage, 101580U (2016)
10. Mimbela, L.Y., et al.: Applications of fiber optic sensors in Weigh In Motion (WIM) systems for monitoring truck weights on pavements and structures. Final report, submitted to U.S. Dept. of Transportation's, FHWA Washington, D.C., Apr 17 2003. www.pooledfund.org/document/download/238. Accessed 5 Feb 2017

11. Batenko, A., et al.: Problems of fibre optic sensor application in Weight-In-Motion (WIM) systems. In: Proceedings of the 11th International Conference RelStat 2011, pp. 311–316 (2011)
12. Iele, A., et al.: Fiber optic sensing system for Weighing In Motion (WIM) and Wheel Flat Detection (WFD) in railways assets: the TWBCS system. In: Proceedings of the 8th EWSHM (2016)
13. Turek, I., et al.: Effect of mechanical stress on optical properties of polydimethylsiloxane. Opt. Mater. **36**(5), 965–970 (2014)
14. Tarjányi, N., Turek, I., Martinček, I.: Effect of mechanical stress on optical properties of polydimethylsiloxane II - Birefringence. Opt. Mater. **37**, 798–803 (2014)
15. Liu, R., et al.: Evaluating innovative sensors and techniques for measuring traffic loads: final report. Technical report FHWA/TX-07/0-4509-1, Univ. of Houston, October 2006. https://d2dtl5nnlpfr0r.cloudfront.net/tti.tamu.edu/documents/0-4509-1.pdf. Accessed 5 Feb 2017
16. Lančarič, M.: Usage of fiber optical sensors in telematics systems. Bc. thesis, Department of Control & Information Systems, Žilina University, Žilina, Slovakia (2014)

Analysis of the Motor Vehicle Dynamics on the Example of a Fish Hook Maneuver Simulation

Jarosław Zalewski[✉]

Faculty of Administration and Social Sciences,
Warsaw University of Technology, plac Politechniki 1,
00-661 Warsaw, Poland
j.zalewski@ans.pw.edu.pl

Abstract. In the paper analysis of the fish hook maneuver simulation is presented, which seems to be an interesting example of a situation, when the vehicle must realise a sudden maneuver.

Research was based on simulation in MSC Admas/Car. However road parameters and vehicle loading were assumed to reflect a real situation, in which the vehicle has to change its direction. The maneuver was realised on both flat and uneven road surface, where the irregularities occurred randomly.

Certain aspects of vehicle dynamics have been taken into account, as well as the influence of the external disturbances on vehicle response. It is necessary to mention, that the correlation coefficient between irregularities for the left and right wheel was 1, which means, that they were identical for both sides of vehicle wheels.

Keywords: Fish hook maneuver · Road irregularities · Motor vehicle dynamics

1 Introduction

Dynamics of the means of transport, particularly considering motion in difficult conditions, seems to be an important issue, especially towards the modern, interdisciplinary attitude to connect different branches of knowledge. Of course, such attitude is essential nowadays, as scientists face the unavoidable fact, that different technologies can be used in order to obtain one goal.

As for transport telematics, three main directions should be taken into account: information technology, telecommunications and automatics. Although these are mainly used to easier the process of transport, considerations related directly or indirectly should also refer to both infrastructure and means. It could then be concluded, that telematics interacts with the means of transport and their infrastructure. So, the motion conditions are one of the main problems included in the wide scope of research on the proper, yet safe and economic transport organisation.

© Springer International Publishing AG 2017
J. Mikulski (Ed.): TST 2017, CCIS 715, pp. 248–259, 2017.
DOI: 10.1007/978-3-319-66251-0_20

One of the most popular and available means of transport is a motor vehicle, for which both road conditions and steering are of primary importance, when considering safety and dynamics.

Certain issues, concerning dynamics and safety in motor vehicles, have already been analysed, such as:

- the problem of wheel – road cooperation during vehicle motion (e.g. [1–4]);
- the problem of vertical and tangential reaction forces of the road, acting on the vehicle wheels (e.g. [5, 6]);
- the problem of road irregularities in realisation of the specified maneuvers (e.g. [7, 8]), causing different vehicle behaviour;
- the problem of vehicle stability, also as a result of the modelling and simulation (e.g. [9, 10]);
- modelling and simulation of vehicle dynamics (e.g. [11, 12]).

In this paper the issue of a fish hook maneuver of a vehicle model has been considered in relation to both vehicle trajectories, and selected problems resulting from different conditions of motion.

The selected vehicle model, laden as discussed in Sect. 2, realised the so called fish hook maneuver. The simulation was prepared in MSC Adams/Car software and the road conditions were described in Sect. 3.

As a result, certain conclusions have been drawn, as the given maneuver seems to be rather rare in the normal vehicle driving, so it can reflect the neccesity to rapidly change the direction, caused by e.g. sudden obstacle occurring on the road.

2 Assumptions

For studying the dynamics of the vehicle during the presented maneuver, certain assumptions were made. As for the vehicle model, the mass of its body was increased from 995 kg to 1150 kg, by adding two masses, each 73 kg, representing the driver and the passenger. The location of both masses were 0.7 m from the "origo" point along the x axis. The initial coordinates of the center of mass in a vehicle body were (in relation to the "origo" point, which is the origin of a coordinate system moving along with the vehicle, but remaining on the road surface [8]):

$$x_c = 1.5\,\text{m}, \quad y_c = 0, \quad z_c = 0.45\,\text{m},$$

however, after increasing the vehicle body mass, they amounted to:

$$x_c = 1.589\,\text{m}, \quad y_c = 0, \quad z_c = 0.452\,\text{m},$$

which indicates the symmetrical loading of the vehicle with the driver and the passenger.

The vehicle model used in the simulation is described, among others, in [8, 10, 13].

As a tool of research, the fish hook maneuver in Adams/Car was selected in order to examine the selected parameters of the vehicle motion. The initial speed of the vehicle was 80 km/h, while the gearbox was set on the fifth gear. Simulation was prepared for the following configurations:

- laden vehicle moving on a dry, flat road surface;
- laden vehicle moving on an icy, flat road surface;
- laden vehicle moving on a dry, uneven road surface;
- laden vehicle moving on an icy and uneven road surface.

Of course, the road irregularities occurring on the uneven surface have been previously described [8, 13]. Information about the file describing the irregularities are given in further parts of the paper.

Other assumptions were adopted as in [6, 14], i.e. the non-linear characteristics of the elastic – damping elements in vehicle suspension (however the non-linearity was rather weak), regarding the vehicle body as a set of quasi-rigid cuboids, etc. Characteristics of the used suspension have been presented in [8].

In Fig. 1 a spatial view of the analysed vehicle model is presented. It is worth noticing, that the x axis is pointed backwards, i.e. the positive direction points to the back of the vehicle.

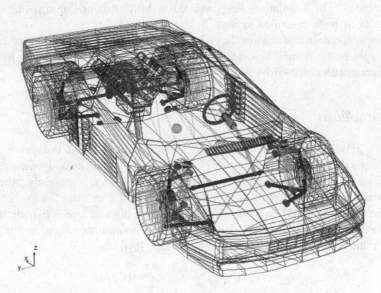

Fig. 1. Full vehicle model used in simulations [own study]

3 Simulation of the Fish Hook

Simulation of the fish hook maneuver, for all four configurations described in p. 2, has been performed. For each vehicle configuration a set of trajectories has been obtained. In Fig. 2 the trajectories of vehicle lateral displacement for the flat road surface is

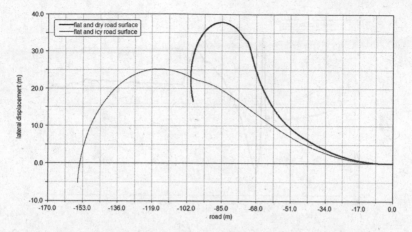

Fig. 2. Lateral displacement of the vehicle versus the covered distance on a flat road [own study]

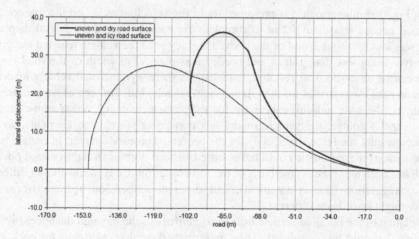

Fig. 3. Lateral displacement of the vehicle versus the covered distance on an uneven road [own study]

presented, meanwhile in Fig. 3 – for the uneven road. In both figures it can be noticed, that the shape of all four trajectories resembles the fish hook, however narrower for dry surface, and wider for the icy one.

In Fig. 2 the vehicle needed only about 102 m to complete the maneuver on the dry road, but it also required about 40 m wide road lanes (lateral displacement). For the icy road the maneuver has been completed using about 160 m of the road, however less than 30 m road width was necessary to perform the whole fish hook.

It is necessary to remember, that the trajectories present the motion of the characteristic point in the vehicle (in this case the center of mass of the vehicle body). So, in order to analyse the area used for the maneuver, vehicle width should be taken into account.

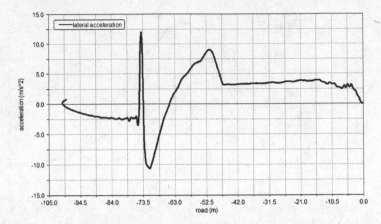

Fig. 4. Lateral acceleration of the vehicle versus the covered distance on a flat and dry road [own study]

In Fig. 3 similar scenario is presented, only the vehicle moving on the uneven and dry road needed less than 102 m to complete the maneuver, and the lateral displacement was less by about 2 m. For the uneven and icy road the vehicle realised the given maneuver using less than 150 m and needed less than 30 m width of the road.

It can be observed, that certain circumstances (here, symmetrical vehicle load and the same road irregularities for the left and right wheels) initially seen as more extreme, can in fact become helpful, especially for the vehicle realising rather non-common maneuver. Of course, the vehicle model has non-linear suspension elements, so the obtained results can be analysed only for the presented case. Moreover, in certain parts of the maneuver the vehicle could have drifted instead normal straightforward course, which has been analysed on the basis of the following figures. It means that in different load configuration (non-symmetrical), and different suspension characteristics, the results might have been different.

Relating to the possible vehicle drift in some parts of the covered distance, Figs. 4, 5, 6 and 7 should be considered. They present the lateral acceleration for each configuration described in p. 2.

In Fig. 4 the highest amplitude of the acceleration value is around 53rd, 70th and 75th meter. It indicates that on the distance between 50 m and 75 m the wheels might have been drifting or at leats sliding, whereas the direction of their resultant velocity could have been unparallel to the direction of vehicle motion. If Fig. 4 is related to Fig. 2, and the distance between 50 m and 75 m is taken into account along with the trajectory for the flat and dry road surface, then it is clearly seen, that for this road period the vehicle realised the first part of the u-turn in the fish hook maneuver. It, of course, means that the vehicle was in the middle of the maneuver and undergone the greatest lateral acceleration, reaching up to 12.5 m/s^2.

The same observation can be made for each of the obtained results. As for the Fig. 5, the changes in lateral acceleration are not so sharp and only for around 100th meter its amplitude increases rapidly to about 7.5 m/s^2. Relating it to Fig. 2, the

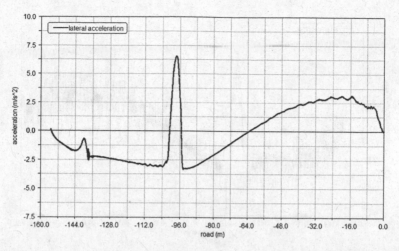

Fig. 5. Lateral acceleration of the vehicle versus the covered distance on a flat and icy road [own study]

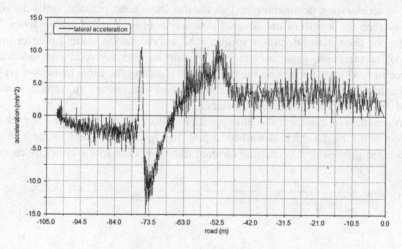

Fig. 6. Lateral acceleration of the vehicle versus the covered distance on an uneven and dry road [own study]

100th meter marks the start of the proper turn, where the momentary sliding or drift could have occurred. Also the shape of the trajectory corresponding to the discussed distance provides such conclusion (the slight deformation of the curve around 100th meter). However on the icy road surface the peak value of lateral acceleration is smaller than this obtained for the dry road, which corresponds to the longer and more rounded curve for these road conditions.

The changes in lateral acceleration for the uneven road are a lot sharper than those for the flat road surface. In Fig. 6 the greatest amplitude of acceleration on the uneven and dry road reaches around 12 m/s², whereas in Fig. 7 – around 8.75 m/s². The shapes of both curves are similar to those from Figs. 4 and 5 respectively. The peak

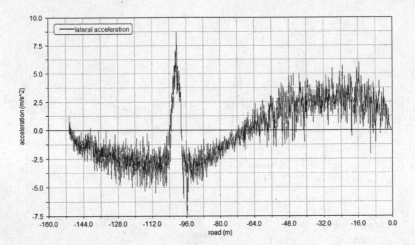

Fig. 7. Lateral acceleration of the vehicle versus the covered distance on an uneven and icy road [own study]

values of acceleration occur more or less for the same road periods, since the differences between the length of the covered distance in Figs. 2 and 3 are by about 3 m.

From the above considerations similar conclusions can be made, however greater values of acceleration, especially in case of uneven and icy road can indicate the influence of road irregularities on the difficulty in realising the same maneuver with the same vehicle, especially in this critical moment of the change in vehicle direction (around 100th meter of the covered distance).

Further analysis contains the changes in longitudinal acceleration for each configuration of the road-vehicle described in p. 2. In Figs. 8, 9, 10 and 11 these changes are presented graphically, versus the covered distance. Like in the conclusions drawn

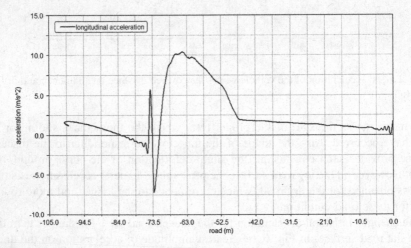

Fig. 8. Longitudinal acceleration of the vehicle versus the covered distance on a flat and dry road [own study]

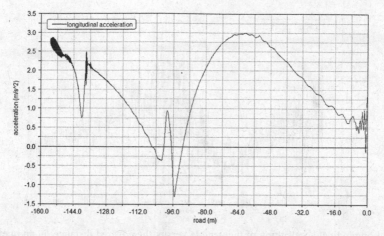

Fig. 9. Longitudinal acceleration of the vehicle versus the covered distance on a flat and icy road. [own study]

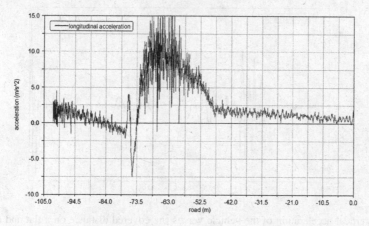

Fig. 10. Longitudinal acceleration of the vehicle versus the covered distance on an uneven and dry road [own study]

for lateral phenomena, certain similarities can be noticed for the motion on flat and uneven road. However, for the irregularities occurring on the surface, the observed changes are more rapid.

The highest value of longitudinal acceleration for the flat and icy road (Fig. 8) occurred after about 65 m of the covered distance, which correspond with the part of the trajectory in Fig. 2, where the vehicle has temporarily been realising straightforward motion before the u-turn.

In Fig. 9 the highest values of acceleration occurred for 64th and 155th meter of the distance, which also corresponds with the trajectory in Fig. 2, where the vehicle has been temporarily moving along the straight line in the above mentioned sections of the covered distance.

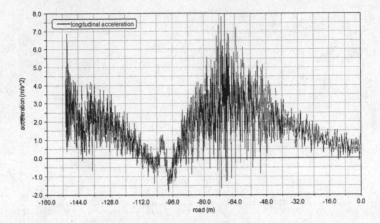

Fig. 11. Longitudinal acceleration of the vehicle versus the covered distance on an uneven and icy road [own study]

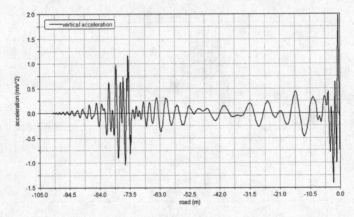

Fig. 12. Vertical acceleration of the vehicle versus the covered distance on a flat and dry road. [own study]

The courses of longitudinal acceleration obtained for the uneven road (Figs. 10 and 11) may look similar to those above (Figs. 8 and 9), although the amplitudes of acceleration are much higher, especially for the icy road with random irregularities, reaching as much as twice the highest value obtained for the flat road surface (Fig. 11).

As for the vertical acceleration, the conclusions have turned to be different than previously, since the obtained results do not indicate any resemblance between each other. As for the vehicle motion along the flat and dry road (Fig. 12), acceleration amplitudes have reached as much as 2 m/s², at the beginning of the maneuver and around 1.25 m/s² for the 74th meter of the covered distance. Meanwhile, on the uneven

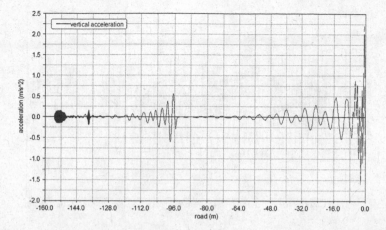

Fig. 13. Vertical acceleration of the vehicle versus the covered distance on a flat and icy road [own study]

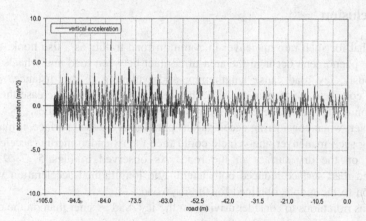

Fig. 14. Vertical acceleration of the vehicle versus the covered distance on an uneven and dry road [own study]

and dry road (Fig. 14) its value is about 2.5 m/s² at the beginning of the maneuver and as much as nearly 5 m/s², with one peak up to 7.5 m/s² for the 84th meter of the distance.

It seems that icy road surface has reduced the influence of irregularities on the vertical phenomena, because on the flat road covered with ice (Fig. 13), the highest amplitude of vertical acceleration (excluding the first 10 m similar to the dry, flat road) was only 0.5 m/s², whereas for the uneven and icy road it amounted to nearly 5 m/s² for the 48th meter of the covered distance. For the rest of the course, the vertical acceleration remained on the level below 4 m/s² (Fig. 15).

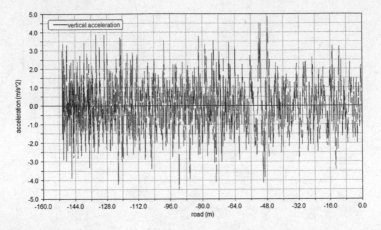

Fig. 15. Vertical acceleration of the vehicle versus the covered distance on an uneven and icy road [own study]

4 Conclusion

It seems that for such rare maneuver in common road traffic, as a fish hook, the most interesting phenomena occur in the area of contact between road and wheels.

Obviously, icy road cause lengthening and widening of the distance, which is needed to complete the maneuver. Also, the ice on the road may decrease the vertical acceleration of the vehicle, even on the road with random irregularities.

However, when considering accelerations parallel to the road surface, it may occur, that during such maneuver, the vehicle could face the extended inertia (accelerations), especially on the dry surface of the road. As observed in Figs. 5, 7, 9 and 11 ice-covered road surface reduced both lateral and longitudinal acceleration values in comparison to the graphs obtained for the dry road.

It seems ridiculous to consider driving on the icy road as safer than on the dry road, considering, say, Figs. 2 and 3, where the vehicle needed wider area to complete the fish hook, comparing to the trajectories obtained for the dry road. Random irregularities may be dangerous in so far as the amplitudes of the discussed longitudinal acceleration increased momentarily even by 5. No such phenomena have been observed in case of lateral acceleration.

Nevertheless, it should be stressed that the changes in all discussed accelerations have been more rapid and turbulent, when the vehicle has been moving on the road with irregularities.

Further considerations, concerning such extreme maneuvers, will contain different speeds and unbalanced vehicle loading.

References

1. Cebon, D.: Handbook of Vehicle-Road Interaction. Taylor & Francis, Abingdon (2000)
2. Guiggiani, M.: The Science of Vehicle Dynamics, Handling, Braking and Ride of Road and Race Cars. Springer Science+Business Media, Dordrecht (2014)
3. Pacejka, H.B.: Tyre and Vehicle Dynamics, 3rd edn. Butterworth-Heinemann, Oxford (2012)
4. Pacejka, H.B.: Tyre Models for Vehicle Dynamics Analysis. Taylor & Francis, Abingdon (1993)
5. Reński, A., Sar, H.: Wyznaczanie dynamicznych charakterystyk bocznego znoszenia opon na podstawie badań drogowych. In: Proceedings of the Institute of Vehicles, no. 4(67) (2007). (in Polish)
6. Zalewski, J.: Impact of road conditions on the normal reaction forces on the wheels of a motor vehicle performing a straightforward braking maneuver. In: Mikulski, J. (ed.) TST 2015. CCIS, vol. 531, pp. 24–33. Springer, Cham (2015). doi:10.1007/978-3-319-24577-5_3
7. Zalewski, J.: Impact of the selected road parameters on the motor vehicle motion and maintenance. In: Studies & Proceedings of Polish Association for Knowledge Management, no. 80 (2016). (in Polish)
8. Zalewski, J.: Influence of road conditions on the stability of a laden vehicle mathematical model, realising a single lane change maneuver. In: Mikulski, J. (ed.) TST 2014. CCIS, vol. 471, pp. 174–184. Springer, Heidelberg (2014). doi:10.1007/978-3-662-45317-9_19
9. Abe, M.: Vehicle Handling Dynamics. Theory and Application, 2nd edn. Butterworth Heinemann, Elsevier, Oxford (2015)
10. Kisilowski, J., Zalewski, J.: Certain results of examination of technical stochastic stability of a car after accident repair. In: Proceedings of the Institute of Vehicles, no. 5(81) (2010)
11. Genta, G.: Motor Vehicle Dynamics, Modeling and Simulation. World Scientific, Singapore (2006)
12. Rill, G.: Road Vehicle Dynamics: Fundamentals and Modeling. CRC Press, Boca Raton (2011)
13. Using Adams: MSC Software Corporation
14. Zalewski, J.: Analysis of the course of tangent reaction forces on the wheels of a motor vehicle performing a straightforward braking maneuver. Arch. Transp. Syst. Telemat. 9(1), 52–56 (2016)

Visibility of Satellites and Their Geometry for Different Numbers of Satellites of Global Navigation Systems

Jacek Januszewski[✉]

Gdynia Maritime University, al. Jana Pawla II 3, 81-345 Gdynia, Poland
jacekjot@am.gdynia.pl

Abstract. At the time of this writing (January 2017) two global satellite navigation systems (SNS), GPS and GLONASS, are fully operational, two next, Galileo and BeiDou, are under construction. The number of operational satellites in nominal constellation of these four SNSs is equal 31, 24, 24 and 27 (three MEO orbits only), respectively. As the user's position accuracy is a function of DOP (Dilution of Precision) coefficient and UERE (User Equivalent Range Error) the decreasing of the number of satellites visible by the user at different latitudes and the increasing of DOP coefficient in the case of one or two satellites out of service are analysed for each SNS in this paper. The additional analysis were made for GPS system and its UERE because this error depends among other things on the number of civil frequencies transmitted by the satellites; because some GPS satellites transmit two civil frequencies, some only one. The knowledge of the consequences of less than nominal number of satellites is for each SNS necessary for the users in all modes of transport in restricted area in particular.

Keywords: Satellite navigation system · Satellite visibility · System geometry

1 Introduction

At the time of this writing (January 2017) two global satellite navigation systems (SNS), American GPS and Russian GLONASS, are fully operational, two next, also global, Galileo in Europe and BeiDou in China, are under construction.

In the case of GPS the number of operational satellites – 31, is, since few months, greater than 24, i.e. the number of satellites in nominal constellation. In the case of GLONASS the number of operational satellites cannot be greater than nominal number 24 meanwhile the number of satellites on the orbits with different status is since few years greater than 24, e.g. January 28, 2017 – it was 3 satellites, under check by the Satellite Prime Contractor (1), spares (1), in flight tests phase (1) [1].

The Galileo constellation design was originally planned based on a three-plane constellation with a minimum of nine operational satellites in each plane and three active spares, one per orbital plane; total number of satellites – 30 (9 × 3 + 1 × 3). Over the course of time the planned Galileo constellation changed and in 2014 these changes became evident. Currently the reference space segment will have 24 operational

J. Mikulski (Ed.): TST 2017, CCIS 715, pp. 260–270, 2017.
DOI: 10.1007/978-3-319-66251-0_21

satellites only with up to six operating spares – two in each plane. The eight satellites in each plane are equally spaced; locations of the spares will be determined [2–4].

As the BeiDou constellation of 3 IGSO and 5 GEO satellites can be used in limited area, China and Asia–Pacific region, the constellation of 27 MEO satellites only will be taken into account. Currently (January 2017) there are 20 BeiDou satellites in orbits and healthy and one with status in commissioning and 11 Galileo usable satellites, 4 with status in commissioning, 2 under tests and one partially unavailable. Full operational Capability (FOC) of BeiDou and Galileo systems is planned in 2020 [5–7].

That's why all calculations were made for constellation of 31 GPS satellites, 24 GLONASS, 24 Galileo and 27 BeiDou MEO satellites. However in the case of each SNS at any moment one or more satellites can be for different raisons (e.g. in commissioning phase, in maintenance, in flights tests phase, under check by the Satellite Prime Contractor) out of service. The consequences of this fact if one or two satellites are out of service are the subject of this paper.

2 Satellite Geometry and Dilution of Precision Coefficient

In each SNS fix position can be calculated only from satellites of which elevation angle in user's receiver at the moment of measurements is higher than the masking elevation angle H_{min}. There is need of at least four satellites to calculate latitude, longitude, altitude and time. The accuracy of the position solution determined by SNS is ultimately expressed as the product of a geometry factor and a pseudorange error factor [8–10]:

$$(\text{error in SNS solution}) = (\text{geometry factor}) \times (\text{pseudorange error factor}) \quad (1)$$

As the error solution can be expressed by σ_ρ – the standard deviation of the positioning accuracy, geometry factor by the dilution of precision (DOP) coefficient and pseudorange error factor by σ_{UERE} (UERE – User Equivalent Range Error), this relation can be defined as:

$$\sigma_\rho = DOP \cdot \sigma_{UERE} \quad (2)$$

If we can obtain all four coordinates of the user's position (latitude, longitude, altitude, time – φ, λ, h, t), geometry factor DOP is expressed by GDOP (Geometric Dilution of Precision), if we want obtain one coordinate, e.g. altitude h only, geometry factor DOP is expressed by VDOP (Vertical Dilution of Precision).

3 Test Methods

In order to know the number of satellites visible by the user above angle H_{min} and the distributions of GDOP coefficient values for different numbers of operational satellites for each SNS at different user's latitudes author's simulating program were used.

The parameters of spatial segment, the number of MEO satellites and time interval of nominal constellation repeatability of all four SNSs are presented in the Table 1. The geographical longitude of ascending node and argument of latitude of all GPS and

GLONASS satellites were taken from current constellations of these operational systems, in the case of Galileo and BeiDou it was nominal future constellation.

Table 1. Global satellite navigation systems, parameters of spatial segment and time interval of nominal constellation repeatability [1, 2, 6, 11]

System	Orbit altitude [km]	Orbit inclination [°]	Number of orbital periods/time interval [min]	Number of MEO satellites
BeiDou	21,500	55	13/10,091.48	27
Galileo	23,222	56	17/14,360.75	24
GLONASS	19,100	64.8	17/11,488.44	24
GPS	20,183	55	2/1,435.94	31

The interval of the latitude of the observer between 0° and 90° was divided into 9 user zones, each 10° wide. In the user's receiver masking elevation angle H_{min} was assumed to be 5° (the most frequently used value of H_{min}) and 25°. This last angle is representative for the positioning in restricted area where the visibility of satellites can be limited. This problem is very important in two modes of transport − road (urban canyon) and maritime. All calculations were based on reference ellipsoid WGS−84.

For each zone of latitude, for each SNS, for each number of operational satellites and for each angle H_{min} one thousand (1,000) geographic-time coordinates of the observer were generated by random−number generator with uniform distribution:

- latitude interval 0–600 min (10°),
- longitude interval 0–21,600 min (360°),
- time interval in minutes equal time of constellation repeatability (Table 1).

For each geographic-time coordinates the number of satellites ls visible by the user and GDOP coefficient values were calculated. Depending on the system and user's latitudes this value (v) was divided at latitudes 0–10° and 50–60° into 6 intervals ($v < 2$, $2 \leq v < 3$, $3 \leq v < 4$, $4 \leq v < 5$, $5 \leq v < 8$, $v \geq 8$) for BeiDou and GPS systems, into 6 intervals ($v < 3$, $3 \leq v < 4$, $4 \leq v < 5$, $5 \leq v < 6$, $6 \leq v < 8$, $v \geq 8$) for Galileo and GLONASS systems and at latitudes 80–90° into 5 intervals ($4 \leq v < 5$, $5 \leq v < 6$, $6 \leq v < 8$, $8 \leq v20$, $v \geq 20$) for all four systems. If the number ls is less than 4 it means that 3D position cannot be obtained and GDOP value calculated. That's why each time next parameter called No Fix was calculated and expressed in per cent.

As BeiDou, Galileo and GLONASS systems have three orbit planes and all theirs satellites on each orbit are evenly distributed the calculations for these three systems were made for nominal constellation, 27 MEO, 24 and 24 satellites, respectively, for one non unhealthy satellite (in each system number one) and three cases of two satellites out of service:

- two satellites adjacent of the same orbit (satellites number 1 and 2),
- two satellites of the same orbit for which theirs latitude arguments differ in 180° (Galileo and GLONASS systems, satellites number 1 and 5) or are the nearest to this value (BeiDou system, satellites number 1 and 5),

– one satellite of the first orbit (number 1) and one satellite of the second orbit for which the difference of theirs latitude arguments is the lowest (BeiDou system satellite number 10, Galileo and GLONASS satellite number 9).

GPS system has six orbit planes and its satellites on each orbit are not evenly distributed. That's why the calculations for this system were made for current (31 satellites) constellation, 30 satellites (without satellite number 1) and 29 satellites (without two satellites, one of the first orbit and one of the adjacent orbit for which the difference of theirs latitude arguments is the lowest).

4 The Distributions of GDOP Coefficient Value

The distributions of GDOP coefficient values and the lack of 3D position (mentioned above No Fix) both in per cent for two latitude zones, 0–10° and 50–60°, for different masking elevation angles H_{min} for different numbers of operational satellites ls for BeiDou system (27 satellites and less) are presented in the Table 2, for Galileo system (24 satellites or less) in the Table 3, for GLONASS system (24 satellites or less) in the Table 4 and for GPS system (31 satellites or less) in the Table 5. The same distributions for latitude zone 80–90° for all four systems are showed in the Tables 6 and 7. Zones 0–10°, 50–60° and 80–90° are representative for low, middle and high latitude, respectively.

4.1 BeiDou System

In the case of BeiDou system (Tables 2 and 6) No Fix is greater than 0 for $H_{min} = 25°$ only. Its value depends on user's latitude. If one only satellite is out of service, the greatest values are at middle latitudes, the lowest at high latitudes and this difference is considerable. It two satellites are non operational (ls = 25) the lowest value of No Fix is for each user's latitude if these satellites are on the same orbit and the difference of theirs latitude arguments is the nearest to 180°.

BeiDou GDOP coefficient value depends on user's latitude and angle H_{min}. This value v is for $H_{min} = 5°$ less than for $H_{min} = 25°$ considerably; for the first angle for each constellation of satellites (27, 26 and 25) the lowest value v_{lo} can be less than 2 (0.8% or less) in zone 0–10°, greater than 2 but less than 3 (about 60%) in zone 50–60° and greater than 4 but less than 5 (about 9%) in zone 80–90°. For $H_{min} = 25°$ this value can be greater than 3 but less than 4 at latitudes 0–10° (about 20%) and 50–60° (about 20%) and greater than 6 but less than 8 (about 9%) in zone 80–90°.

If $H_{min} = 5°$ for all 5 constellations the greatest value of GDOP v_{gr} is in zone 0–10° less than 8, in zone 50–60° less than 8 in almost 100% and in zone 80–90° greater than 20 approximately in 22%. If $H_{min} = 25°$ for the same constellations v_{gr} can be greater than 8 in dozen or so per cent and in 20 or more per cent in zone 0–10° and 50–60° respectively and in zone 80–90° greater than 20 in 30% or more.

Table 2. The distribution of GDOP coefficient values and the lack of 3D position both in per cent for different observer's latitudes for different masking elevation angle H_{min} for different numbers of operational satellites ls (27 or less) for BeiDou system [own study]

φ [°]	H_{min} [°]	ls without	No Fix [%]	GDOP coefficient value – v					
				<2	$2 < v \leq 3$	$3 < v \leq 4$	$4 < v \leq 5$	$5 < v \leq 8$	>8
0–10	5	27	0	0.8	89.9	9.3	–	–	–
		1	0	0.7	87.5	11.6	–	0.2	–
		1 & 2	0	0.7	82.8	16.0	0.3	0.2	–
		1 & 5	0	0.6	84.5	14.5	0.1	0.3	–
		1 & 10	0	0.5	85.0	14.0	0.1	0.4	–
	25	27	2.7	–	–	23.1	34.0	25.3	14.9
		1	5.0	–	–	20.2	30.3	26.8	17.7
		1 & 2	9.6	–	–	17.4	27.2	26.5	19.3
		1 & 5	8.4	–	–	18.0	26.7	27.6	19.2
		1 & 10	6.9	–	–	17.3	27.8	28.5	19.5
50–60	5	27	0	–	66.9	30.9	2.0	0.2	–
		1	0	–	61.8	35.6	2.2	0.3	0.1
		1 & 2	0	–	55.6	37.7	4.3	2.3	0.1
		1 & 5	0	–	56.1	40.1	2.9	0.8	0.1
		1 & 10	0	–	58.2	37.1	3.2	1.4	0.1
	25	27	2.8	–	–	12.3	35.0	23.5	26.4
		1	8.6	–	–	10.4	29.2	26.6	25.2
		1 & 2	16.9	–	–	9.1	23.7	27.6	22.8
		1 & 5	12.2	–	–	8.9	23.5	30.3	25.1
		1 & 10	16.9	–	–	9.7	26.2	26.3	20.9

4.2 Galileo System

For Galileo system (Tables 3 and 6) No Fix is greater than 0 for $H_{min} = 25°$ only except for zone 80–90° and ls = 24. For all 5 constellations the values of No Fix are in zone 0–10° and 50–60° almost the same. If ls = 24 No Fix is less than 1%, if ls = 23 less than 10%, if ls = 22 No Fix increases to dozen or so per cent.

For $H_{min} = 5°$ the lowest value v_{lo} of GDOP can be less than 3 in zone 0–10° (about 90%) and in 50–60° (about 60%) and greater than 4 but less than 5 (less than 1%) in zone 80–90°. The greatest value v_{gr} of GDOP can be in zone 0–10° less than 5 (almost in 100%), in zone 50–60° less than 6 (almost in 100%) and greater than 20 (20% or more) for all 5 constellations.

For $H_{min} = 25°$ for all constellations the v_{lo} can be in zone 0–10° and zone 50–60° less than 4 but greater than 3 in dozen or so per cent, in zone 80–90° greater than 20 in 30% or more.

Table 3. The distribution of GDOP coefficient values and the lack of 3D position both in per cent for different observer's latitudes for different masking elevation angle H_{min} for different numbers of operational satellites ls (24 or less) for Galileo system [own study]

φ [°]	H_{min} [°]	ls without	No Fix [%]	GDOP coefficient value – v					
				<3	$3 < v \leq 4$	$4 < v \leq 5$	$5 < v \leq 6$	$6 < v \leq 8$	>8
0–10	5	24	0	93.5	5.2	1.3	–	–	–
		1	0	89.0	8.0	3.0	–	–	–
		1 & 2	0	83.2	12.7	4.1	–	–	–
		1 & 5	0	86.3	9.3	4.3	0.1	–	–
		1 & 9	0	84.0	12.2	3.8	–	–	–
	25	24	0.9	–	17.6	27.5	5.6	10.1	38.3
		1	8.0	–	14.6	24.8	6.6	10.2	35.8
		1 & 2	13.8	–	13.4	21.2	6.0	10.4	35.2
		1 & 5	15.5	–	12.9	20.3	7.5	9.3	34.5
		1 & 9	15.6	–	12.3	21.2	7.5	9.9	33.5
50–60	5	24	0	69.6	30.1	0.3	–	–	–
		1	0	63.2	34.7	2.1	–	–	–
		1 & 2	0	55.0	40.2	4.5	0.3	–	–
		1 & 5	0	57.6	39.2	3.2	–	–	–
		1 & 9	0	57.5	37.3	4.8	0.1	–	0.3
	25	24	0.3	–	17.7	45.0	0.3	2.8	33.9
		1	7.3	–	15.1	39.2	1.3	3.5	33.6
		1 & 2	17.6	–	12.2	34.2	1.2	2.8	32.0
		1 & 5	12.9	–	13.0	36.4	2.2	3.9	31.6
		1 & 9	15.4	–	11.3	36.2	2.7	4.2	30.2

4.3   GLONASS System

For GLONASS system (Tables 4 and 7) No Fix is greater than 0 for $H_{min} = 5°$ only, its value is in zone 0–10° greater than 30% and in zone 50–60° less than 10% for all constellations and in zone 80–90° for two constellations with 22 satellites equal 0.2% only.

For $H_{min} = 5°$ the v_{lo} can be for all constellations less than 3 in zone 0–10° (about 80%) and in zone 50–60° (about 55%) and greater than 4 and less than 5 (less than 1%) in zone 80–90°. The v_{gr} is in zone 0–10° and zone 50–60° for ls = 24 less than 5, for all other constellations can be greater than 6 but less than 8 (less than 1% in zone 0–10°) or greater than 8 (less than 1% in zone 50–60°). In zone 80–90° for all constellations the v_{gr} can be greater than 20 (20% or more).

For $H_{min} = 25°$ the v_{lo} can be for all constellations greater than 3 but less than 4 in zone 0–10° (about 10%) and in zone 50–60° (about 3%) and greater than 6 but less than 8 in zone 80–90° (several %). The v_{gr} can be for all constellations in zone 0–10° and 50–60° greater than 8 (about 30%) and in zone 80–90° greater than 20 (32% or more).

Table 4. The distribution of GDOP coefficient values and the lack of 3D position both in per cent for different observer's latitudes for different masking elevation angle H_{min} for different numbers of operational satellites ls (24 or less) for GLONASS system [own study]

φ [°]	H_{min} [°]	ls without	No Fix [%]	GDOP coefficient value – v					
				<3	$3 < v \le 4$	$4 < v \le 5$	$5 < v \le 6$	$6 < v \le 8$	>8
0–10	5	24	0	84.1	15.7	0.2	–	–	–
		1	0	79.2	19.3	1.3	0.1	–	0.1
		1 & 2	0	72.1	23.0	3.3	0.5	0.2	0.9
		1 & 5	0	74.0	21.7	3.6	0.3	0.2	0.2
		1 & 9	0	73.7	22.5	3.3	0.3	0.1	0.1
	25	24	30.7	–	13.1	10.6	4.8	7.9	32.8
		1	38.5	–	11.3	9.4	4.3	7.1	29.4
		1 & 2	43.9	–	9.8	8.3	4.1	6.4	27.5
		1 & 5	48.9	–	9.3	8.3	4.0	5.4	24.1
		1 & 9	46.2	–	9.6	7.7	3.7	6.5	26.3
50–60	5	24	–	61.1	38.5	0.4	–	–	–
		1	–	55.1	41.7	3.2	–	–	–
		1 & 2	–	51.3	42.7	6.0	–	–	–
		1 & 5	–	48.6	46.5	4.6	0.2	0.1	–
		1 & 9	–	49.3	45.9	4.6	0.1	0.1	–
	25	24	0.1	–	3.5	50.7	15.0	2.8	27.9
		1	3.3	–	3.4	43.2	15.1	4.5	30.5
		1 & 2	9.8	–	3.4	37.3	14.3	4.6	30.6
		1 & 5	9.2	–	2.4	37.7	15.2	5.6	29.9
		1 & 9	8.3	–	3.2	37.7	12.3	6.7	31.8

4.4   GPS System

For GPS system (Tables 5 and 7) No Fix is greater than 0 for $H_{min} = 25°$ only in zone 0–10° for all three constellations and in zone 50–60° (the greatest value 12%) and for constellation with 29 satellites (9.5%).

If $H_{min} = 5°$ the v_{lo} can be for all constellations less than 3 in zone 0–10° (89% or more) and zone 50–60° (85% or more) and greater than 4 but less than 5 in zone 80–90° (2% or less). The v_{gr} is for all constellations less than 4 in zone 0–10° and 50–60° and can be greater than 20 in zone 80–90° (20% or more).

If $H_{min} = 25°$ the v_{lo} for all constellations can be in zone 0–10° greater than 2 but less than 3 (less than 1%), in zone 50–60° greater than 3 but less than 4 (about 10%) and in zone 80–90° greater than 6 but less than 8 (several %). The v_{gr} for all constellations can be greater than 8 in zone 0–10° (about 30%), in zone 50–60° (about 30%) and greater than 20 in zone 80–90° (about 35%).

Table 5. The distribution of GDOP coefficient values and the lack of 3D position both in per cent for different observer's latitudes for different masking elevation angle H_{min} for different numbers of operational satellites ls for GPS system [own study]

φ [°]	H_{min} [°]	ls	No Fix [%]	GDOP coefficient value – v					
				<2	$2 < v \leq 3$	$3 < v \leq 4$	$4 < v \leq 5$	$5 < v \leq 8$	>8
0–10	5	31	0	0.4	89.6	10.0	–	–	–
		30	0	–	89.0	11.0	–	–	–
		29	0	–	89.0	10.0	–	–	–
	25	31	0.2	–	0.8	14.7	22.6	26.6	35.1
		30	0.9	–	0.6	14.0	20.2	28.9	35.4
		29	12.0	–	0.5	10.3	18.2	27.6	35.4
50–60	5	31	0	–	87.3	12.7	–	–	–
		30	0	–	85.0	15.0	–	–	–
		29	0	–	83.7	16.3	–	–	–
	25	31	0	–	–	16.3	14.8	48.5	20.4
		30	0	–	–	7.8	10.0	48.7	33.5
		29	9.5	–	–	6.0	11.0	49.0	24.5

Table 6. The distribution of GDOP coefficient values and the lack of 3D position both in per cent for latitudes 80–90° for different masking elevation angle H_{min} for different numbers of operational satellites ls (27 or less) for BeiDou system and Galileo (24 or less) system [own study]

System	H_{min} [°]	ls without	No Fix [%]	GDOP coefficient value – v				
				$4 < v \leq 5$	$5 < v \leq 6$	$6 < v \leq 8$	$8 < v \leq 20$	$v > 20$
BeiDou	5	27	0	0.5	12.8	27.5	37.2	22.0
		1	0	0.3	12.1	27.8	37.6	22.2
		1 & 2	0	0.3	11.2	27.2	38.6	22.7
		1 & 5	0	0.3	11.2	28.5	37.8	22.2
		1 & 10	0	0.3	11.5	27.7	38.2	22.3
	25	27	0	–	–	10.2	59.0	30.8
		1	0.2	–	–	8.9	57.0	33.9
		1 & 2	3.5	–	–	8.0	52.6	35.9
		1 & 5	0.2	–	–	7.6	55.8	36.4
		1 & 10	7.2	–	–	8.1	53.9	30.8
Galileo	5	24	0	0.6	11.5	26.6	39.0	22.3
		1	0	0.6	10.2	27.2	39.4	22.6
		1 & 2	0	0.6	9.7	27.2	39.4	23.1
		1 & 5	0	0.6	8.8	27.6	40.2	22.8
		1 & 9	0	0.5	9.8	27.0	39.5	23.2
	25	24	0	–	–	7.2	61.8	31.0
		1	0.2	–	–	6.4	59.7	33.7
		1 & 2	1.0	–	–	5.4	55.6	38.0
		1 & 5	0.4	–	–	5.6	58.1	35.9
		1 & 9	2.3	–	–	5.6	55.4	36.7

Table 7. The distribution of GDOP coefficient values and the lack of 3D position both in per cent for latitudes 80–90° for different masking elevation angle H_{min} for different numbers of operational satellites ls (24 or less) for GLONASS system and 31, 30 and 29 for GPS system [own study]

System	H_{min} [°]	ls without	No Fix [%]	GDOP coefficient value – v				
				$4 < v \leq 5$	$5 < v \leq 6$	$6 < v \leq 8$	$8 < v \leq 20$	$v > 20$
GLONASS	5	24	0	0.6	9.0	27.7	39.2	23.5
		1	0	0.6	7.8	28.5	39.4	23.7
		1 & 2	0	0.5	7.3	28.0	40.1	24.1
		1 & 5	0	0.6	7.3	27.7	40.4	24.0
		1 & 9	0	0.5	7.2	28.1	40.3	23.9
	25	24	0	–	–	8.8	58.6	32.6
		1	0	–	–	7.4	58.2	34.4
		1 & 2	0.2	–	–	6.8	54.9	38.1
		1 & 5	0	–	–	6.6	57.1	36.3
		1 & 9	0.2	–	–	6.7	55.2	37.9
GPS	5	31	0	1.7	14.5	24.5	38.2	21.1
		30	0	1.4	11.6	27.3	39.4	21.3
		29	0	1.2	11.3	26.4	39.7	21.4
	25	31	0	–	–	4.4	67.7	27.9
		30	0	–	–	3.3	65.7	31.0
		29	0.1	–	–	2.0	55.0	43.0

5   GPS Signals and Frequencies

The number of frequencies and signals transmitted by satellite depends on the SNS and in the case of GPS satellite block also. Currently in this system there are three blocks – IIR, IIR-M and IIF. Satellites block IIR transmit three signals on two frequencies (L1 and L2), block IIR-M – 6 signals (L1 and L2 also), block IIF – 7 signals on three frequencies (L1, L2 and L5). For all civil users the most interesting are civil signals, L1 C/A (all blocks, all 31 satellites), L2C (block IIR-M and IIF, 19 satellites) and L5C (block IIF only, 12 satellites). The number of satellites, their block and PRN (Pseudo Random Noise) number located on six orbits are presented in the Table 8.

Table 8. GPS system, the number of satellites, their blocks and PRN number (January 2017) [own study]

Orbit	Block			Total number
	IIR	IIR-M	IIF	
A	–	7, 31	24, 30	4
B	16, 28	12	25, 26	5
C	19	17, 29	8, 27	5
D	2, 11, 21	–	1, 6	5
E	18, 20, 22	5	3, 10	6
F	13, 14, 23	15	9, 32	6
Total	12	7	12	31

In the case of all three other SNS the frequencies and signals transmitted by satellites of each system are the same. Two and more frequencies permit the user's receiver to calculate the ionospheric correction thus the position's accuracy is greater.

Currently almost all GPS receivers used in transport are the receiver L1 C/A; i.e. without the possibility of calculations ionospheric correction. That's why we can say that for these users the knowledge of satellite block in position determination is an unimportant matter.

6 Conclusion

The user of each satellite navigation system, in particular the user of each mode of transport must take into account the possibility that the current number of operational satellites can at any moment decrease and finally in some cases the position will not be calculated. In open area for all global SNSs (BeiDou, Galileo, GLONASS and GPS) it is no problem if user's receiver masking elevation angle $H_{min} = 5°$.

If one or even two operational satellites are for any raison out of service the 3D position can be in each SNS obtained always, at any moment and at any point on the Earth, (No Fix = 0%).

The percentage of the lack of the user's position expressed by parameter No Fix depends on the parameters of the SNS's spatial segment, angle H_{min}, user's latitude and time.

If $H_{min} = 25°$ and all satellites of nominal constellation are healthy No Fix is at low and middle latitudes greater than 0 for all SNSs except for GPS and at high latitudes equal 0 for all SNSs.

If $H_{min} = 25°$ and one satellite of nominal constellation is unhealthy No Fix is at low latitudes equal for GLONASS almost 40% and for all other three SNSs several per cent, at middle latitudes several per cent, except for GPS (0%) and at high latitudes for BeiDou and Galileo less than 1% and for GLONASS and GPS zero per cent.

If $H_{min} = 25°$ and two satellites of nominal constellation are unhealthy the lowest values of No Fix are at high latitudes for all SNS, for GPS it is zero per cent, for GLONASS less than 1% for some constellations, for Galileo less than 3% and for BeiDou several per cent. The greatest values of No Fix are for GLONASS, almost 50%, at low latitudes, for GPS, Galileo and BeiDou dozen or so at low and middle latitudes.

If two operational satellites are out of service and the 3D position cannot be obtained the knowledge of PRN numbers of these satellites, theirs orbits and slots is very important for the user of the receiver of one SNS only because these parameters decide on the reducing of position accuracy.

References

1. www.glonass-ianc.rsa.ru. Accessed 28 Jan 2017
2. European GNSS Service Centre. Constellation Information (2016)
3. www.gsc.europa.eu. Accessed 28 Jan 2017

4. www.gpsworld.com. Accessed 26 Jan 2017
5. Munich Satellite Navigation Summit, Munich (2016)
6. www.beidou.gov.cn. Accessed 20 Jan 2017
7. www.insidegnss.com. Accessed 26 Jan 2017
8. Beitz, J.W.: Engineering Satellite-Based Navigation and Timing Global Navigation Satellite Systems, Signals and Receivers. Wiley, Hoboken (2016)
9. Hofmann-Wellenhof, B., et al.: GNSS – Global Navigation Satellite Systems GPS, GLONASS, Galileo, and More. Springer, New York (2008)
10. Kaplan, E.D., Hegarty, C.J.: Understanding GPS Principles and Applications. Artech House, Boston, London (2006). doi:10.1007/978-3-211-73017-1
11. www.gps.gov. Accessed 28 Jan 2017

Political and Market Challenges in Relation to Services Using Intelligent Transport Systems

Elżbieta Załoga$^{(\boxtimes)}$ and Władysław Wojan

University of Szczecin, Szczecin, Poland
elzbieta.zaloga@wzieu.pl, w.wojan@gmail.com

Abstract. Development of intelligent transport systems (ITS) is one of the tools used to achieve long-term objectives of the EU transport policy. The new EU strategy sets the course of action towards a cleaner, environment friendly and efficient transport, promotes intelligent mobility in order to change standards of mobility patterns. Development of technologies concerning mobile devices and common access to the Internet (at any time and place) forms a basis for the creation of new services also in the transport sector. Although, their further development requires integration in the institutional, technological and operational areas. Examples of actions taken in such direction are C-ITS Platform, an EU initiative, and a new service called Mobility as a Service (MaaS). Rationale and effects of these innovative solutions are this article's main subject for considerations.

Keywords: EU transport policy · ITS · New transport services

1 Introduction

Intelligent transport systems (ITS) as applications of the information and communication technologies are increasingly reflected in the EU agendas and regulations. Particularly important for the support of development of the ITS in the European Union in the past decade was the 2001 White Paper on transport [1], which contained many indications for the development and use of modern technologies as an alternative to the growing – in a broad sense – costs of the transport sector. Also the currently in force 2011 White Paper [2] recognizes innovation as the essence of a new EU strategy focused on decarbonisation of transport, process optimization, resource saving, efficient use of the modes of transport and infrastructure, and implementation of intelligent mobility systems. Concerning the scope of modern technologies, there are three key factors that are conditioning the achievement of these objectives: higher efficiency of vehicles - due to modern engines, materials and construction designs, use of greener energy – due to the use of new fuels and propulsion systems, better, safer and more reliable use of the Internet – thanks to information and communication systems.

Intelligent transport systems are also of interest to many other EU development programmes. In the 2016 European strategy for low-emission mobility [3] one of the priorities is to increase the efficiency of transport system through the optimal use of digital technologies (ITS), smart pricing, and further encouraging the shift towards low-emission transport modes. Intelligent and interoperable technologies are also the basis for development of the multimodal Trans-European transport network (TEN_T).

J. Mikulski (Ed.): TST 2017, CCIS 715, pp. 271–281, 2017.
DOI: 10.1007/978-3-319-66251-0_22

They support the European Rail Traffic Management System (ERTMS) and the telematics applications for freight service (TAF-TSI), air traffic management for the single European air space (SESAR), river information services system (RIS), maritime safety system (SafeSeaNet), and vessel traffic monitoring and information system (VTMIS). There are also other systems under implementation, like automatic identification system (AIS) and remote identification and vessel tracking system (LRIT). The strategy adopted by the European Commission in 2016, concerning the digitization of European industry [4], is also a major support for the development of ITS-based services, as it provides funding of research for innovative solutions concerning large data sets (data platforms), cybersecurity, and systems for steering a connected vehicle. In road transport, the most common examples of intelligent transport systems are systems for control and management of urban and highway traffic, electronic tolling systems, and navigation systems. The pan-European eCall system (now known as the emergency number 112) will also take advantage of ITS elements.

The above examples are proving that ITS are becoming an increasingly common tool for achieving the EU's strategic objectives, as they optimize transport processes, serve the better use of transport infrastructure, increase transport safety, improve transport accessibility, and reduce resource consumption (Fig. 1). They also contribute to changing mobility patterns. ITS can facilitate the delivery of a wide range of policy objectives, beyond those directly associated with transport, bringing significant benefits to transport users and those who live and work within the area [5]. Many ITS solutions provide significant environmental benefits.

The most relevant contemporary challenge is to create a legal framework for the integration of intelligent systems at the institutional, technological and operational level, as well as to ensure the ability of transport sector to adapt innovative solutions and social acceptance of so called *intelligent services*.

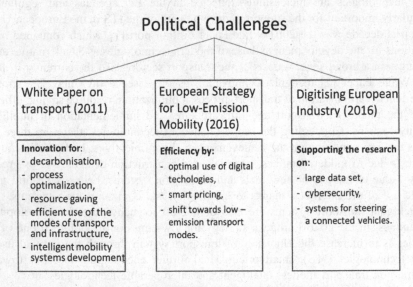

Fig. 1. Political challenges for ITS [own study]

1.1 Progress in the Implementation of EU Regulations in the ITS Area

Essential for the development of ITS in the EU is Directive 2010/40/EU of the European Parliament and of the Council of 7 July 2010 on the framework for the deployment of Intelligent Transport Systems in the field of road transport and for interfaces with other modes of transport. It contains six priority actions covering as follows [6]:

- the provision of EU-wide multimodal travel information services;
- the provision of EU-wide real-time traffic information services;
- data and procedures for the provision, where possible, of road safety related minimum universal traffic information free of charge to users;
- the harmonised provision for an interoperable EU-wide eCall;
- the provision of information services for safe and secure parking places for trucks and commercial vehicles;
- the provision of reservation services for safe and secure parking places for trucks and commercial vehicles.

According to the abovementioned directive the European Commission is empowered to adopt, by means of delegated acts, the necessary specifications governing the technical, organizational, functional and service aspects of ITS. So far there have been adopted specifications concerning (in chronological order): road safety (priority 3), eCall (priority 4), information on parking places for trucks (priority 5) and real time information services (priority 2).

The period during which EC entitlements are valid expires in 2018, but the EC declares to extend it.

A 2016 EU regulation [7] on the protection of natural persons with regard to the processing of personal data and on the free movement of such data (entry into force on 25 May 2018) is the EU's significant achievement concerning creation of grounds for intelligent mobility. It is considered that the protection of personal data and privacy is a decisive factor in the successful implementation of intelligent mobility services. This is due to the fact that technological progress and globalization have led to an increase in the amount of personal data exchanges between public and private entities (also between natural persons and enterprises). At the same time public administration authorities require more stable and coherent framework for data protection in order to perform the functions, carrying out which they were established for.

1.2 Rationale and Directions for the Development of Intelligent Mobility

The EU's common transport policy within the scope of the priorities adopted until 2050 will require significant updates regarding functioning of the changing market of transport services, as well as the extension and verification of the existing scope of regulation. These changes are resulting from the following rationale:

- it is predicted that by 2050 cargo traffic in the EU will increase by 80%, passenger traffic by more than 50% [8];
- the number of road accidents should be reduced, knowing that 93% of accidents correspond to the human factor;

- the development of ITS technology (including telematics) is increasingly utilized in the implemented solutions, in the infrastructure of transport corridors as well as in the civic communication systems;
- the market of new vehicles provides a wide range of wireless and touchless on-board solutions that enhance traveling comfort, and sensor-based solutions for safer road movement in various conditions and environments;
- services like "mobility-as-a-service" (MaaS) that are based on mobile solutions (aka: "Transportation-as-a-Service"- TaaS) are being developed;
- the phenomenon of urbanization is progressive (in 2010 half of the people in the world lived in cities, in 2050 this will increase to nearly 70%);
- there is increasing pressure to regard environmental considerations into account in every activity;
- the EU population is aging (in 2050 people aged 65+ will account for nearly one third of the population).

By positively assessing the system approach of the EC to the use of ITS in all modes of transport, based on currently available solutions and implementations in land transport, it can be assumed that ITS will play a major role in two of its areas [9]:

- in agglomeration transport and urban mobility management;
- in inter-agglomeration and international transport - in public road and highway information systems, road tolling systems, as well as in interoperability of railways and development of inland waterways transport.

The future of intelligent transport systems, like the future of the transport system, will not rely solely on modernization of transport infrastructure and modes, but will result from a gradual shift of the mobility paradigm and the transport process itself. The development of many technologies (including mechatronics, nanoelectronics, intelligent software) will play a significant role in generating changes in many areas of socio-economic life, will primarily affect people's attitudes and behaviours, interactions between technology and its users, and will also help to solve contemporary social problems [10]. The main market challenges are shown at Fig. 2. Data from previous

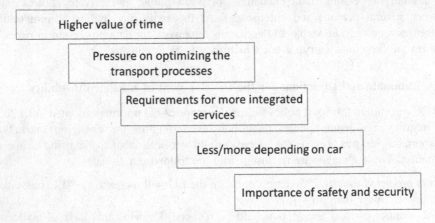

Fig. 2. Market challenges for ITS [own study]

years indicate [11] that in the EU 40 million people use satellite navigation, which relates in annual savings of 10 million traffic hours and more than 100 million km. This is one of the tangible effects of the implementation of ITS, but there are others, such as increased safety and reliability of transport, improved travel comfort and reduced congestion.

1.3 Strategy for Development of Cooperative Intelligent Transport Systems

At the end of 2016, the European Commission adopted a Strategy on Cooperative Intelligent Transport Systems (C-ITS), which is considered as a milestone for cooperative, connected and automated mobility [12]. Actions in this regard were preceded in 2014 by the launch of the C-ITS Deployment Platform and the C-Roads Platform. The second phase of the C-ITS Platform is currently underway, and the legal framework for this initiative is set to be determined before 2018. The coordinated implementation of C-ITS concerning the services specified in the foundation package (Release 1) is scheduled for 2019. The automotive industry has confirmed the readiness of the vehicles (cooperative, connected and automated) functioning within C-ITS to participate.

European Committee for Standardization (CEN), in cooperation with European Telecommunications Standards Institute ETSI, in 2014 defined C-ITS as [13] "*a subset of the overall ITS that communicates and shares information between ITS stations to give advice or facilitate actions with the objective of improving safety, sustainability, efficiency and comfort beyond the scope of stand-alone systems*". Work on standardization packages for C-ITS services is underway. Release 1. package was completed in 2014. Currently work on Release 1.5 package is commenced. The scope of services included in the individual packages is shown in Table 1.

Cooperative intelligent transport systems C-ITS are using technology allowing communication between one vehicle to another (V2V), between vehicles and infrastructure (V2I), and/or between different infrastructures (I2I). Cooperation means that vehicles warn each other about potentially dangerous situations and communicate with local road infrastructure. It also means a two-way communication between the vehicle and traffic control centres. C-ITS digital technologies will be based on static data (maps, traffic regulations) and dynamic data (e.g. real-time traffic information). Utilising them in C-ITS requires creation of appropriate legal framework. By 2018, European Commission is considering issuing, as a delegated act to the directive 2010/40UE, regulations regarding the following [14]:

- providing continuity of C-ITS services,
- providing safety of communication of C-ITS,
- data protection within C-ITS,
- providing a forward-thinking approach based on hybrid communication (combination of complementary communication technologies),
- interoperability of system components,
- conformity assessment processes.

Table 1. The scope of C-ITS services [14]

List of C-ITS release 1 services	List of C-ITS release 1.5 services
Notification of dangerous location – warning about slow or immobilised vehicles and traffic ahead; – warning about road works; – weather conditions; – emergency braking light; – approaching of an emergency vehicle; – other dangers. **Applications of marking** – on-board marking of the vehicle; – on-board speed limiters; – violation of a signal/intersection safety; – request of marked vehicles for the traffic signalling priority; – information on optimal speed on the green light; – data from vehicle probes; – shockwave damping (falls into the "local danger warning" category)	– information on gas and charging stations; – protection for unprotected road users; – street parking management and information; – information on parking outside the street; – information on "park and ride" systems; – navigation for connected and cooperative vehicles in and out of the city (first and last section of transport, parking, route information, coordinated traffic lights); – traffic information and intelligent route planning.

The subject literature indicates that these systems will create great potential for improving road safety, traveling comfort and transport efficiency. Considering the favourable [14] benefit-cost ratio of implementing these solutions in the EU (3:1 - cumulative values over the period of 2018–2030), a pressure exist to implement C-ITS services as soon as possible, and to give them network character at European level.

Launched in December 2016, the initiating operating phase of GNSS – the EU Satellite Navigation System (there are 18 satellites operating within the CNSS system, and the number will increase to 24 by 2020) will become the support for positioning, navigation and on-line data access, and the development of services based on the use of satellite feed (establishment of location and time). It will become a tool for the development of new services. The financial instruments for implementing C-ITS services are "Connecting Europe Facility", structural and investment funds, and the European Fund for Strategic Investments.

1.4   Mobility-as-a-Service (MaaS) – an Example of Intelligent Mobility Service

Mobility-as-a-service (MaaS) is an innovative platform for mobility services, concept of which was presented at the ITS Europe Congress in Helsinki in 2014. This cutting-edge service is defined [15] as *"using a digital interface to source and manage the provision of a transport related service(s) which meets the mobility requirements of a customer"*. The first use of MaaS took place in Helsinki in 2016, but the promotion of this type of service as a new business model is increasing, especially in the UK and Finland.

It is anticipated that MaaS (which also takes a form of uberization [16]) will particularly become of interest to the younger generation of transport users, i.e. generation Y [17], which cares less about ownership (owning a car) but more about experiences. They prefer the model of occasional car access and related services, seeking solutions meeting their private needs (*tailored travel options*). The literature [15] describes this phenomenon as a new paradigm shift - from the ownership model to the service model, emphasizing that this could lead to a "shift away from public transport" model, which would not be conforming to the EU transport development strategy.

The idea of MaaS assumes focusing on the user (who and why seeks the solution?), and searching for solutions meeting his/her mobility needs, at the same time guaranteeing (alongside providing on-line information) the ease of concluding transaction and various forms of payment (pre-paid, pay-as-you-go, post-pay, monthly passes). Development of this form of service is supported by the common use of smartphones, smartwatches and smartcards, as well as by the policy of lowering prices for these devices and roaming calls. The global smartphone market is still growing at a steady pace due to more widespread adoption in emerging markets. The global smartphone report estimate that global market will hit about 2.1 billion units shipped in 2021 [18]. Even in the literature appears term the 'mobile phone culture', which shows the transformations from communication technology to a key social tool, and important aspect of people's everyday life, e.g. human behavior, interaction, and way of cognition. Mobile phones have altered the way people live, communicate, interact, and connect with others as well as access and use information and media [19].

Innovation of MaaS is reflected by the fact that the service provider must be able to aggregate the transport operator services using a digital platform. There are [15] two main strengths of the MaaS business model:

– Servitization, when the service provider creates an expedient, innovative proposal that combines packets of various services, mainly concerning mobility, what triggers competition between transport operators;
– Data sharing, when the MaaS operator uses data concerning customer mobility needs, and shared infrastructure and available fleet data, to improve the quality of provided services.

Among the main benefits for transport operators (including public transport operators [20]) resulting from participation in the MaaS system, apart from the above-mentioned ones, worth mentioning are increased profitability of the business, and the achievement of the company's mission objectives by providing services of higher added value to the users.

The scope of integration of the services created by the MaaS operator, and his/her functions within the system (Fig. 3) pertain to following areas: fee and financial settlement management, infrastructure and travel information management, services concerning individual travel planning and transport mode management.

Solutions of MaaS translate into a number of benefits not just for the users but also for those who manage the infrastructure. Many of these benefits, such as increased efficiency and effectiveness of transport infrastructure, enhanced safety, and optimized infrastructure or services usage fee charging systems, are directly attributable to the ITS

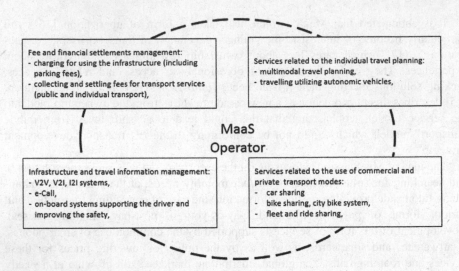

Fee and financial settlements management:
- charging for using the infrastructure (including parking fees),
- collecting and settling fees for transport services (public and individual transport),

Services related to the individual travel planning:
- multimodal travel planning,
- travelling utilizing autonomic vehicles,

MaaS Operator

Infrastructure and travel information management:
- V2V, V2I, I2I systems,
- e-Call,
- on-board systems supporting the driver and improving the safety,

Services related to the use of commercial and private transport modes:
- car sharing
- bike sharing, city bike system,
- fleet and ride sharing,

Fig. 3. Areas of integration of MaaS data [own study]

characteristics, but the relevant effect of change in approach to the mobility requires integration of processes in terms of creation of the services within the scope of broadly defined intelligent mobility, which will combine already available solutions and the development of mobile computing technologies.

In order to provide further development to the services within MaaS, the ITS will require integration in three basic areas:

- Institutional - by implementing solutions integrating public sector institutions (both regional and international) as well as the sector of enterprises and individual users. In this area it will still be important to fund the research and to test the implementations within Cooperative Intelligent Transport Systems (C-ITS).
- Technological - by unifying standards for ITS solutions and ranges of data exchange between the systems and the users. For certain systems in the EU there are standards in force, concerning, for example, the range of telematic data shared with all users, regarding electronic fee collection for infrastructure use, or - in case of rail transport - the signal cabin system (ERTMS), the functioning of the international emergency number "112". Standards are systematically implemented in the introduced solutions, but new ones are emerging (e.g. MaaS), which require decisions regarding standardization, since only these solutions will become fully interoperable within the EU.
- Operational - by adjusting and/or clarifying requirements for systems operating in different transport sectors, in order to guarantee multimodal interoperability and utilization of the capacity of TEN-T network.

Furthermore, it must considered that MaaS services can make use of fully automated vehicles. They will find the most extensive use in urban mobility, creating a new package of transport services in door-to-door system. Such solutions are designed to improve road safety, reduce pollution and congestion, but above all, they can change the approach to car ownership. According to KPMG research, conducted among production managers from automotive industry [21], more than a half of current vehicle owners in 2025 will not be interested in owning a car. Introduction of services utilizing fully automated vehicles requires specific regulations as the current legislation does not solve the key issues related to the use of these vehicles, such as: authority to use a fully automated vehicle by those who are not authorized to drive motor vehicles, civil liability and insurance coverage concerning damages resulting from the use of fully automated vehicles, or criminal liability resulting from possible collisions or accidents involving fully automated vehicles. The biggest challenge, however, lies with vehicle manufacturers that need to develop proper vehicle software that works with different road infrastructure technologies in different urban infrastructure conditions. A completely different aspect is the level of social acceptance of solutions related to fully automated vehicles, which so far have not raised public trust. It is necessary to undertake a wide-ranging debate with the participation of various representatives of the public in order to know the public concerns and reservations related to these solutions.

Another aspect affecting the change in approach to mobility is the implementation of solutions that improve passenger safety and broadly defined travel comfort through vehicle-to-infrastructure and vehicle-to-vehicle data exchange technologies, which are becoming increasingly popular.

2   Conclusion

Intelligent Transport Systems (ITS) are becoming the foundation for creation of new services in the cargo and passenger transport, both mass and individual, in different its branches and modes. Development of technology related to mobile devices and common Internet access are also playing an important role, and are considered as relevant factors in the development of society's needs in the coming decades. They will also be the basis for the creation of new services improving the mobility, thus increasing the added value on the account of the use of transport. It is not without significance for the future of mobility process to develop fully automated vehicle technology, which first series, according to most car manufacturers, are likely to hit the market between 2020 and 2025, and the use of these vehicles will not be possible without extending the scope of ITS functioning.

Development of the market for services, transport processes and modes of transport, also requires on-going adjustment of the legislative solutions. It is not enough to introduce technology guidelines that are applicable to the various systems within ITS. Development of MaaS solutions shows the direction in which the solutions in agglomeration areas are heading. New understanding of mobility as a service has forced a change in the approach to the services provided in this area and their integration. Future provision of mobility services must be tailored to the individual needs of users, with the use of public transport, car sharing, car renting, taxi, city bike, etc., all being able to accessed with one application installed on a mobile device, such as smartphone. The change in approach to provision of services requires appropriate standards and legal framework so that intelligent mobility services were safe and did not lead to fraud, and at the same time encouraged actions based on the "blue ocean" principle, i.e. creating free market space and capturing new demand through cooperation and coopetition.

References

1. White Paper. Common Transport Policy Till 2010—Time to Decide. Commission of the European Communities, Brussels (2001)
2. White Paper. Roadmap to a Single European Transport Area—Towards a Competitive and Resource Efficient Transport System. European Council, Brussels 29.3.2011, COM 144 (2011)
3. A European Strategy for Low-Emission Mobility, SWD(2016)244 final, EC, Brussels (2016)
4. Digitising European Industry. Reaping the Full Benefits of a Digital Single Market {SWD (2016) 110 final}, Brussels (2016)
5. Understanding the benefits and costs of Intelligent Transport Systems—A Toolkit Approach. www.its.uk.org.uk, July (2005)
6. Report from the Commission to the European Parliament and the Council. Implementation of Directive 2010/40/EU of the European Parliament and of the Council of 7 July 2010 on the framework for the deployment of Intelligent Transport Systems in the field of road transport and for interfaces with other modes of transport, Brussels 2014, COM (2014)624 final, p. 5, www.europa.eu. Accessed 15 Jan 2017
7. Regulation (EU) 2016/679 of the European Parliament and of the Council of 27 April 2016 on the protection of natural persons with regard to the processing of personal data and on the free movement of such data, and repealing Directive 95/46/WE
8. http://europa.eu/rapid/press-release_MEMO-13-897_pl.htm. Accessed 16 Jan 2017
9. Załoga, E.: Trendy w transporcie lądowym Unii Europejskiej, p. 216. Uniwersytet Szczeciński, Szczecin (2013)
10. van Voorsttot Voorst, M.P.: Future of the super intelligent transport systems, EFP Brief 2011, no. 202
11. Innovation to compete. Europe is a leader in transport innovation. www.transport-research. info. Accessed 28 Dec 2012
12. European Strategy on Cooperative Intelligent Transport Systems (C-ITS), COM (2016) 766. Accessed 17 Jan 2017
13. Cooperative Intelligent Transport Systems, EU, Brussels 2016, www.transport-research.info, p. 2. Accessed 20 Feb 2017
14. Communication from the Commission to the European Parliament, the Council, the European Economic and Social Committee and the Committee of the Regions. A European strategy on Cooperative Intelligent Transport Systems, a milestone towards cooperative, connected and automated mobility, COM(2016)766 final, European Commission, Brussels, 30 Nov 2016
15. Mobility as a service. Exploring the opportunity for mobility as a service in the UK. CEO Transport Systems Catapult, July 2016. Accessed 12 Jan 2017
16. Uberization is a concept from sharing economy, also known as on—Demand economy. It originates form the name of a company Uber Technologies Inc. established in 2009 in San Francisco. Uber is a mobile application that offers on-demand transport carried out by non-licensed carriers (private individuals)
17. Generation Y, aka. Millenials, are people born between 1980 and 1995 (some close the Y generation period before the year 2000). It is predicted that the Y generation, representing 1/4 of the population, in a few years will shape the reality. Members of this generation trust in mobile applications, but they also believe in evaluation of the offered service from the two perspectives: service recipient and service provider
18. http://www.businessinsider.com/global-smartphone-market-forecast-2016-3?IR=T. Accessed 15 Jan 2017

19. Lee, S.-H.: Mobile phone culture: the impacts of mobile phone use. In: Encyclopedia of Mobile Phone Behavior, USA, 2015

20. According to art. 4.1 section 8 of the Urban Mass Transportation Act dated 16.12.2010, public transport operator is a budgetary body of the local government or a business entity authorized to carry out passenger carriage, who has entered into a contract with the public transport organizer regarding provision of public transport services, on the communication line specified in the contract

21. https://www.theguardian.com/business/2017/jan/09/fewer-car-owners-more-driverless-vehicles-future-survey-reveals. Accessed 15 Jan 2017

Implementation of New Solutions of Intelligent Transport Systems in Railway Transport in Poland

Mieczysław Kornaszewski$^{(\boxtimes)}$, Marcin Chrzan, and Zbigniew Olczykowski

Faculty of Transport and Electrical Engineering, University of Technology and Humanities in Radom, Malczewskiego 29, 26-600 Radom, Poland
{m.kornaszewski, m.chrzan, z.olczykowski}@uthrad.pl

Abstract. There is a need of continuous modernization of Polish railways and the introduction of innovative changes. Such a development strengthens the position of rail transport on the European market, improves the image of this mode of transport in the eyes of society and, above all makes it safer. Implementation of new computer's solutions and microprocessor technology, microcomputers and the programmable controllers (PLC) is particularly important for management systems of train traffic. This is the inspiration for the new control of the rail traffic systems that provide high operational reliability, low power consumption, stability and safety of the trains' movement. Intelligent Transport Systems specify the technical systems, which improve the functioning and development of transport, thanks to the implementation of information advanced technology. Implementation in Poland of modern European Railway Traffic Management System is a significant step in the development of ideas of Intelligent Transport Systems and safety of railway traffic.

Keywords: Rail transport · Intelligent Transport Systems · Rail traffic systems · Safety · ERTMS · ETCS · GSM-R

1 Introduction

Intelligent Transportation Systems (ITS) constitute a wide set of diverse tools, which are based on information technology, wireless communication and vehicle electronics. They allow for an effective management of the transport infrastructure, as well as for an efficient service of passengers. In such systems transport's functioning is mainly supported by integrated measuring (sensors, detectors), telecommunications, information technology, information and automatic control solutions [1].

ITS allow for increasing security, traffic flow (traffic capacity) and for reducing environmental pollution. Well-designed ITS become part of an integrated land transport infrastructure and they fulfil a variety of responsible tasks [2, 3]:

- dynamic alerts to dangerous situations (hold-ups, fog, black ice, strong wind, damages to the surface and the like),
- influencing driving speed,

© Springer International Publishing AG 2017
J. Mikulski (Ed.): TST 2017, CCIS 715, pp. 282–292, 2017.
DOI: 10.1007/978-3-319-66251-0_23

- keeping proper distance between vehicles,
- regulating entering the traffic flow,
- controlling light signalling devices,
- controlling changes of direction,
- automatic control of following the rules and the like.

The presented tasks can also be used in the rail transport in Poland, and can be accomplished by the railway traffic control.

2 ITS Solutions in the Rail Transport

An example of the ITS use in the rail transport is the *Safety Management System* (SMS). Infrastructure administrators and rail operators functioning in European Union need to show for approval own SMS to the national security organ (in Poland: Office for Rail Transport). SMS needs to assure the management of all hazards, including the hazards resulting from the commission to maintain systems and devices, purchase materials and to cooperate with contractors in regards to the matters crucial to security [4, 5].

Among new concepts and innovative technologies implemented in the rail transport in Poland worth mentioning are: track-side light signalling devices which optical systems use LED diodes, photovoltaic cells for powering railway automatic devices, open standards of the radio broadcast for railway traffic control systems' cooperation and the like [6, 7].

Where it comes to the railway traffic control devices, because of the safety priority, it is very important to deliver to the engine driver proper information regarding the way of driving a train. Delivering the information can happen point-wise, meaning in limited time, when a train is driving above the appropriate devices installed in the track or, it can take place continuously – during the uninterrupted communication of the train with the fixed devices. Apart from the old, technically uncomplicated systems of the point-wise effect on a train, systems used nowadays include automatic control of the train drive system and the system of high bit rate information transmission and a low probability of a faulty decision. There is also an access to the systems of the precise location of stationary and mobile objects via the reception and processing of the signals sent by the satellite systems [8].

There are many Intelligent Transport Systems dedicated to only one transport branch. Where it comes to the rail transport, this system is *European Rail Traffic Management System* (ERTMS).

Train traffic control applies to the control of the distance between trains and their speed and to the security against a train driving on the tracks occupied by other vehicles. Assuring security when trains go at 160 km/h requires constant exchange of information between the trains on a given track. At the moment, a modern ERTMS system is being implemented in Poland, it is a safe system, with a constant control of the train's drive. The implementation of the ERTMS will allow to fulfill the requirements of the national and European law where it comes to the interoperability of the rail transport, it will also improve the comfort and safety of travelling. *European Train Control System* (ETCS), as the ERTMS's subsystem, will ensure the realization of the cab signalling and a constant control over the engine driver's work, whereas *Global System for Mobile Communications* (GMS-R) will create and improve vocal communication and track-vehicle data transmission. In Germany, Global System for Mobile Communications technology is used in a half of the railway network [8, 9].

3   ERTMS as an Example of a Modern ITS Solution in Poland

European Union tries to create a unified, modern European railway system, in which owners of the interoperable railway infrastructure will enter traffic with an interoperable rolling stock of various carriers offering people and goods transport. Among technical differences between European railways, one should especially point out control systems (Fig. 1), in which there are different signal images, various systems of the engine driver's control etc. [10, 11].

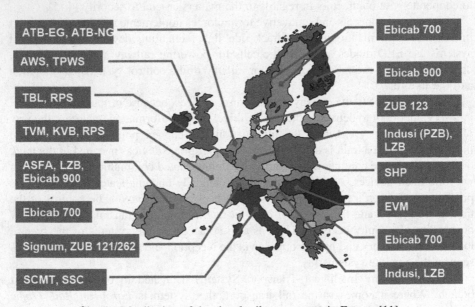

Fig. 1. Distribution of the signal railway systems in Europe [11]

3.1   Development of the Driving Support Systems (SSC)

Railway traffic control devices, together with the railways' development, have improved their technical solutions along with growing needs, requirements and expectations. Throughout the years, railway traffic control systems have undergone a gradual evolution, from mechanical devices, through electromechanical and relay ones, to hybrid and computer devices.

The newest generation of the railway traffic control devices are computer systems, which combine modernity, reliability, they also assure a really high level of the traffic security. An example is the development of points used in the switch railway traffic control systems. According to PKP Polish Railway Lines' Manager, 10% of points used nowadays are computer-controlled devices and 3% are relay-computer devices (hybrid). Most of the railway traffic control devices are relay (41%) and mechanical (40%) technologies, where the latter include central mechanical (31%), key mechanical (9%) and electric slide (6%) technologies – Fig. 2 [9, 12].

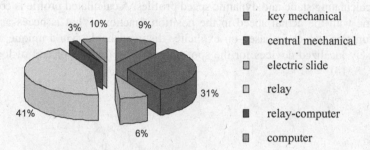

Fig. 2. A percentage breakdown of the points used in the switch railway traffic control systems made in various technologies [12]

A train is driven by an engine driver who chooses the train's speed based on the track-side semaphores' indications, timetable and parameters so that the ride is safe, on schedule and in accordance with the binding regulations. Because of the human's unreliability (for speed exceeding 160 km/h human eye's perception can be fallible), the engine driver needs to be supported in his actions by the appropriate railway traffic control devices. Among the full automation devices for driving a train one can distinguish two main groups of the railway traffic control systems, classes ATC (*Automatic Train Control*) and ATP (*Automatic Train Protection*).

The ATC system is used for constant control over the train's speed and for interference in its ride through automatic speed limitation in case of speeding, on the basis of the parameters of the rolling stock, track or the traffic situation. ATC devices allow for signalling the target speed in the engine driver's cab or for substituting side-track signalling devices.

The ATP system constitutes a supplement to the track signalling and controls if the engine driver reacts correctly to the signals transmitted by the track signalling devices. Normally, when the engine driver acts appropriately, the ATP does not react. If the engine driver makes mistakes, the ATP activates service or emergency braking, which causes the train to stop (before the dangerous spot) [13, 14].

3.2 ERTMS as an Example of the New Attitude Towards the Railway Traffic Management

ERTMS is mostly about safety. Thanks to this system, the engine driver receives a lot of information as they arise, they concern the location of other trains, allowed speed, the closure of the level crossing, as well as an unexpected obstruction on the road. All this data is displayed on the screen of a special monitor inside the engine driver's cab. That is why it is possible to react immediately and to adjust the driving speed to the conditions on the road [3, 5, 15].

3.2.1   European Train Control System ETCS

The idea of the ETCS is based on the digital track-vehicle transmission. The transmission can be accomplished through balises, short, middle or long loops, digital radio or specialized transmission modules (Fig. 3). Data describing the track and the vehicle serve for calculating static and dynamic speed profiles. A calculated profile is compared in real time with the actual speed in the position function. What is necessary is the location function, which is based on explicitly distinguishable (via a unique number) and precisely localized devices for the spot transmission (balises or end of loop indicators) [5, 8, 14].

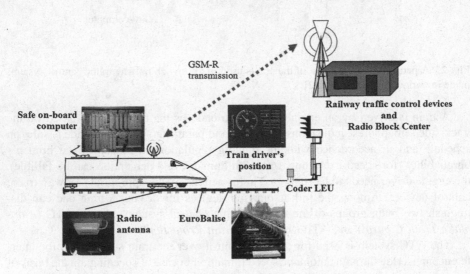

Fig. 3. The rule of functioning of the ERTMS/ETCS system [own study]

The ETCS is a system used for controlling the railway traffic. Because the needs to control depend on the type of the railway, specifications provide for three levels of the ETCS implementation, where for each level it is possible to additionally distinguish various hardware configurations [10, 16].

Level 1 is based on the transmission via balises of permission to move emitted by the light signalling devices. There is a switched balise, attached to the signalling device through a coder, which transmits the permission to move depending on the signalling device's indicator to the on-board ETCS device, which, based on the received information, controls if the engine driver drives the vehicle with accordance to the signalling device's indication. That is why, in order to implement the ETCS on this level, there have to be sections of the railway with already existing stationary light signals.

Level 2 is based on the GSM-R radio communication for giving permission to move and on the conventional technique for the control of tracks' occupation for preparing permission to move, basing on the existing railway traffic control devices of the basic layer.

Level 3 is based on the GSM-R radio communication for giving permission to move and on the substitution of the classic operation for the control of tracks' occupation with a combination of the train's location and the train integrity unit. It allows to prepare permission to move based on the moving block rule [10, 16].

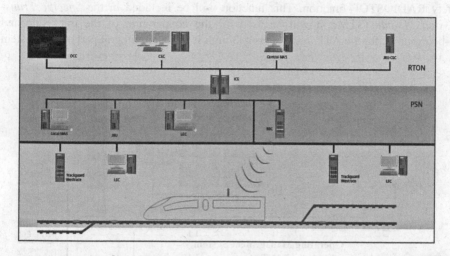

Fig. 4. Architecture of the "Trainguard Futur 2500" system [17]

The "Trainguard Futur 2500" system, presented in the Fig. 4, is a complete solution from Siemens Rail Automation for Level 1 and Level 2 of the European Rail Traffic Management System. Abbreviations used in the figure: RTON – Real-Time Operating Network, CEC – Central ERTMS Control, PSN – Packet Switched Network, LEC – Local ERTMS Control, MAS – Maintenance Assistance System, ICE – Interface Control Equipment, JRU – Juridical Recorder Unit, OCC – Operations Control Center, RBC – Radio Block Center [17].

3.2.2 Global System for Mobile Communications – Railway GSM-R

The GSM-R system works on the 900 MHz band and corresponds to the GSM2+ version. Apart from allowing for communication, it also allows for data transmission, group calls, defining calls' priority, functional addressing (e.g. with the trains' numbers) and other specialized functions devoted to the railway. GSM-R constitutes a channel through which permissions to move, issued by the Radio Block Centre (RBC), are sent to various trains located in area of a given centre. Because of the safety, the GSM-R system was assigned two frequency bands with 25 MHz bandwidth each: $890 \div 915$ MHz used for transmission from the mobile station to the base station and $935 \div 960$ MHz used for an opposite transmission. The location of the GSM-R devices remains a priority strategy in most of the new European Union members [9, 16, 18].

GSM-R system's architecture (Fig. 5) is a typical GSM mobile network and it consists of the Network Switching Subsystem (NSS) and the Operating Management System (OMS) on the main level, and of the Base Stations Subsystem (BSS), consisting of peripheral group of Base Station Controllers (BSC) and of peripheral group of Base Transceiver Stations (BTS) [19].

Contrary to the majority of European radio communications train systems, the Polish GSM-R system includes not only verbal transmission, but also the railway traffic safety RADIOSTOP function. This function will be included in the *Specific Transmission Module* (STM), it will be used for the cooperation of the track-side and on-board units, for the ATP system, even though it is not an integral part of this system [4, 14, 19].

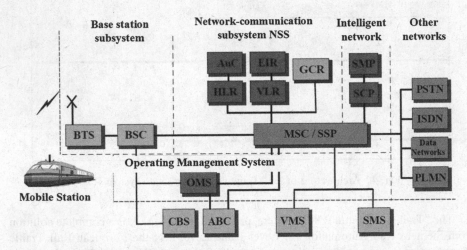

Fig. 5. GSM-R system's architecture [4], which includes the following: ABC – Administration and Billing Centre; AuC – Authentic Centre; BSC – Base Station Controller; BTS – Base Station; CBS – Broadcasting Services; EIR – Equipment Identification Register; HLR – Home Location Register; MSC – Mobile Switching Centre; GCR – Group Call Register; SCP – Service Control Point; SMP – Service Management Platform; SMS – Short Message System; SSP – Service Switching Point; VLR – Visitor Location Register; VMS – Voice Mail Service.

3.2.3   On-Board ETCS Devices

In order for the engine driver to be able to drive the train safely, the train has to have access to the information that is also passed on to the on-board ETCS devices and serves for controlling the correctness of driving the train by the engine driver. In order to achieve that, the on-board devices were equipped with the *Man Machine Interface* (MMI). Proposed solution is based on the touch screen technology. Information is transmitted via icons, which appear always on the same part of the screen (the middle part Fig. 6) [16].

Fig. 6. Subsystems of the cab's signalization in the ERTMS/ETCS system [20]

The MMI device was standardized by the *International Union of Railways* (UIC). The clock on the left side of the MMI indicates types of speed: actual, allowed, allowed in the next track section, whereas the indicator on the right shows the following types of speed: actual, recommended and the maximum brake power of the locomotive and train. On the verge of the casing, there are events icons and buttons controlling e.g. communication via the GSM-R [9].

4 Implementing New Informatics Solutions and PLC Controllers into the Rail Transport

Implementing innovative informatics solutions, as well as microprocessor techniques and programmable PLC controllers is of a special significance for the new rail transport systems, especially for the railway traffic control systems responsible for the safety during running the railway traffic.

Equipment and software used for creating systems responsible for safety, have to conform to high quality and reliability standards. After many years of work of the *European Committee for Electrotechnical Standardization* (CENELEC) normalized algorithms for the creation, checking and allowing to exploitation of the safe railway applications were developed and implemented [8, 21].

Basic methods of achieving the required level of safety of the modern computer systems for the railway traffic control include [9, 21]:

- equipment redundancy – in the controls system one can distinguish two independent channels: feed systems, controllers, input/output (I/O) systems etc. (Fig. 7). Devices installed on a given railway equipment, regardless of their quantity and type, are distributed among the two channels so that in case of failure of one of the control channels, the other channel will assure a sufficient protection of the vehicle;
- variety of control programs in both channels – programs for the channels' control are developed by independent computer programmers teams. In most cases, the programs are written on the level of the used processor's assembler and on the level of the ladder schemes;

- control channels' synchronization – the PLC controllers of both channels are usually connected with a serial interface, through which the exchange of information between devices takes place, in order to assess the conformity of the system's layer deciding about the turning on or off the railway traffic control in one of the channels;
- real-time testing of the operation of modules and devices – control programs additionally include testing procedures for the correct operation of the feed system, input/output (I/O) devices, chosen modules, as well as procedures and mechanisms for the selftesting of the control systems;
- real-time testing of the control program – the control program, apart from the procedures realizing particular system's functions, includes procedures checking the correct operation of each program cycle and procedures checking the correctness of the current railway traffic control system's parameters (e.g. trains'meters).

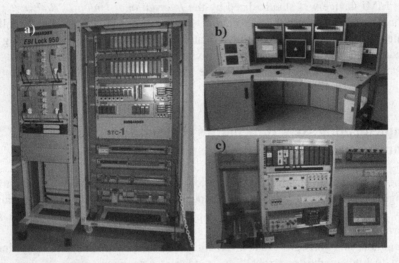

Fig. 7. Chosen laboratory stands in the University of Technology and Humanities in Radom, including modern computer railway traffic control devices: (a) railway traffic control device type EbiLock 950 produced by Bombardier Transportation (ZWUS) Polska; (b) control panel of the ZSB 2000 system produced by Scheidt & Bachmann Polska; (c) system controlling the unoccupancy of railway tracks, type SKZR produced by KOMBUD Radom [own study]

5 Conclusion

All actions taken towards intelligent transport allow for significant increase of safety, assessing more efficient time table, natural environment protection and, most of all, allow for more effective use of the existing railway infrastructure.

Introducing systems using methods and ITS' funds will contribute to: reducing expenditures on the transport infrastructure by $30 \div 35\%$, with the same system's efficiency improvement as when building new roads, raising the transport network's bandwidth by ca. 20% without building new road sections, significant reduction of the transportation accidents and their casualties, saving travel time, reducing emission of the carbon dioxide (e.g. because of the reduced number of detentions at countries' borders) and improving traffic flow.

Efficiency and technological progress of the railway traffic control systems influence directly the safety of rail vehicles. Railway traffic control devices contribute to a complicated, interrelated system, in which even a failure of the smallest element leads to the lack of possibility to lead the trains' movement in an organized way, on secured routes.

The ERTMS fits into the strategy, which aim is to enliven the railway sector. Facilitation of the locomotives' pass through border will contribute to increase of the international railway traffic for the transport of people and goods. Issues related to the reduction of external costs, such as pollution, noise, safety, overload, are very important, and each factor which could lead to the levelling of participation of particular means of transportation on the market will cause reduction of these costs.

Gradual implementation of modern satellite systems is slowly entering the railway traffic control area. However, due to security reasons, the *Global Positioning System* (GPS) system is an auxiliary agent in Poland, serving for locating technical emergency's vehicles, emergency trolleys or technical emergency's trains. The use of satellite communication will allow for a remote transmission of information regarding the location and the vehicle's condition to the proper data bases and subjects directly interested in these information. This will also allow for the implementation of many functions, which will support the work of the engine driver and the staff.

Currently, in Poland, many actions regarding the development of the ITS in the road communication take place, among others car traffic control in big city agglomerations, including delivering information about the traffic condition, faster access to emergency services, electronic tolling service on motorways etc.

Acknowledgement. This material is based upon work supported by Polish National Center for Research and Development under Grant No. PBS3/A6/29/2015.

References

1. Jamroz, K., Krystek, R.: Inteligentne Systemy Transportu – rozwój i struktura. Transport Miejski i Regionalny **5**, 2–11 (2006)
2. Koźlak, A.: Inteligentne systemy transportowe jako instrument poprawy efektywności transportu, Logistyka, vol. 2, CD (2008)
3. Kornaszewski, M., Chrzan, M., Wojciechowski, J.: Intelligent transportation systems example of modern eco-transport solutions in Europe. In: Proceedings of Conference of Globalizácia a Jej Sociálno-Ekonomické Dôsledky 2010, pp. 247–251 (2010)
4. Białoń, A., Gradowski, P.: ERTMS (ETCS i GSM-R) problematyka wdrażania w Polsce. presented at Conf. Najnowsze technologie w transporcie szynowym. SITK, Warszawa, Polska (2011)
5. Ghazel, M.: Formalizing a Subset of ERTMS/ETCS Specifications for Verification Purposes, 2014, vol. 42, pp. 60–75. Pergamon-Elsevier Science Ltd, The Boulevard, Langford Lane, Kidlington, Oxford Ox5 1 GB, England. doi:10.1016/j.trc.2014.02.002

6. Martinelli, G., Morini, A., Tortella, A.: Innovative technologies for public electric transport systems. In: Proceedings of Urban Transport and the Environment in the 21st Century, Algarve, Portugal, pp. 569–578 (2005)
7. Kornaszewski, M., Bojarczak, P., Pniewski, R.: Introduction of world innovative technologies to railway transport in Poland. In: Proceedings of 16th International Scientific Conference Globalization and Its Socio-Economic Consequences, Rajecke Teplice, Slovak Republik, vol. 2, pp. 962–969 (2016)
8. Pricevicius, G., et al.: Railway Segment Management Information System in Open Conference of Electrical, Electronic and Information Sciences (EStream). IEEE, 345 E 47th St, New York, NY 10017 USA (2016)
9. Kornaszewski, M.: Współczesne systemy sterowania ruchem kolejowym w Polsce. Logistyka 3, 3083–3093 (2014)
10. Kornaszewski, M.: Integracja europejskich systemów kolejowych na przykładzie systemu ERTMS. Transport i Komunikacja 1, 52–55 (2009)
11. Kregelin, V.: Bombardier transportation. In: Rail Control Solutions, Presented of LST Symposium Transport Logistics 2011, Munich, Germany (2011)
12. PKP Polskie Linie Kolejowe S.A. Raport roczny 2015 (edition). http://www.plk-sa.pl/files/public/raport_roczny/RR_za_2015_rok_-13_grudnia.pdf. Accessed 25 Jan 2017
13. Branishtov, S.A., et al.: Automated traffic control system in railways. In: International Conference on Mechanical Engineering, Automation and Control Systems (MEACS), pp. 1–5. IEEE, 345 E 47th St, New York, NY 10017 USA (2015)
14. Dyduch, J., Pawlik, M.: Systemy automatycznej kontroli jazdy pociągu, pp. 16–49. Wydawnictwo Politechniki Radomskiej, Radom (2002)
15. Bombardier Transportation (Rail Engineering) Polska Sp. z o.o. Bombardier Transportation w Polsce. Sprawdzony Partner (edition). http://pl.bombardier.com/Bombardier_Transportation_Poland.pdf. Accessed 25 Jan 2017
16. Dvornik, D., Mrvelj, S.: Intelligent transport systems (ITS) application using GSM-R system in railway traffic. In: International Conference on Traffic in Transitional Conditions: Intelligent Transport Systems and Their Interfaces. TTC-ITSI 1999 Proceedings, pp. 107–113, Zagreb, Croatia (1999)
17. Siemens, A.G.: Mobility Division. Trainguard Futur 2500. Level 2 ERTMS solution (edition). https://www.mobility.siemens.com/mobility/global/SiteCollectionDocuments/en/rail-solutions/rail-automation/train-control-systems/trainguard-futur-2500-en.pdf. Accessed 16 Jan 2017
18. D'Amore, P., Tedesco, A.: Technologies for the implementation of a security system on rail transportation infrastructures. In: Setola, R., Sforza, A., Vittorini, V., Pragliola, C. (eds.) Railway Infrastructure Security. TSRRQ, vol. 27, pp. 123–141. Springer, Cham (2015). doi:10.1007/978-3-319-04426-2_7
19. Ruscelli, A.L., Cecchetti, G., Castoldi, P.: Cloud networks for ERTMS railways systems. In: 5th IEEE International Conference on Cloud Networking (IEEE CloudNet), pp. 238–241. IEEE, 345 E 47th St, New York, NY 10017 USA (2016)
20. Bombardier Transportation. ERTMS Onboard Components (edition). http://www.bombardier.com/en/transportation/products-services/rail-control-solutions/ertms/onboard-components.html. Accessed 16 Jan 2017
21. Costea, I.M., et al.: Technologies used in the security and safety of intelligent transport systems. In: International Conference on Applied and Theoretical Electricity (ICATE), pp. 1–6. IEEE, 345 E 47th St, New York, NY 10017, USA (2014)

Simulation of Road Capacity with Loading/Unloading Bays Based on Cellular Automaton Model

Krzysztof Małecki[✉]

West Pomeranian University of Technology,
Żołnierska Str. 52, 71-210 Szczecin, Poland
kmalecki@wi.zut.edu.pl

Abstract. Unloading bay is one of the most popular and simple solutions to implement to promote the development of sustainable freight transport system in cities. The idea is to reduce traffic congestion in the busy city streets, which are often caused by the van parked directly on the lanes to perform the loading/unloading. Computer simulation can help make a difference in this area. The main objective of this paper is to present the application to simulate the events of position unloading bays on the road lane or outside road lane. The application implements a mathematical model based on cellular automaton model. This article presents examples of the results of the application.

Keywords: Loading/unloading bays · Cellular automata · Microscopic simulator

1 Introduction

Organization of freight deliveries in urban areas is not an easy task and needs to face with many challenges, e.g. a large number of entities that require delivery at the same time in large quantities. Narrow streets and necessity of parking of commercial vehicles on the streets give rise to congestion and thus increasing the local pollution and increased noise. Analysing the impact of urban freight transport (UFT) measures is particularly important, since improving the situation for freight deliveries more often than not will be at the expense of the citizens. For instance, designated spaces for loading/unloading operations in city centres will usually be at the expense of public parking spaces. Therefore, UFT solutions are to a considerable extent resisted by the public, and for this reason, local governments often give up on implementing such solutions.

In this article the possibilities of computer simulation is used to determine the selected dependencies for urban freight transport, in particular the loading/unloading bays. Computer simulation executes a specific mathematical model and allows to consider input parameters to obtain a certain results as system outputs [1]. Numerous publications on this subject are focused especially on traffic research and modelling, developing theory and traffic models as well as the research on behaviour of traffic participants or the possibility of using telematics [2–16]. There are also works in which computer simulation is carried out on the basis of hardware [17–19].

J. Mikulski (Ed.): TST 2017, CCIS 715, pp. 293–306, 2017.
DOI: 10.1007/978-3-319-66251-0_24

The aim of the article was to present the possibilities of the model for the analysis of road capacity with unloading bays located at.

The article is organized as follows: in the next section there is an analysis of the issues, then there is introduced the concept of cellular automata and assumptions of the model to the unloading bays, briefly discussed the model and developed simulator. At the end, the results of research are presented and a discussion was held.

2   Related Work

Unloading bays increasingly appear on the streets. One of the main advantages of this solution is a significant reduction in traffic congestion. This translates into a noticeable reduction in emissions. When truck drivers know where unloading bays are located, they lost less time searching for a parking space [20]. In [21] there was developed a model for determining the optimal location of loading-unloading space. Important from the point of view of determining the needs of different stakeholders, research conducted in [22], highlighted the different factors that would need to be taken into account when evaluating a managed loading bay system, from the standpoints of the various actors involved, including the traffic authority, freight operator, driver, retailer and other road users. Planning and managing delivery spaces requires consideration of the roles and objectives of various stakeholders, including carriers, shippers and receivers [23].

Assessment of factors influencing the reduction of congestion related to parking of commercial vehicles on the street has been concern in [24]. In [25] described a case of reducing CO_2 emissions by 40 kg per day when using unloading bays. In [26] a discrete choice model to estimate the demand of the use of loading/unloading bays was presented.

3   Research Environment

In this section there is presented the concept of the cellular automata, assumptions for the presented "loading/unloading bays" model, the model and the computer application where the model is implemented.

3.1   The Concept of Cellular Automata

The simplest cellular automaton is a mathematical model consisting of three elements [27, 28]:

- a discrete n-dimensional space (grid) divided into equal cells,
- a finite k-element set of states for a single cell,
- F rule defining state of the cell at time t + 1, depending on its state and state of the neighbour cells at the moment of time t.

N-dimensional grid of identical cells evolves in discrete moments of time. Each cell in the grid can take one k state defined by the rule:

$$S_i(t+1) = F(s_j(t)), \quad j \text{ belongs to } O(i), \tag{1}$$

where: $O(i)$ – neighbourhood of i-th cell, S_i – state of cell, F – transition rule.

Grid of automata is the n-dimensional discrete space composed of identical cells with the same parameters, i.e. with the same number of neighbours, the same shape if the state space. Its structure is influenced by the dimension of automata space, the shape of cells and the connected number of neighbours.

3.2 Assumptions for the "Loading/Unloading Bays" Model

The cellular automaton model applied at the research work was used to simulate physical phenomena determined by some defined rules, connected with events occurring within the systems [29]. Road traffic organisation is a phenomenon of this type. It involves events such as e.g. movement of vehicles, braking, accelerating etc. The simulator applied in the analysis of unloading bays functioning was based on the Nagel-Schreckenberg model [30]. This model is one of the basic models of cellular automata that simulate vehicle traffic. It was developed in 1992 by Karl Nagel and Michael Schreckenberg. The model describes single lane traffic of vehicles and constitutes the basis for testing various scenarios in road traffic [31]. The basic assumptions of the model are:

- the length of a single cell side: ca. 7.5 m (which corresponds to a mean length of a vehicle),
- one-way and single lane road,
- the cell may be in one of the two states: "free" or "occupied" by a vehicle moving with velocity v_i,
- one cell may be occupied by max. one vehicle,
- a cell has a recorded numeric value that corresponds to the current velocity of the vehicle,
- the vehicle's velocity is a discrete value, which means it may have the following values: 0, 1, 2, ... v_{max}.

In this model, the transition rule of the automaton consists of four fundamental steps:

- acceleration – if the vehicle has not yet achieved its maximum velocity (v_{max}), it may increase its velocity (v_i) as per the formula:

$$v_i < v_{max} \rightarrow v_i = v_i + 1 \tag{2}$$

- braking – if the number of subsequent free cells (distance from the preceding vehicle d_i) is smaller than the current velocity (v_i), the vehicle is braking, which is described by the relationship:

$$v_i > d_i \rightarrow v_i = d_i \tag{3}$$

- random event – from time to time, as per the specified probability, the vehicle velocity is decreased by 1 (which makes it possible to simulate phenomena such as pedestrians suddenly running into the street or drivers getting lost in thought):

$$v_i > 0 \cap P < p \rightarrow v_i = v_i - 1 \tag{4}$$

- shifting – apart from the relationships described in the previous steps, a vehicle may be shifted by the number of cells equalling the current velocity, and the time variable is increased by 1 (from t to $t + 1$), which is reflected by the following formula:

$$x_i = x_i + v_i, \quad t = t + 1 \tag{5}$$

The model applied in the simulation has several features which significantly distinguish it from the Nagel-Schreckenberg model:

- it is a two-lane road, therefore the grid of the cellular automaton takes the form of a table with two rows and n columns;
- a single cell, apart from the value specifying whether it is occupied or free (i.e. whether there is a vehicle in the given place or not), may be assigned an additional value specifying whether the given section of the road is used for unloading operations in any given unit of time (occupied or free);
- an occupied cell shows a numeric value depending on the kind of vehicle ($V_{j,i} = 0, 1, 2, \ldots, V_{MAX1}$ or $V_{j,i} = 0, 1, 2, \ldots, V_{MAX2}$, where: j - road number, i - cell number, V_{MAX1} – max. velocity of a passenger car, V_{MAX2} – max. velocity of a delivery vehicle);
- additionally, the cell has a value that specifies whether it is a road cell or an unloading bay cell;
- the transition function of the automaton consists of six parts: four of them as per the Nagel-Schreckenberg model (acceleration, braking, random events and shifts) and two additional – lane changing and parking in a bay;
- the driver courage parameter has been taken into account ($c_{j,i}$ – courage) which is a distance between the vehicle located in the ith cell of the road and the vehicle moving behind it, on the other lane, located in cell j.

3.3 "Loading/Unloading Bays" Model Based on Cellular Automaton

Two models that allow to analyze the impact of deliveries on the congestion level have been developed as an example of the use of cellular automata to evaluate the efficiency of urban freight transport solutions:

- "2-roads-ub-inside", in which delivery requires the car to stop directly at the lane of two-lane road (Fig. 1a);
- "2-roads-ub-outside" model, which involves simulation of locating one or several loading/unloading bays in a given road stretch bays or more loading/unloading bays, so that the delivery does not require the car to stop at the lane (Fig. 1b).

Both of these models are based on Nagel-Schreckenberg model [30, 31]. However, they are different in the following aspects:

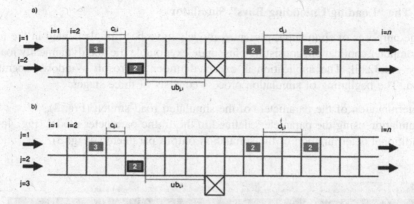

Fig. 1. Two-lane road stretch with delivery stop marked with X: a) inside the lane, b) outside the lane, using the loading/unloading bay. A rectangle with a contour symbolizes the truck (it has a lower maximum velocity) [own study]

- the road is not a one-lane but two-lane road, so net cellular automaton is an array with two rows and n columns.
- a single cell, in addition to the value specifying whether it is occupied or empty (is there a car at a given location) can be assigned additional value specifying whether a given road stretch is the loading/unloading bay (occupied or empty).
- occupied cell has a numeric value ($V_{j,i} = 0, 1, 2, \ldots, V_{MAX1}$ or $V_{j,i} = 0, 1, 2, \ldots, V_{MAX2}$, where: j – road number, i – cell number, V_{MAX1} – max. velocity of a passenger car, V_{MAX2} – max. velocity of a truck), depending on the type of vehicle. In this manner passenger cars and trucks are specified.
- in addition, cell has a value specifying whether it is a cell of road or loading/unloading bay.
- automatic transition rule consists of six steps: four in accordance with the
- N-Sch model (acceleration, braking, randomization and car motion), and two additional (lane change and parking in the bay).
- it includes driver's courage parameter ($c_{j,i}$ – courage), which is the distance between the car located in the i-th cell of the road number j and car behind it, on the adjacent lane.

The model takes into account the existence of two types of cars with different characteristics - passenger cars and commercial vehicles. A rectangle with a contour in Fig. 1 symbolizes commercial vehicle, which has a lower maximum velocity compared to the passenger car. The numbers on cars indicate the current velocity, j – road number, i – cell number, $d_{j,i}$ – the distance to the nearest preceding car, $c_{j,i}$ – the distance between the car located in the i-th cell of the road number j and car behind it, on the adjacent lane, $ub_{j,i}$ – the distance of the car from the nearest loading/unloading bay.

3.4   The "Loading/Unloading Bays" Simulator

For the purposes of simulation, presented models have been implemented in the form of a computer application, consisting of a single executable file and dynamically loaded library glut32.dll. The application is executed under Microsoft Windows operating system. The beginning of simulation process consists of three stages:

- determination of the parameters of the simulated road stretch (Fig. 2),
- simulation using the parameters defined in the traffic parameter model (Fig. 4a–c),
- additional determination of the simulation output parameters (Fig. 3).

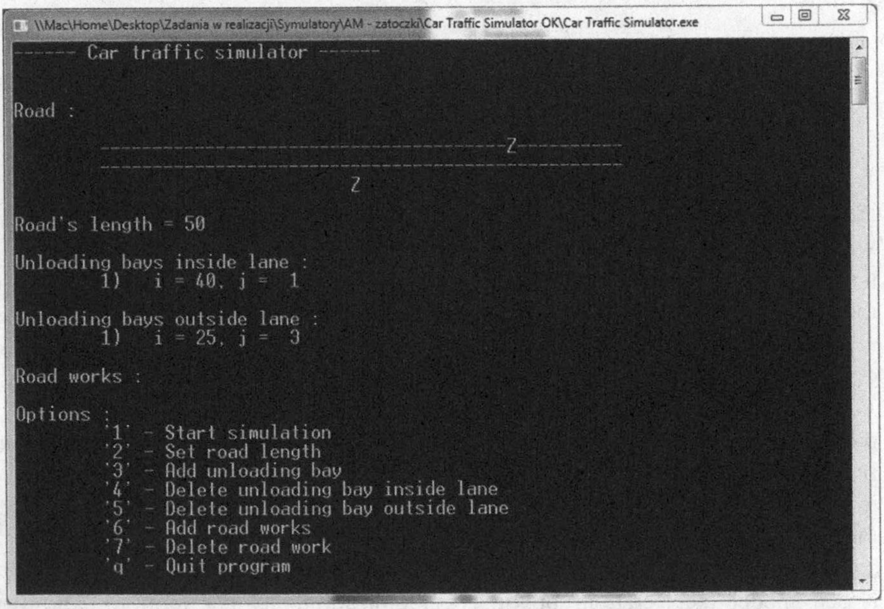

Fig. 2. The dialog windows for setting parameters for road and loading bays [own study].

The first step uses a dialog window presented in Fig. 2, which enables to establish the length of the analyzed road stretch, to add or remove parking spaces for delivery or loading/unloading bays, as well as to add or remove additional, permanent road obstructions (roadblocks), which allows to simulate traffic obstacles caused by e.g. the ongoing roadworks.

Adding unloading bay (in-side or out-side) requires two parameters:

- i – specifying stretch, where the parking space or loading/unloading bay is to be added,
- j – specifying the type of element added with the assigned value: 0 - in the case of adding loading/unloading bay at the left lane, 1 - in the case of adding the parking space at the left lane, 2 - in the case of adding the parking space at the right lane, 3 - in the case of adding the loading/unloading bay at the right lane.

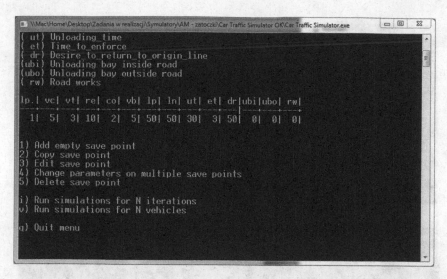

Fig. 3. The dialog windows for setting additional output parameters of the simulation [own study] (Color figure online)

The next stage of operation of the simulator is the process simulation on the basis of the settings in the dialog box presented in Fig. 2. This step is described in more detail after the Fig. 4.

The simulator was developed to allow running simulations for the following settings:

- the simulation for N iterations,
- the simulation for N vehicles.

Making these settings is possible in the window presented in Fig. 3, which is displayed when you press the blue icon (second icon from right at the top) located in the simulation window. This solution allows to study, the effect of which is to save the resulting data files. Observation of the simulation process is possible in the main window of the simulation (Fig. 4). In addition, this dialog box gives you the opportunity to correct the existing parameters. At the top of this window are presented parameter settings to aid the user in making changes.

The main window of the simulator is opened after setting the parameters of the studied road stretch (Fig. 4a–c). The center of the screen shows the individual automaton cells, representing the studied road stretch.

Light blue cells represent the parking spaces on the lanes (ub-inside) or loading/unloading bays located on the side of the road (ub-outside). Inside the cells there may be two types of cuboids: green, representing passenger cars, and red, representing the commercial vehicles. At the bottom of the panel there are sliders, which are responsible for setting the key parameters of the simulation:

- the maximum passenger car velocity,
- the maximum commercial vehicle velocity,

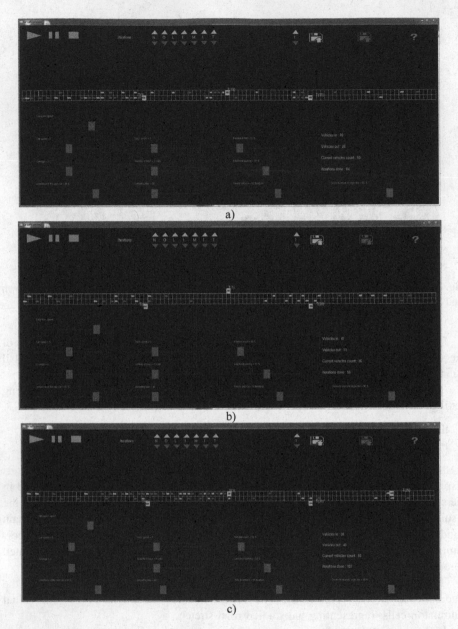

Fig. 4. The main window of the simulator for (a) simulation for 3 ub-inside, (b) simulation for 3 ub-outside, (c) simulation for 2 ub-inside and 2 ub-outside [own study] (Color figure online)

- the probability of a random event, corresponding to lowering car velocity by 1 (according to the N-Sch model),
- the probability that the driver will want to return to the previous lane,

- the tendency of the driver to change lane, visibility of loading/unloading bays, linked to the step "Lane change",
- the frequency of commercial vehicles stopping for delivery,
- the probability of a new car appearing in the first empty cell from left at random lane,
- time of delivery, corresponding to the number of iterations, during which commercial vehicle will remain in the cell representing the parking space or bay.

The application allows to save the results in CSV format. The report takes into account the parameters set in the simulation window and the number of car appearing on the input and the output of the model.

4   Experimental Results

In order to verify the proper operation of the model, several experimental have been done. A measure of the capacity of the road in all the presented experiments is the number of iterations of the developed cellular automaton model with the assumed number of 10,000 vehicles passing through a certain stretch of road. The greater the number of iterations of the cellular automaton, the road capacity is smaller, what means that a mathematical model performs more computing operations to handle the assumed number of vehicles moving through the fixed road. The path length was set at 100 cells cellular automaton, which, under the premise of the Nagel-Schreckenberg model gives the actual length of 750 m.

The first study involved to determine the effect of unloading time of goods in unloading bays located in the lane on the road. The results of the experiment are shown in Fig. 5. The simulation was carried out for the following input parameters:

- the maximum velocity of the passenger car: 5;
- the maximum velocity of the commercial vehicle: 3;
- the probability of random event: 0.1;
- the timidity of the driver: 2 (the higher the value is, the greater number of free cells in cellular automaton (between the vehicle in unloading bays and a vehicle approaching the unloading bay) is required to a vehicle parked in that bay could leave the unloading bay; it is a simulation of the phenomena associated with the skills of the driver - the higher the skill are, the driver can leave unloading bay faster without causing unnecessary congestion);
- the visibility of bays: 5 (the number of cells of cellular automaton - the distance from which the unloading bay can be seen, which allows you to simulate the phenomena of parking in the unloading bay);
- the probability of the new car: 0.9;
- the unloading time: 20–100 iterations of cellular automaton;
- the number of unloading bays (ub-inside) – for the first and third experiment: 0–4;
- the number of unloading bays (ub-outside) – for the second and third experiment: 0–4;

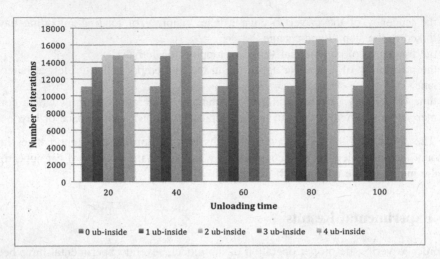

Fig. 5. The impact of the number of unloading-inside bays and unloading time on the number of iterations of cellular automaton [own study]

The second study involved to determine the effect of unloading time of goods in unloading bays located outside the lane of the road. The results of the experiment are shown in Fig. 6 and the input parameters were the same as in the previous study.

The parameter observed in both experiments was the number of iterations of the cellular automaton.

The third experiment was also concerned to determine the effect of unloading time unloading of goods in unloading bays on the road capacity. However, in this experiment on the fixed road they have been deployed both unloading bays situated in the lane of the road and bays located outside the lane of the road (in the amounts indicated in Fig. 7).

Analysis of the results leads to the conclusion that the time of unloading of goods has a significant impact on the capacity of the road, where there are unloading bays. This parameter becomes crucial when unloading bays are located in the line of a road. In the case of the four ub-inside bays arranged evenly on the fixed road (as shown in Fig. 5 relative to the road without bays the capacity of the road decreased by 52%. If the bays are located outside the main road the difference in capacity is much smaller and amounts to little more than 1%, which is mainly related to parking of vehicles in the bays and departuring from bays (scooping speed for commercial vehicles is time-consuming). This means that locating unloading bays outside the main road is the right urban trend, despite some consequences associated with it, e.g. to reduce the number of parking spaces. This consequence seems to be less painful for drivers going by this road.

A last experiment carried out in this research work was whether there is a correlation between the skills of the driver regarding the departure of unloading bay and the capacity of the road. Skills referred to above was defined in the model as courage in meaning uncertainty (shyness) drivers expressed in numbers of cells, which are the distance from the approaching vehicle to bay, allowing to leave bay by the vehicle

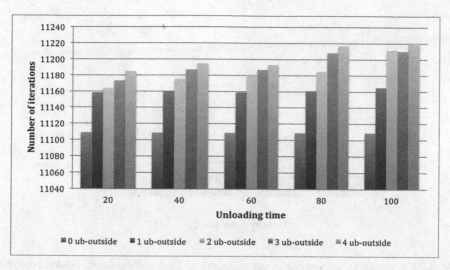

Fig. 6. The impact of the number of unloading-outside bays and unloading time on the number of iterations of cellular automaton [own study]

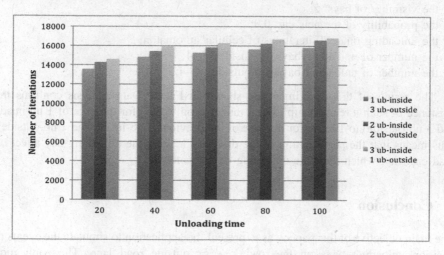

Fig. 7. The impact of the number of unloading-inside bays and unloading-outside bays and unloading time on the number of iterations of cellular automaton [own study]

parked. The assumption is that this manoeuvre should not lead to a rapid deceleration of the vehicle approaching to the unloading bay. The simulation was carried out for the following input parameters:

- the maximum velocity of the passenger car: 5;
- the maximum velocity of the commercial vehicle: 3;
- the probability of random event: 0.1;
- the timidity of the driver: 2–5;

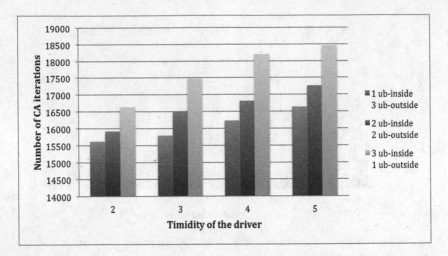

Fig. 8. The impact of the number of unloading-inside bays and unloading-outside bays and timidity of the driver on the number of iterations of cellular automaton [own study]

- the visibility of bays: 5;
- the probability of the new car: 0.9;
- the unloading time: 80 iterations of cellular automaton;
- the number of unloading bays (ub-inside): 0–4;
- the number of unloading bays (ub-outside): 0–4;

The results of this experiment are shown in Fig. 8. Data analysis confirms the existence of such a relationship, which ranged from 6.5% (for the case of 1 ub-inside and 3 ub-outside) to 10.9% (for the case of the previous 3 ub-inside and 1 ub-outside). This means that the parking skill of the driver or leaving the unloading bay affect on road capacity, which there are unloading bays located at.

5   Conclusion

The main objective of this paper was to present the application to simulate the events of position unloading bays on the road lane or outside road lane. The application implements a mathematical model based on cellular automaton model.

Several experiments were carried out: study of the effects of time unloading of goods on the road capacity and study of assessing the skills of the driver parked and leaving the unloading bay on the road capacity. The study were conducted in a simulation environment with parameters similar to the actual road conditions prevailing in some streets of the city of Szczecin. The results indicate the need to organize the unloading bays outside the main road and the awareness of drivers that their actions have a significant impact on the road capacity.

References

1. Gordon, G.: The Simulation of Systems. WNT, Warsaw (1974)
2. Chowdhury, D., Santen, L., Schadschneider, A.: Statistical physics of vehicular traffic and some related systems. Phys. Rep. **329**, 199–329 (2000)
3. Helbing, D.: Vehrkers-dynamik: Neue Physikalische Model lierungskonzepte. Rev. Mod. Phys. **73**, 1067 (1997)
4. Chopard, B., Luthu, P.O., Queloz, P.A.: Cellular automata model of car traffic in a two-dimensional street network. J. Phys. A **29**, 2325–2336 (1996)
5. Chowdhury, D., Schadschneider, A.: Self-organization of traffic jams in cities: effects of stochastic dynamics and signal periods. Phys. Rev. E **59**, 1311–1314 (1999)
6. Kerner, B.S.: Complexity of synchronized flow and related problems for basic assumptions of traffic flow theories. Netw. Spacial Econ. **1**, 35–76 (2001)
7. May, A.D.: Traffic Flow Fundamentals. Prentice Hall, Upper Saddle River (1990)
8. Daganzo, C.F.: Transportation and Traffic Theory. Elsevier, Amsterdam (1993)
9. Wolf, D.E., Schreckenberg, M.: Traffic and Granular Flow. Springer, Singapore (1998). doi:10.1007/978-3-662-10583-2
10. Małecki, K., Iwan, S.: Development of cellular automata for simulation of the crossroads model with a traffic detection system. In: Mikulski, J. (ed.) TST 2012. CCIS, vol. 329, pp. 276–283. Springer, Heidelberg (2012). doi:10.1007/978-3-642-34050-5_31
11. Cernicky, L., Kalasova, A., Mikulski, J.: Simulation software as a calculation tool for traffic capacity assessment. Komunikacie **18**(2), 99–103 (2016)
12. Mikulski, J.: The possibility of using telematics in urban transportation. In: Mikulski, J. (ed.) TST 2011. CCIS, vol. 239, pp. 54–69. Springer, Heidelberg (2011). doi:10.1007/978-3-642-24660-9_7
13. Mikulski, J.: Using telematics in transport. In: Mikulski, J. (ed.) TST 2010. CCIS, vol. 104, pp. 175–182. Springer, Heidelberg (2010). doi:10.1007/978-3-642-16472-9_19
14. Kalašová, A., Faith, P., Mikulski, J.: Telematics applications, an important basis for improving the road safety. In: Mikulski, J. (ed.) TST 2015. CCIS, vol. 531, pp. 292–299. Springer, Cham (2015). doi:10.1007/978-3-319-24577-5_29
15. Iwan, S., Małecki, K.: Data flows in an integrated urban freight transport telematic system. In: Mikulski, J. (ed.) TST 2012. CCIS, vol. 329, pp. 79–86. Springer, Heidelberg (2012). doi:10.1007/978-3-642-34050-5_10
16. Iwan, S., Małecki, K., Korczak, J.: Impact of telematics on efficiency of urban freight transport. In: Mikulski, J. (ed.) TST 2013. CCIS, vol. 395, pp. 50–57. Springer, Heidelberg (2013). doi:10.1007/978-3-642-41647-7_7
17. Jaszczak, S., Małecki, K.: Hardware and software synthesis of exemplary crossroads in a modular programmable controller. Prz. Elektrotech. **89**(11), 121–124 (2013)
18. Kołopieńczyk, M., Andrzejewski, G., Zając, W.: Block programming technique in traffic control. In: Mikulski, J. (ed.) TST 2013. CCIS, vol. 395, pp. 75–80. Springer, Heidelberg (2013). doi:10.1007/978-3-642-41647-7_10
19. Andrzejewski, G., Zając, W., Kołopieńczyk, M.: Time dependencies modelling in traffic control algorithms. In: Mikulski, J. (ed.) TST 2013. CCIS, vol. 395, pp. 1–6. Springer, Heidelberg (2013). doi:10.1007/978-3-642-41647-7_1
20. Odani, M., Tsuji, T.: An experiment to demonstrate the effectiveness of on-street parking facilities for delivery vehicles. In: Taniguchi, E., Thompson, R.G. (eds.) Proceedings of 4th International Conference on City Logistics. Institute for City Logistics, Kyoto, pp. 107–116 (2001)

21. Auira, N., Taniguchi, E.: Planning on-street loading-unloading spaces considering behavior of pickup-delivery vehicles and parking enforcement. In: Taniguchi, E., Thompson, R.G. (eds.) Recent Advances in City Logistics. Elsevier, pp. 107–116 (2006)
22. McLeod, F., Cherrett, T.: Loading bay booking and control for urban freight. Int. J. Logist. Res. Appl. 14(6), 385–397 (2011)
23. Johansen, B.G., Andersen, J., Levin, T.: Better use of delivery spaces in Oslo. Procedia Soc. Behav. Sci. 151, 112–121 (2014)
24. Alho, A., de Abreu e Silva, J., Pinho de Sousa, J.: A state-of-the-art modeling framework to improve congestion by changing the configuration/enforcement of urban logistics loading/unloading bays. Procedia Soc. Behav. Sci. 111, 360–369 (2014)
25. Roche-Cerasi, I.: State of the art report. Urban logistics practices. Green Urban Distribution, Deliverable 2.1. SINTEF Teknologi og samfunn (2012)
26. Ma, N., Shi, X., Zhang, T.: Urban freight transport in Shanghai: the demand for a commercial loading/unloading bay. In: Proceedings of the 15th COTA International Conference of Transportation Professionals, pp. 788–797 (2015)
27. Kułakowski, K.: Cellular automata. Akademia-Górniczo-Hutnicza, Kraków (2000). (in Polish)
28. Malarz, K.: Cellular automata. Akademia-Górniczo-Hutnicza, Kraków (2008). (in Polish)
29. Wolfram, S.: Universality and complexity in cellular automata. Physica D 10, 1–35 (1984)
30. Nagel, K., Schreckenberg, M.: A cellular automata model for freeway traffic. J. de Physique I 2, 2221–2229 (1992)
31. Nagel, K., Wolf, D.E., Wagner, P., Simon, P.M.: Two-lane traffic rules for cellular automata: a systematic approach. Phys. Rev. E 58(2), 1425–1437 (1998)

System for Monitoring and Guarding Vehicles on Parking Areas

Dušan Nemec$^{(\boxtimes)}$, Aleš Janota, and Rastislav Pirník

Faculty of Electrical Engineering, University of Zilina,
Univerzitna 1, 010 26 Zilina, Slovak Republic
{dusan.nemec,ales.janota,
rastislav.pirnik}@fel.uniza.sk

Abstract. The paper deals with design and realization of the system monitoring vehicles at the open type of parking area. Design and realization of such a system, with the use of video-cameras and relevant SW and HW equipment, help to warn the vehicle's owner about movement of his/her vehicle (in an adverse case of its theft). Information about movement is sent by email or SMS message. The system analyzing images from cameras has been implemented on the basis of the approved utility model PUV 96-2015 "Automated system for monitoring and guarding vehicles on parking areas" utilizing knowledge about image processing and computer vision. The system has been implemented and tested within the University of Žilina Campus. The system is available to the end users in the form of Internet web-page which makes it accessible from any ICT device.

Keywords: Monitoring · Guarding · Vehicle · ICT device · Parking area

1 Introduction

One of the issues of the static transportation problem is to reduce the area and time which is taken by parked vehicles. Solution of this problem is provided by system Park & Ride (P&R) demonstrated on the example of Vienna city (see Fig. 1). System P&R (or P&GO) is a smart solution for the local government which allows drivers to park their vehicles on the car park away from the centrum of the city for a reasonable fee. Such car parks are situated at the edge of the city next to the highway and are well connected to the public transportation. The public transportation (a subway in the ideal case) then transports passengers directly to/from the city centrum in a reasonably short time [1].

Car park is not just some dedicated clean area where vehicles can be stored while inactive. It is a complex system which has to allow comfortable and secure way to park a vehicle. Besides standard social services it has to provide additional services especially for drivers. Such Smart services require the P&R system management to introduce access control systems, management of free parking places and especially security of the parked vehicles.

Increasing number of local governments in large cities of EU introduces paid parking zones in the centrum of the city in order to suppress negative influence of an uncontrolled traffic, optimize usage of the parking places and also collect

© Springer International Publishing AG 2017
J. Mikulski (Ed.): TST 2017, CCIS 715, pp. 307–319, 2017.
DOI: 10.1007/978-3-319-66251-0_25

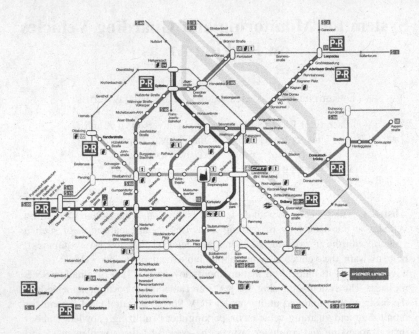

Fig. 1. P&R system utilizing subway in the city of Vienna [1]

non-negligible income to the city budget. In the past the only required action was to find a free parking place and park the vehicle. Nowadays the "SmartCity" customer (driver) expects some counter-value for the parking fee – in the form of some kind of service or security benefit. The price for a vehicle can reach up to 50 000€ and more. Only in the Slovak republic there were 1932 vehicles stolen during 2015, which caused loss of 16.5 million of euro. According to the statistical institute of EU (Eurostat) thieves stole more than 762 400 vehicles in the whole EU during 2012. Despite the decreasing tendency of car thefts, their number is still high. Theft of a vehicle also decreases mobility of its owner and causes emotional stress to him. Level of security of the more expensive vehicles is quite high – various security systems cooperating with localization system are installed. These systems are capable to inform a vehicle's owner about any movement of the vehicle during standard usage but mainly after the theft. However, the technical equipment and skills of professional thieves are also high – e.g. high power GSM jammer is capable to cut off the information flow about vehicle's position which should be sent to the vehicle's owner.

2 Smart Parking Concept in the Parking Zones Inside the Cities

The system proposes independent service of guarding of the vehicles parked in the zone which is completely or at least partially covered by security camera system. According to the current knowledge it is possible to use existing camera systems installed in the centrums of the cities. Concept has been developed for an open-air type of car parks which does not provide such functionality yet. In Fig. 2 there is coverage of one street in Prague (yellow rectangle marks the detection zone above the parking place).

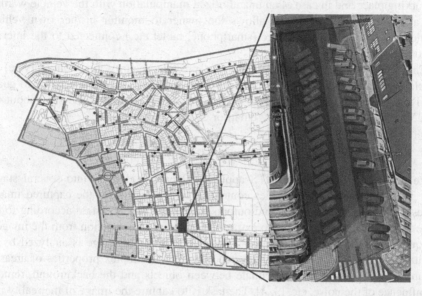

Fig. 2. Concept of the secure parking system inside centrum of Prague [2] (Color figure online)

During collaboration with organization participating in the project which supported this contribution we have designed and developed a concept of guarding system for lorry trucks. The entrance and exit of the system is equipped with weight detector Measure-In-Motion® which allows not just detection of unauthorized exit (possible theft) of the vehicle but also detection of changes in vehicle's weight caused by theft of transported goods or boarding of an unauthorized person seeking for an illegal transportation.

3 System Design

Currently the camera systems are used for monitoring of public areas, streets, parking areas and vehicles. These systems are interconnected with a communication subsystem which is centralized and the response time is not always sufficient (not in real-time) and flexibility of the output signaling is low (the output device is given without chances to adopt it to different scenarios). Also the widely used approach for monitoring vehicles in the parking areas is a security person.

Proposed technical solution in the form of automated system for monitoring and guarding of vehicles in parking areas allows additional flexibility and variability to the user enabling variety of output display terminals (smartphone, tablet, notebook etc.).

The connection of the output terminals is provided by data network in real time while specialized software automatically monitors and guards the vehicles and in the case of unauthorized manipulation with vehicle alerts the owner of the vehicle. The proposed system means improvement of the standard approach which uses a security service officer. System with the aid of cameras and processing software analyzes situation in the parking place and in case of an unauthorized manipulation with the vehicle warns a user about the situation. It also allows the owner to monitor his/her own vehicle remotely by mobile display device (smartphone, tablet etc.) connected to the internet network.

The goal of the proposed system is to provide security function (service) of vehicle's guarding to the end user (owner). This security function is capable to guard the given vehicle using video camera system; its output is an alert message about car theft. This message can be sent by email or SMS.

3.1   Tool for Capturing and Processing of the Video Image

Image processing (not only for ITS application) can be divided into several stages (Fig. 3). The highest level provides contextual understanding of the captured image, processing of the obtained facts and outputs the results of the system according to the requirements. In order to be able to extract the necessary information from the image it is required to process it first at lower levels where the raw image is analyzed by its brightness, contrast, color, its changes and other more complex properties of areas in the image which allows to differentiate between objects and the background, remove the influence of the noise, etc. [3, 4]. The task is to capture the image of the reality in a suitable way in real time, process it and evaluate the important properties.

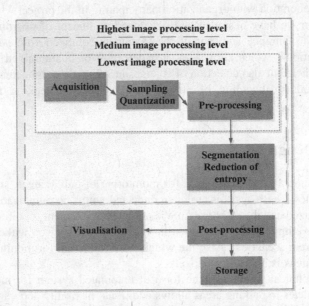

Fig. 3. Block diagram of digital image processing [3]

Lowest image processing level

At the lowest level we use processes that do not require making any specific decisions by the system. Each function works automatically according to the predefined algorithm. Those basic steps are as follows:

- Image capturing – this is the very first step in the whole process. It requires equipment capable of capturing and digitalizing image data such as video camera (TV or industrial), scanner (single line or rotary) or digital camera. In case of legacy devices which provide analog output the signal has to be digitalized by suitable analog to digital converter (ADC).
- Image preprocessing – the raw image obtained in the previous step is usually unsuitable for usage in higher levels of processing. Key role of the preprocessing is to improve the digital image in order to highlight certain features in the image and increase the overall reliability of methods applied in the following steps. Commonly used techniques use adjustment of contrast, removal of optical discrepancies, noise or blur.

Medium image processing level

Medium level extracts and characterizes items represented as areas (segments) in the image. Compared to the lowest processing level this level requires much higher flexibility. It can be composed of the following steps:

- Segmentation – in general the goal of segmentation is to separate input image into captured objects. According to the theory of systems we consider the term "object" as the part of the image which we are interested in. In many cases this is one of the most demanding tasks during analysis of the digital image. During segmentation the image semantics (knowledge what should the image contain) is used. Output of the segmentation is usually a set of points (elementary image elements) representing either borders of the object or pixels within the borders.
- Representation and description – obtained sets of pixels needs to be converted to another form more suitable for next processing. It is possible to describe them quantitatively by numerical characteristics and also qualitatively by relations between them. Therefore it is needed to decide which representation of objects (borders or pixels) is better for the given problem (whether the object is characterized by its shape or its internal properties). The result of description is some quantitative information about the object or its property which can be used for classification of the object to different classes.

The highest image processing level

The last and highest level of image processing implements the image content recognition and interpretation which is required for automated analysis of the digital image. The analysis consists of discovering, identification and understanding of meaning of patterns obtained from analyzed image without any previous knowledge. Understanding of the image is based on the knowledge, goals and methods and feedbacks between various levels of processing. Basic task of these steps is to equip the machine (PC) with certain human-like abilities of vision and sensing. For successful recognition and interpretation knowledge from geometry, statistics, neural networks, artificial intelligence or fuzzy systems are being used.

Methods of image processing can be deployed in wide area of intelligent transportation control system (ITS systems). Every video image processing (VIP) system used to observe and collect information from traffic process has to meet the following requirements:

- automated identification of objects and background allowing to capture every object of interest,
- correct recognition of object type – lorry trucks, cars, bicycles, pedestrians, railway vehicles etc.
- functionality has to be provided in large number of possible situations which can occur in the traffic process,
- insensitivity to light or temperature conditions (thermal vision) of the captured scene. The system has to work during both day and night, summer and winter etc.
- real time execution.

The specific level of requirements depends on the application. Telematic image processing systems can be usually found in the road traffic applications.

The software of the system has been developed in the IDE Microsoft Visual Studio which supports multiple programming languages allowing usage of frameworks for static and dynamic image processing. In our case we have used C++ language with OpenCV framework which is an open source, multiplatform library focused on computer vision and image processing in real time. OpenCV library supports C++, C, Python, Java and MATLAB programming languages [5]. It is supported by Windows, Linux, Android and Mac OC operating systems.

3.2 Basic Tasks of the System

The automated system (1) shown in Fig. 4 allows monitoring and guarding of vehicles (8) parked on the parking places (6) and locating of free parking places (7) [5].

User (2) (owner of the vehicle, driver or another privileged person) is capable to connect its imaging device (9) (smartphone, tablet, notebook, etc.) to the server (4) using external communication subsystem (5a). External communication subsystem is based on intranet or internet network. After the connection is established the user is able to monitor his/her vehicle on the parking place and also search for unoccupied parking places. Necessary data are obtained by cameras (3) installed in site. The system is designed as decentralized, therefore it is not needed to transfer image data from all cameras using internal communication subsystem into one server.

3.3 Placement of the Camera

In order to obtain as much as possible usable image information from the camera, the placement of the camera is very important. Every car park requires specific approach for solving this task. In general, there are the following requirements:

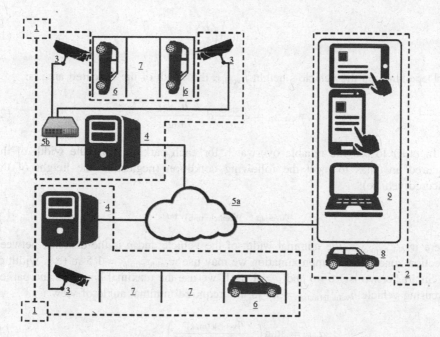

Fig. 4. Proposed conception of the system [6]

- placement with the direct view to maximum number of parking places,
- the angle of view which avoids obscuring of one vehicle by another (e.g. supply van hides smaller car).

Two approaches can be considered:

First approach expects that the experimental car park is viewed longitudinally in order to capture all parking places. The problem is the height of the camera's placement (see Fig. 5) which allows satisfying above conditions. More distant parking places are more critical since a distant large vehicle (e.g. a supply van) obscures more from the view of the camera.

It is important to keep the ratio between the height of the camera h and the distance d as high as possible in order to increase the angle of view α:

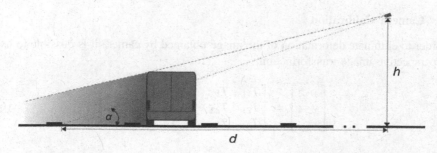

Fig. 5. Obscuring of one parking place by a nearby vehicle [own study]

$$\alpha = \operatorname{atan}\left(\frac{h}{d}\right). \tag{1}$$

Depending on the vehicle's height h_{vehicle} the width of the obscured area is:

$$w_{\text{overlap}} = \frac{h_{\text{vehicle}}}{h} d = \frac{h_{vehicle}}{\tan \alpha}. \tag{2}$$

In order to provide reliable overwatch for each parking place the width of the obscured area has to meet the following condition (neglecting the height of the obscured vehicle):

$$w_{\text{overlap}} < w_{\text{vehicle(min)}} + \delta, \tag{3}$$

where $w_{\text{vehicle(min)}}$ is the minimal width of the vehicle and δ is the distance between detection zones. For an approximation we may use $w_{\text{vehicle(min)}} = 1.5$ m (e.g. width of the Smart Fortwo is 1.6 m) and $\delta = 1$ m. If we use the maximal height of the parked obscuring vehicle $h_{\text{vehicle(max)}} = 3.0$ m, the required minimal angle of view is:

$$\alpha_{\min} = \operatorname{atan}\left(\frac{h_{\text{vehicle(max)}}}{w_{\text{vehicle(min)}} + \delta}\right) \approx 50°, \tag{4}$$

The problem can be eliminated by installing the camera in sufficient height (not on the lamp post but in the level of 4–5th floor). If the height of the camera $h = 15$ m the system can reliably work up to the distance $d_{\max}$ from the camera:

$$d_{\max} = \frac{h}{\tan \alpha_{\min}} \approx 12.5 \text{ m}. \tag{5}$$

In order to cover larger working area it is more suitable (if possible) to capture the car park laterally (see Fig. 6). Horizontal angle of the camera's view does not allow to monitor all parking places of a wider car park. This can be solved by adding additional cameras to the system and place them optimally in order to cover the whole area of the car park. The proposed and implemented system experimentally evaluates both (longitudinal and lateral) orientations.

3.4   Camera Calibration

In order to eliminate deformation of the image obtained by camera it is possible to use 2D perspective image transformation:

$$\begin{bmatrix} x \cdot z \\ y \cdot z \\ z \end{bmatrix} = \begin{bmatrix} T_{11} & T_{12} & T_{13} \\ T_{21} & T_{22} & T_{23} \\ T_{31} & T_{32} & T_{33} \end{bmatrix} \cdot \begin{bmatrix} x' \\ y' \\ 1 \end{bmatrix}, \tag{6}$$

Fig. 6. Concept of the secured parking system in UNIZA campus [own study]

where x, y are coordinates of the pixel in the source image, x' and y' are coordinates of the pixel in the resulting (undistorted) image and z is a scaled distance between the camera and point represented by the given pixel. T_{ij} are constant calibration coefficients related only to the position and orientation of the camera. In order to keep the transformation normalized the coefficient $T_{33} = 1$, therefore there are 8 unknown coefficients. If we know 4 pairs of corresponding points in source and transformed image denoted $[x_k \quad y_k]$ and $[x'_k \, y'_k]$ respectively, it is possible to determine all calibration parameters.

The Eq. (6) can be rewritten as follows:

$$T_{11}x'_k + T_{12}y'_k + T_{13} - T_{31}x_kx'_k - T_{32}x_ky'_k = 1, \tag{7}$$

Fig. 7. Original (left) and transformed (right) image with displayed calibration points and detection zones [own study]

$$T_{21}x'_k + T_{22}y'_k + T_{23} - T_{31}y_kx'_k - T_{32}y_ky'_k = 1, \tag{8}$$

which can be expressed in the matrix form as:

$$\begin{bmatrix} x'_1 & y'_1 & 1 & 0 & 0 & 0 & -x_1x'_1 & -x_1y'_1 \\ x'_2 & y'_2 & 1 & 0 & 0 & 0 & -x_2x'_2 & -x_2y'_2 \\ x'_3 & y'_3 & 1 & 0 & 0 & 0 & -x_3x'_3 & -x_3y'_3 \\ x'_4 & y'_4 & 1 & 0 & 0 & 0 & -x_4x'_4 & -x_4y'_4 \\ 0 & 0 & 0 & x'_1 & y'_1 & 1 & -y_1x'_1 & -y'_1 \\ 0 & 0 & 0 & x'_2 & y'_2 & 1 & -y_2x'_2 & -y_2y'_2 \\ 0 & 0 & 0 & x'_3 & y'_3 & 1 & -y_3x'_3 & -y_3y'_3 \\ 0 & 0 & 0 & x'_4 & y'_4 & 1 & -y_4x'_4 & -y_4y'_4 \end{bmatrix} \cdot \begin{bmatrix} T_{11} \\ T_{12} \\ T_{13} \\ T_{21} \\ T_{22} \\ T_{23} \\ T_{31} \\ T_{32} \end{bmatrix} = \begin{bmatrix} 1 \\ 1 \\ 1 \\ 1 \\ 1 \\ 1 \\ 1 \\ 1 \end{bmatrix}, \tag{9}$$

If selected calibration points are not collinear, the matrix (9) is invertible. The effect of the transformation is shown in Fig. 7. As can be seen, the detection zones in the transformed image are aligned and have equal size.

3.5   Functionality of the RUNIZA System

User connects to the web server via web client (see Figs. 8 and 9).

First he/she enters his/her credentials, and then he/she selects the number of the camera which provides the monitoring of his/her vehicle. After confirmation of the selection the system sets variables NAME, CAMERA, ZONE, EMAIL and TELE-PHONE. Then the flag GUARD is set and VIDEOGUARD subprogram will be started for the given parking place.

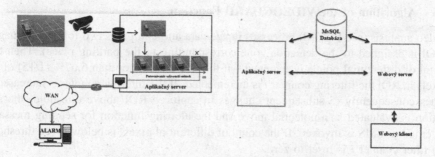

Fig. 8. Scheme of the RUNIZA system [7]

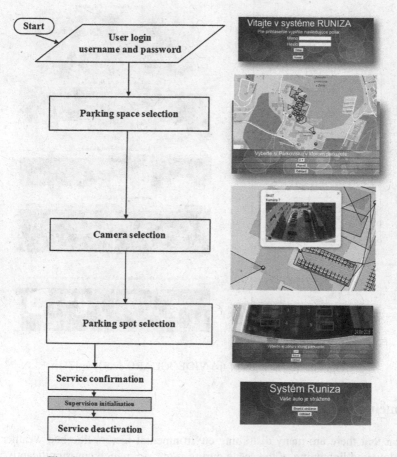

Fig. 9. Functionality of the implemented RUNIZA system [7]

3.6   Algorithm of the VIDEOGUARD Function

Video analysis (see Fig. 10) processes image data and compares two following images (ROI is assigned to the detection zone corresponding to the parking place). Then the count of differential points is achieved – if the count is higher than 6.67% (1/15) of all pixels in ROI the filtering counter i is incremented. If the filtering counter is increased 5 times consequently (5 subsequent changes in captured ROI above threshold) the situation is evaluated as a potential move and the alerting function for sending message by email or SMS is invoked. If the count of differential pixels is below given threshold the filter counter i is reset to zero.

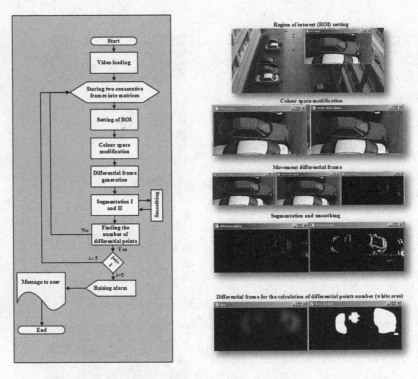

Fig. 10. Explanation of the VIDEOGUARD function [7]

4   Conclusion

It is clear that there are many disturbing environmental factors like bad weather conditions, lowered lightening of the scene during night etc. which can significantly affect the system based on video-detection. In our case we examine the dependency of the system reliability on the resolution of the camera and data throughput (at given data compression) which is necessary to provide raw video stream in online mode. At higher count of ROI (above 40 parking places) in one shot of the camera the lower resolution of the camera will not provide reliable detection of the vehicle's movement (yellow

Table 1. System reliability as a function of the resolution and the throughput of the network [own study]

Count of ROI = 40	Data throughput [kb/s]					
Resolution	500	1000	1500	2000	5000	10000
480p (768x432)						
720p (1280x720)						
1080p (1920x1080)						
5MP (2992x1680)						

fields in Table 1). Higher resolution and high data throughput caused HW and SW failures (overload) of the used application server (red fields in Table 1).

The main advantage of the proposed and implemented automated system for monitoring and guarding of vehicles is its flexibility and variety of the output display terminals, decentralization and remote access into the communication subsystem in real time which decreases the need for physical (personal) guarding service in car parks. This work was supported by Slovak Research and Development Agency under the project No. APVV-15-0441.

References

1. Pulka, R.: U-Bahn Plan. http://www.parkandride.at/U-Bahn-Plan.html. Accessed 14 Feb 2017
2. Zóny placeného stání Praha. http://www.praha.cz/zony-placeneho-stani. Accessed 14 Feb 2017
3. Bubeníková, E., Pirník, R., Holečko, P.: Optimisation of video-data transmission in telematic system. Adv. Electr. Electron. Eng. **11**(2), 123–134 (2013)
4. Bubeníková, E., Franeková, M., Holečko, P.: Evaluation of unwanted road marking crossing detection using real-traffic data for intelligent transportation systems. In: Mikulski, J. (ed.) Telematics-Support for Transport. CCIS 471, pp. 137–145. Springer, Berlin, Heidelberg (2014)
5. opencv.org. http://opencv.org/about.html. Accessed 14 Feb 2017
6. Pirník, R., et al.: Automatizovaný systém monitorovania a stráženia dopravných prostriedkov na odstavných plochách, Slovak utility model PUV 96-2015, Nov 16 (2016)
7. Bakša, P., Dohľad nad vozidlami v systémoch statickej dopravy, M.S. thesis, Department of Control and Information Systems, University of Žilina, Žilina, Slovakia (2016)

Multispectral Data Acquisition in the Assessment of Driver's Fatigue

Krzysztof Małecki[✉], Adam Nowosielski, and Paweł Forczmański

West Pomeranian University of Technology,
Żołnierska Str. 52, 71-210 Szczecin, Poland
{kmalecki, anowosielski, pforczmanski}@wi.zut.edu.pl

Abstract. Many factors contribute for the occurrence of the road accidents. The most important are the behaviour of drivers and the level of their fatigue. Appropriate recognition of driver's fatigue is now becoming an important research issue, the results of which are beginning to be implemented in automotive driver assistant systems. In the article the authors present the characteristics of selected multispectral data (visual image, depth map, thermal image) used for automatic assessment of driver fatigue and the station for their acquisition. For the study a simulator station has been proposed and developed. It reflects the driver's cabin (based on physical measurements of a wide range of vehicles), is equipped with the appropriate video sensors (including depth and thermal recorders) and monitors showing real driving situations. The acquired data streams can be used for research on the development of non-invasive methods for assessing the degree of driver fatigue.

Keywords: Driver fatigue · Multispectral data acquisition

1 Introduction

Many studies in the field of road safety show that driver behaviour and errors committed by them are the cause of most accidents. For example, the US research on the causes of road accidents conducted on more than 2000 cases have been identified that man is responsible for about 92% of accidents [1]. In turn, [2] determined that fatigue and drowsiness are the two main causes of road accidents. Similar statements appear in a study by the Virginia Tech Transportation Institute in 2013 [3], which showed that the reason for 20% of accidents was driver fatigue. Although many times there is no possibility to determine whether the fatigue was a factor in an accident afterward and what level of fatigue the driver was suffering, the tiredness is a considerable contributory factor in road accidents. The fatigue causes loss of concentration and prevents the driver from making quick and correct decisions [4, 5].

There are various methods for measuring fatigue in terms of driving: cognitive methods, subjective approach by using the Karolinska Sleeping Scale (KSS) and the physiological approach. The first is the measurement of the performance of neuro-behavioral characteristics (cognitive methods) with the use of the Psychomotor Vigilance Test (PVT). PVT measures the rate of continuous attention span. This measurement has become a major standard in detecting fatigue [6]. Individuals will

© Springer International Publishing AG 2017
J. Mikulski (Ed.): TST 2017, CCIS 715, pp. 320–332, 2017.
DOI: 10.1007/978-3-319-66251-0_26

provide a response to stimuli or visual stimuli by pressing a button on the computer screen during a period of 5–10 min. PVT measures reaction time. However, the PVT performance may not fit poor driving performance, due to the large individual differences in poor driving performance, which skew the test results. Thus, PVT is better at predicting fatigue-related accidents [7]. Another approach is the subjective method based on Karolinska Sleeping Scale (KSS). KSS is based on EEG (electroencephalographic) and testing the activity and eye movement EOG (electrooculographic) [8]. In addition, it is used also the physiological approach, e.g. measurement of heart rate and blood pressure test. It was found that the longer time driving times resulted in faster heartbeats because the male drivers experienced fatigue [9]. Another possibility is the measurement of physiological driver parameters like brain activity. However, this technique requires some uncomfortable sensors to be placed on the body that renders it inconvenient.

Researches on the drivers' fatigue are conducted both during the real driving [10] and in the simulation environment [11].

The vehicle safety can be improved by monitoring the drowsiness of the driver [4]. Some inference can be conducted from lane monitoring system (attentiveness to the lane-keeping task [12]). After the detection of the fatigue the driver assistant system is able to stimulate the driver with some auditory signals like playing lively music. In critical situation, after the sleep detection, such system may even stop the vehicle.

A broader look at the problem of the estimation of driver's physical conditions unveils, that it can be estimated using certain, independent and complementary means. Selected sources of data used in this problem are presented in Fig. 1. These sources belong to three main groups: visual data associated with the observation of drivers head/face area, non-visual data related to the behavioral information like pulse or brain activity, together with the characteristic of the steering wheel movements or pressure on the driver's seat, and finally, data taken from car's built-in systems, like GPS, accelerometer, speed- or G-sensors.

Most of these sources are associated with non-invasive acquisition; hence they are acceptable for a typical driver. The first and the third groups of data are also quite easy to capture and process in real-time, which is an important requirement for a successful implementation. On the other hand, visual sources require, in some cases, good imaging conditions, e.g. proper lighting, adequate perspective, scale and geometrical orientation. Therefore, they are supplemented with data from other sources.

As it was mentioned, the vision-based driver monitoring systems are the most convenient and noninvasive solution, hence they are in the scope of the research presented in this paper. Computer vision techniques allow the constant observation of the driver. Drowsy drivers exhibit some observable behaviors in head and eyelid movement or eye gaze [4]. When the driver is drowsy his eyes begin to close. Long lasting closure or downward diverted sight are derogations indicating the increased risk situation. Such case may be detected automatically. It is possible to detect the state of the eyes, to discover the face or gaze direction or to detect recurrent yawning. Because the environmental conditions are not always favorable, and not always state of the driver might be discovered in visible spectrum, it is crucial to analyze multispectral data acquisition.

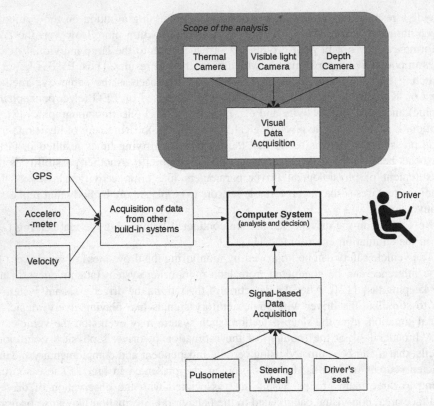

Fig. 1. Schematic concept of data acquisition in advanced driver assistance system oriented at the estimation of driver's physical conditions [own study]

The aim of this publication is to present the process of obtaining the data used to recognize fatigue and weariness of the driver, using vision-based systems.

The article is organized as follows: in the next part of the article the analysis of the subject is presented, then there is the theoretical background related to the selected multispectral imaging. In the next part there is the concept of position to acquire video sequences and experimental results.

2   Related Work

In general, data acquisition on driver fatigue is performed based on different techniques. In [13] the technique of the questionnaire was presented. The study was conducted for 45 randomly selected municipal bus drivers, riders along the route "easy" (outside the city centre, little traffic) and "difficult" (the centre of the city, heavy traffic). The authors used the fatigue test questionnaire, based on the list of symptoms of fatigue prepared by the Research Committee of Japan Fatigue. Distinguished four groups of symptoms that describe physical fatigue, deterioration of efficiency of movements, mental fatigue and discomfort of vision and visual fatigue (symptoms of ocular). In [14] the authors took up of the registration and evaluation of biometric parameters of the driver to determine the emotional state of the driver. For this purpose, a biomedical system concept based on three different mechanisms of measurement: recording vehicle speed, recording changes in the heartbeat of the driver and recording the driver's face.

Vision-based solutions provide an excellent mean for fatigue detection. The initial step in vision-based driver fatigue detection systems consist of detection of face and facial features. Detected features are subsequently tracked to gather important temporal characteristics from which the appropriate conclusion of driver's fatigue can be drawn.

Detection of face and facial features are classical face recognition problems. The technology has reached its maturity, and although a number of issues still exist for consideration, it is regarded as solved (under controlled conditions). There are well-functioning procedures for the face detection task like the Viola & Jones method [15, 16]. It is an universal solution and the leading choice option for researchers opting for the ready to use algorithm. The method takes advantage form statistical boosting and uses Haar-like features. The implementation is available for example in the common use OpenCV library (the Open Source Computer Vision library released under a BSD license, available at http://opencv.org/). The OpenCV implementation is equipped with the ready-to-use eye detector constructed upon the same algorithm. By employing existing algorithms and image processing techniques it is possible to create an individual solution for driver fatigue/drowsiness detection based on eyes state. An example is presented in [17] where the OpenCV face and eye detectors are supported with the simple feature extractor based on the two dimensional Discrete Fourier Transform (DFT) to represent an eye region.

Many similar solutions are found in the scientific literature. In [4] the OpenCV implementation of face and eye detectors are also used. The detection is supported by frame difference mechanism. Its aim is to localize the eyes by blinking. The method also employs template matching for tracking purposes. The fatigue of the driver is determined through the duration of the blink.

The duration of the closed eyes is also employed for inference of the driver drowsiness and fatigue in [5]. It is a more sophisticated solution and operates in the visual and near infra-red (NIR) spectra. The infrared spectrum is invisible for the driver and allows to use the additional infrared light source (NIR LED) for the night conditions and poor visibility. Two cascade of classifiers are used, one for the detection of the eye and the other for the verification stage (in each spectrum).

Interesting multimodal platform to identify driver fatigue and interference detection is presented in [18]. The authors present a driving simulator equipped with several sensors. They have designed a framework to acquire sensor data, process and extract features related to fatigue and distraction. In this work, they extracted audio, video colour, depth maps, heart rate, steering wheel and pedals positions. The platform processes the signals in accordance with the three modules: a module perspective, an audio module and the module other signals. The modules are independent of each other and can be enabled or disabled at any time. Each module brings out the essential features and on the basis of hidden Markov models, produces its own estimates of driver fatigue and distraction. The authors detect fatigue accuracy of 98.4% and 90.5%

of the dispersion. The experimental results show that the authors are able to detect fatigue with 98.4% accuracy and distraction with 90.5%.

There are solutions based on mobile devices, especially smartphones and tablets, or based on dedicated hardware. Such solutions, very often, outside the data acquisition driver (based on camera in the device) have built-in mechanisms for identifying fatigue. An example might be the work of [19–21]. Among these works one of the recognized symptoms is yawning. In [22] the authors recognize the act of yawning using a simple webcam. However, in the [23] proposed a dynamic fatigue detection model based on Hidden Markov Model (HMM). This model can estimate driver fatigue in a probabilistic way using various physiological and contextual information. Electroencephalogram, electromyogram and respiration signals were simultaneously recorded by wearable sensors and sent to computer by Bluetooth during the real driving.

Many different technologies have been combined and used in [24]: the image topological analysis technology, Haar features and extreme learning machine algorithm. The authors propose a new detection method of the intelligent fatigue driving. They demonstrated through simulation studies, that their method gives more accurate monitoring of drivers than methods based on the traditional eye tracking technology and traditional artificial neural networks. In a subsequent work [25] authors monitor information about the eyes and mouth of the driver. Then, this information is transmitted to the Fuzzy Expert System, which classifies the true state of the driver. The system has been tested using real data from different sequences recorded during the day and at night for users belonging to different races and genders. The authors claim that their system gives an average recognition accuracy of fatigue close to 100% for the tested video sequences.

The above analysis shows that many of current works is focused on the problem of recognizing driver's fatigue. The results of this work are presented on their own video sequences or exploration is carried out in real conditions on the basis of installed cameras, the most common web cameras. Thus, the acquisition of data that will be used by different authors to compare the solutions seems to be an important research issue.

3 Selected Multispectral Imaging

The human eye is not a perfect sensor and is limited to the detection of electromagnetic waves from about 380–780 nm of wavelengths. To imitate the human vision, the technical sensors are combined with UV and infrared filters which limits the recorded frequency range. Interestingly, most common digital sensors (CMOS and CCD) can only measure the intensity of the light wave and the colour information is obtained through the use of colour RGB filters arranged in a mosaic array (e.g. Bayer filter).

Going beyond the visible spectrum offers a new perspective on many problems and imaging technologies like X-ray, infrared, millimeter or submillimeter wave are examples here. Despite, however, overcoming some of the limitations of visible imaging, not all imaging technologies beyond the visible spectrum are suited as a companion to man on a daily basis. It would be incomprehensible to measure the driver's fatigue with the use of X-ray. For that reason, in the following sections a review on perspective imaging technologies beyond the visible spectrum is provided.

3.1 Visible Lighting Image

Traditional imaging technique, namely capturing image in the visible lighting is the most straightforward method of visual data acquisition. It is easy to introduce to the problem of driver's condition estimation. Required hardware is not expensive and its operational parameters can be very high, in terms of spatial resolution, dynamic range and sensitivity. On the other hand, it should be remembered, that such devices can work only in good lighting conditions, namely during day. It would be impossible to light driver's face are during driving with any sort of additional light source, since it could disturb his/her functioning. Therefore, it is reasonable to equip the system with other capturing devices, working in different lighting spectra.

3.2 Depth Map

A depth map is an image which pixels represent the distance information from the scene objects to the camera. This representation has a number of uses and among others is employed in advanced driver assistance systems or robot navigation. Depth information can be obtained applying the following techniques:

- stereo vision,
- structured-light,
- time-of-flight.

Stereo vision approach requires a pair of calibrated cameras or a single stereoscopic camera. The current price level of the single housing solution can be very low. The proprietary Minoru 3D camera is a good example here (http://www.minoru3d.com/). This is a low-end consumer 3D-webcam, costing about 50$, offering two VGA images (resolution of 640×480 pixels). From these images (called a stereo pair) the 3-D information of a scene is extracted. The corresponding points in the stereo pair are matched and based on mutual displacement relative depth is calculated and final disparity map is generated. Figure 2 presents an example of depth image calculation based on the stereo pair. The result is depicted as a typical 2-D disparity map and as a 3-D projection.

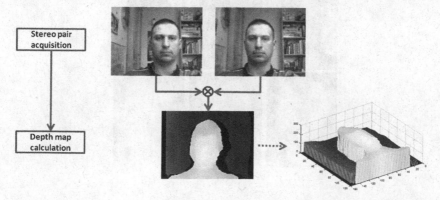

Fig. 2. An example of a depth image calculation based on the stereo pair [own study].

Another concept form the structured light approach. Here, the invisible (for the human eye, i.e. infrared, namely Near-IR) light pattern is projected on the scene. This pattern is distorted on 3-D objects present in the scene. The distortion is registered by the camera, sensitive to the invisible light spectrum, and based on geometric reconstruction the 3-D information about the observed scene is obtained.

The newest of the three approaches of depth information acquisition is the time-of-flight which measures the time required for a light signal to travel from the source to the scene and back to the sensor. The measurements are carried out for each point of the image producing the final depth image. Both concepts, the structured light approach and the-time-of-flight approach, are actually present in the inexpensive consumer electronic. The proprietary Kinect sensor from Microsoft is the widely known example with the version 1 operating on the structured light and the version 2 operating on the time-of-flight approach. The cost of the newest sensor in the official distribution channel is about 100$.

In our study for a simulator station the Intel Real Sense RS300 camera has been employed. Its base parameters are 720p on 60 fps and 1080p on 30 fps. Intel recommends that the camera was controlled under MS Windows 10 and connected to a PC with a processor Intel 6th generation.

3.3   Thermal Image

Data from infrared sensors allows face detection algorithms to disregard the necessity of visible light properly illuminating the subject [26]. In most cases, thermal imaging is performed in the Long-Wavelength Infrared range. Several such problems are presented in Fig. 3. As it can be seen, thermal imaging does not depend on the lighting conditions, which is important fact when dealing with an observation of a driver in uncontrolled environment. In such case, car's interior is often illuminated by directional lighting, coming from the windows (sun, street and other cars' lamps).

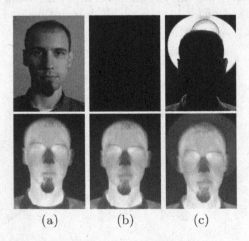

(a)   (b)   (c)

Fig. 3. Examples of problems resolved by using thermal imaging in comparison to the visible lighting imaging: (a) normal conditions, (b) insufficient lighting, (c) backlighting [26]

Fig. 4. Examples of different thermographic imaging effects depending on the temperature range applied: 25.4–35.7 °C, 25.4–36.6 °C, and 26–36.9 °C, respectively [27]

The main problem with thermal imagery is the calibration of the temperature range captured by the camera. The same object, depending on the environment temperature, can be visualized with different appearance. Such problems are presented in Fig. 4. It is a very important issue, especially in case of the problem presented in this paper. Such appearance variations negatively influence the detection of specific face landmarks (in the areas of eyes, mouth, and nose) and pose evaluation. In order to resolve it, a proper calibration should be performed, in order to provide stable imaging conditions.

4 Experimental Setup

In this section there is presented the concept of the position in relation to the actual distance determined by measurements of several cars.

4.1 The Concept of Position to Acquire Video Sequences

The position to be acquiring video sequences showing the driver's fatigue includes the following elements:

- The IR camera model FLIR SC325 with 16-bit matrix 320 × 240 px (no. 1 in Fig. 5),
- The SONY HDR CX-550 video camera with 12 MPix (no. 2 in Fig. 5),
- The Intel SR300 camera (working in visible lighting and infrared) mounted on the steering wheel (no. 3 in Fig. 5),
- Four fluorescent lamps CW 25WT mounted on a stand W806 Light Stand, height 118 cm,
- Five monitors imitating the vehicle windows,
- Three desktops (PC),
- Car rearview mirror,
- Steering wheel,
- Four desks (three desks were used to place monitors imitating the windows of the vehicle; the fourth desk was a center position),
- Chair with a backrest back, imitating the driver's seat,
- Background in the form of a green screen with a size of 275 cm × 200 cm, covering heating pipes on the wall and allow unhindered acquisition of an image of the IR camera.

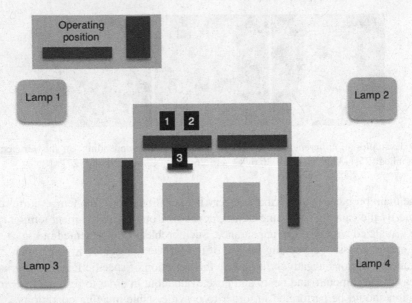

Fig. 5. Explanatory figure position to acquire video sequences (squares marked with numbers 1, 2, 3 indicate the deployment of cameras) [own study]

Table 1. The objects distance [cm] from driver's head in the exemplary car [own study]

Objects in respect of which the distance was measured	VW caddy	VW golf plus	Peugeot 206	The average distance
Head - steering wheel	53.4	48.5	50.8	50.9
Head - mirror	60.5	68.2	56.0	61.6
Head - windshield	92.6	90.5	82.4	88.5
Head - left window	41.5	41.5	48.4	43.8
Head - right window	117.3	119.0	110.0	115.4

The position shown in Fig. 5 reflects the dimensions obtained from the cars indicated in Table 1.

The position was supported by 3 PCs (Table 2). Two computers used to display images on monitors imitating the vehicle window. The third machine was designed to record images from the cameras used in the research position.

During the research experiment multiple video sequences were obtained for the following behaviors:

- Blinking eyes (opening and closing),
- Squinting eyes,
- Rubbing eyes,
- Yawning,

Table 2. Computers used in the research position [own study]

	Computers displaying images	Computer recording images
Processor	Intel® CoreTM2 Duo CPU E8400 @ 3.00 GHz (2 CPU's), 3.0 GHz,	Intel® CoreTM i5-6500 @ 3.60GHz (4 CPU's), 3.2 GHz,
RAM	4096 MB	8192 MB
Graphics Card	NVIDIA GeForce 8800GT	NVIDIA GeForce 760GTX
Operating System	Windows 10 Enterprise 64-bit (6.1, Build 7601)	Windows 10 Enterprise 64-bit (6.1, Build 7601)
Screen resolution	1920 × 1200 px	1920 × 1200 px
USB port	2.0	2.0
Quantity	2	1

- Lowering the head,
- Shaking the head.

They were selected as the most representative actions leading to the successful recognition of driver's fatigue level [5, 18–21].

5 Exemplary Results

In order to test the assumptions taken at the preliminary stage of our research, we captured several video sequences. They present simulated behaviour of various drivers. In each case, four cameras observed driver's head and a fragment of his torso in three spectra: visible (VIS), near-infrared (NIR) and long-wavelength infrared (LWIR). It led to the five video streams: two normal (visible) sequences, thermal sequence, point cloud and depth map. Four selected views are presented below. In the following figures one can find three typical situations: normal pose, with eyes directed to the road, rubbing eyes (associated with medium fatigue level) and lowering the head (associated with high fatigue level).

As it can be seen from Fig. 6 the neutral head position is visible in all views. The glasses can be clearly seen in the thermal representation.

The following figure (Fig. 7) shows an action, which can be associated with medium driver's fatigue level. Depicted person rubs his eyes. In each view, part of the face is covered with a hand.

The last of the figures (Fig. 8) shows an action, which can be associated with high driver's fatigue level. Depicted person drops his head and closes his eyes. In each view, the head pose is changed. In the distance maps (both cloud of points and depth map) upper part of the head are lost.

In all of the above-presented scenes, thanks to the thermal and distance imaging, the background can be ignored, which may improve the detection and recognition

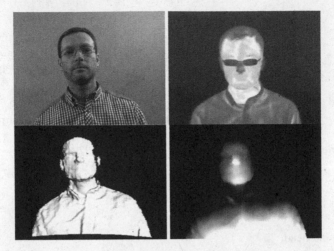

Fig. 6. Selected views of the drivers head and torso areas, visible, thermal (from LWIR), cloud of points and depth map (both from NIR).

Fig. 7. Selected views of the drivers head and torso areas, visible, thermal (from LWIR), cloud of points and depth map (both from NIR).

results. Moreover, thanks to the properties of infrared imaging (both NIR and LWIR), the lighting conditions do not negatively influence such procedures. Of course, when the lighting conditions are good, the detection algorithm, in the firs row, can be based on the visible light spectrum.

Fig. 8. Selected views of the driver's head and torso areas, visible, thermal (from LWIR), cloud of points and depth map (both from NIR).

6 Conclusion

The main purpose of the article was to present the achievements in terms of the creation of the position of the measurement to acquire video sequences showing the state of driver fatigue, weariness, sleepiness and yawning. The article presents the early work in this area and discussed various imaging techniques, depending on the ambient conditions.

As it was shown, proper selection of adequate imaging means can lead to the successful acquisition of visual data, which can later be employed in the algorithms of driver's fatigue estimation. In the paper, three selected imaging technologies were analysed together with the presentation of exemplary recordings.

References

1. Weller, G., Schlag, B.: Road user behavior model. Deliverable D8 project RIPCORD-ISERET, 6 Framework Programme of the European Union. http://ripcord.bast.de/ (2007)
2. Smolensky, M.H., et al.: Sleep disorders, medical conditions, and road accident risk. Accid. Anal. Prev. **43**(2), 533–548 (2011)
3. Virginia Tech Transportation Institute: Day or Night, Driving while Tired a Leading Cause of Accidents. http://www.vtnews.vt.edu/articles/2013/04/041513-vtti-fatigue.html. Accessed 12 Feb 2017
4. Krishnasree, V., Balaji, N., Rao, P.S.: A real time improved driver fatigue monitoring system. WSEAS Trans. Signal Process. **10**, 146–155 (2014)
5. Cyganek, B., Gruszczynski, S.: Hybrid computer vision system for drivers' eye recognition and fatigue monitoring. Neurocomputing **126**, 78–94 (2014)

6. Dinges, D.F., Powell, J.W.: Microcomputer analyses of performance on a portable, simple visual RT task during sustained operations. Behav. Res. Methods Instrum. Comput. **17**, 652–655 (1985)

7. Baulk, S.D., et al.: Chasing the silver bullet: measuring driver fatigue using simple and complex tasks. Accid. Anal. Prev. **40**(1), 396–402 (2008)

8. Kaida, K., et al.: Validation of the Karolinska sleepiness scale against performance and EEG variables. Clin. Neurophysiol. **117**(7), 1574–1581 (2006)

9. Egelund, N.: Spectral analysis of heart rate variability as an indicator of driver fatigue. Ergonomics **25**(7), 663–672 (1982)

10. Philip, P., et al.: Fatigue, sleep restriction and driving performance. Accid. Anal. Prev. **37**, 473–478 (2005)

11. Jagannath, M., Balasubramanian, V.: Assessment of early onset of driver fatigue using multimodal fatigue measures in a static simulator. Appl. Ergon. **45**(4), 1140–1147 (2014)

12. McCall, J.C., Trivedi, M.M.: Video-based lane estimation and tracking for driver assistance: survey, system, and evaluation. IEEE Trans. Intell. Transp. Syst. **7**(1), 20–37 (2014)

13. Makowiec-Dąbrowska, T., et al.: The work fatigue for drivers of city buses. Medycyna Pracy **66**(5), 661–677 (2015). (in Polish)

14. Mitas, A. et al.: Registration and evaluation of biometric parameters of the driver to improve road safety. Scientific Papers of Transport, Silesian University of Technology, pp. 71–79 (2010) (in Polish)

15. Viola, P., Jones, M.: Rapid object detection using a boosted cascade of simple features. In: Computer Vision and Pattern Recognition, pp. 511–518 (2001)

16. Viola, P., Jones, M.J.: Robust real-time face detection. Int. J. Comput. Vis. **57**(2), 137–154 (2004)

17. Nowosielski, A.: Vision-based solutions for driver assistance. J. Theor. Appl. Comput. Sci. **8**(4), 35–44 (2014)

18. Craye, C., et al.: A multi-modal driver fatigue and distraction assessment system. Int. J. Intell. Transp. Syst. Res. **14**(3), 173–194 (2016)

19. Kong, W., et al.: A system of driving fatigue detection based on machine vision and its application on smart device. J. Sens. **2015**, 11 p. (2015). Article ID 548602. doi:10.1155/2015/548602

20. Jo, J., et al.: Detecting driver drowsiness using feature-level fusion and user-specific classification. Exp. Syst. Appl. **41**(4), 1139–1152 (2014)

21. Zhang, Y., Hua, C.: Driver fatigue recognition based on facial expression analysis using local binary patterns. Optik Int. J. Light Electron Opt. **126**(23), 4501–4505 (2015)

22. Alioua, N., Amine, A., Rziza, M.: Driver's fatigue detection based on yawning extraction. Int. J. Veh. Technol. **2014**, 7 p. (2014). Article ID 678786. doi:10.1155/2014/678786

23. Fu, R., Wang, H., Zhao, W.: Dynamic driver fatigue detection using hidden Markov model in real driving condition. Exp. Syst. Appl. **63**, 397–411 (2016)

24. Zheng, C., Xiaojuan, B., Yu, W.: Fatigue driving detection based on Haar feature and extreme learning machine. J. China Univ. Posts Telecommun. **23**(4), 91–100 (2016)

25. Azim, T., Jaffar, M.A., Mirza, A.M.: Fully automated real time fatigue detection of drivers through fuzzy expert systems. Appl. Soft Comput. **18**, 25–38 (2014)

26. Jasiński, P., Forczmański, P.: Combined imaging system for taking facial portraits in visible and thermal spectra. In: Proceedings of the International Conference on Image Processing and Communications - IP&C2015, Image Processing and Communications Challenges 7. Advances in Intelligent Systems and Computing, vol. 389, pp. 63–71 (2016)

27. Hermans-Killam, L.: Cool Cosmos/IPAC website. Infrared Processing and Analysis Center. http://coolcosmos.ipac.caltech.edu/image_galleries/ir_portraits.html. Accessed 10 May 2016

Analysis of Expanded Possibilities of ITS Systems Augmented with New Vision System Elements: The Case of Lodz

Remigiusz Kozlowski[✉], Anna Palczewska, and Lukasz Borowiecki

University of Lodz, ul. Matejki 22/26, 90-237 Lodz, Poland
rjk5511@gmail.com, anna.m.palczewska@gmail.com,
lsb.lsb.017@gmail.com

Abstract. Vision systems are effective tools for optimization and control of numerous processes. They serve as a platform for the development of autonomic, smart solutions both for vehicles and intelligent transportation systems (ITS). The data acquired by the former are critical to the functioning of the latter. Vision systems enhance the capacity of ITS to effectively manage city logistics. This paper aims to identify new possibilities offered by the addition of more cameras to an existing ITS based on analysis of the largest ITS in Poland, which was deployed in Łódź, the country's third most populous city, in December 2015.

Keywords: Intelligent transportation systems · ITS · Use of cameras · Vision systems

1 Introduction

The functionalities of intelligent transportation systems (ITS) have been constantly expanding due to increasing data processing potential at the control centers and server facilities as well as due to improved capacity of the systems supplying external data, such as devices installed directly in the roads (inductive loops) or on elements of vertical traffic infrastructure, such as light posts (cameras and radars). These two areas are crucial both to the potential offered by ITS solutions and their quality and reliability. Data acquisition and processing constitute critical elements which should be carefully incorporated in the process of ITS design and expansion.

In addition to the development of individual ITS, one should also take into consideration the need to integrate them with one another. The questions of how this should be done and what development and integration strategies should be adopted remain open [1, 2], but it seems that such processes are inevitable in the near future. ITS solutions are elements of transport infrastructure, being of great significance to economic development [3].

The city of Lodz has the largest and most advanced ITS in Poland, which has been expanded by the addition of new cameras. The objective of this paper is to identify the resulting new potential offered by this system.

© Springer International Publishing AG 2017
J. Mikulski (Ed.): TST 2017, CCIS 715, pp. 333–343, 2017.
DOI: 10.1007/978-3-319-66251-0_27

2 Sources of Data Used by ITS

Intelligent transportation systems (ITS) comprise of a variety of tools based on information technology and wireless communication [4], which enable efficient traffic services. Wireless networks can be divided into three broad categories: point-to-point networks, terrestrial broadcasting networks, and satellite systems [5]. The term "intelligent transportation systems" was adopted at the first annual World Congress of International Association of Financial Executives Institutes (IAFEI) in Paris in 1994. ITS include telecommunications, information, automatic, and measuring systems, as well as management technologies used in transport to ensure safety to road users, increase the efficiency of transport systems, and protect natural resources. ITS are primarily used to manage [6]:

- road traffic,
- public transport,
- freight transport and fleets of vehicles,
- accident detection and emergency services,
- traffic safety and monitoring of road offenses.

Due to their construction (architecture) ITS may also offer [7]:

- IT systems for road users,
- electronic payments and toll charging,
- advanced vehicle technologies.

To function properly, ITS should be equipped with suitable infrastructure. An example of architecture of the data collection subsystem for ITS is shown below (Fig. 1).

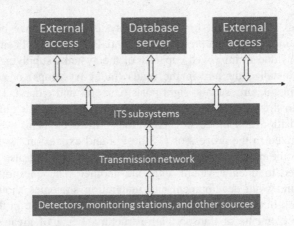

Fig. 1. Architecture of the data collection subsystem for ITS [8]

For ITS to exchange and update data, they must be connected to data storage and transmission devices, which enable:

- organization of data acquisition and exchange,
- management of data collected and generated by the system,

- organization of data collection (tables, charts, statistics),
- data storage for subsequent analysis,
- client–server technologies.

The application of ITS may generate numerous benefits, such as [9]:

- throughput of the road network increased by 20–25%,
- improved road safety and number of accidents decreased by 40–80%,
- travel time and energy consumption reduced by 45–70%,
- quality of the natural environment improved due to reducing exhaust emissions by 30–50%,
- enhanced travel comfort and better traffic conditions for drivers, public transport users, and pedestrians,
- reduced costs of management of rolling stock,
- reduced costs of road maintenance and repairs,
- enhanced economic performance of the region.

As can be seen, the opportunities offered by ITS go well beyond mere traffic management. Increased throughput of the road infrastructure is not the only objective that can be achieved by means of ITS (Table 1).

Table 1. Effects of the application of ITS [10]

Effects of ITS	Type of ITS	Effect size – up to
Increased throughput of road network	Traffic management systems for motorways	25%
	Directing traffic to alternative routes by using variable message signs	22%
	Electronic toll charging	200–300%
Time saving on the road network	Application of traffic lights	48%
	Traffic management at motorway entry points	48%
	Traffic accident management systems	45%
	Electronic toll charging	71%
	Traffic lights prioritizing public transport vehicle (saves time and increases punctuality by up to 59%)	54%
Increased road safety (fewer accidents)	Speed cameras	80%
	Traffic management at motorway entry points	50%
	Advanced traffic management systems	80%
	Road accident management systems	50%
Improved effectiveness of emergency services	Application of accident and emergency services management systems leads to:	
	(a) Shorter accident detection time	66%
	(b) Prompter arrival of emergency services	43%
	Automatic location systems for emergency vehicles and navigation to accident sites, reducing arrival time	40%
Environmental effects	Demand management systems – reduced exhaust emissions	50%
	Traffic management of motorways – reduced fuel use	42%
	Urban traffic management systems – reduced exhaust emissions	30%

The results given in the above table are very ambitious and to achieve them it is necessary to select the right ITS for the needs of a given city and to skillfully manage that system. Indispensable elements of each ITS are tools for collecting external data, such as, sensors, and cameras.

Cameras integrated with an ITS may be useful for all road users, including drivers, cyclists, and pedestrians. For instance, ITS can assess the length of a line of vehicles, its composition, and speed [11].

To date, most vision monitoring systems have been developed as self-contained entities, not connected to other systems. In addition, some of them utilize expensive operational systems due to their popularity and technical support options [12]. Such solutions pose obstacles to the use of vision systems to increase safety, not to mention enabling traffic management, assistance for emergency services, or contribution to environmental protection.

Vision monitoring centers have access to a large number of cameras distributed over different parts of cities and are often equipped with recording devices to more effectively detect events in the field of vision. Moreover, despite the application of many state-of-the-art technical solutions and increasing the number of operators, a large number of events may still escape human attention. This can be avoided by recording and post-processing video material to be analyzed at a later time [13].

3 Deploying Vision Systems as Part of ITS

Technological advances in vision systems, reflected in improved camera parameters, as well as the unification and standardization of communication and data transmission signals, have afforded new application possibilities. Currently, such systems may serve as sources of input data for ITS.

Based on real-time external signals, an ITS processes data and may take effective actions appropriate for a given situation. The subsystem of cameras being part of an ITS should enable at least [14]:

- monitoring intersections or road segments,
- real-time access to images, and also access to images stored at the traffic management center by all system operators,
- video material recording and storage on servers.

In the case of the Lodz ITS, it was assumed that the system must be equipped with an intelligent video analysis module for each camera to ensure the highest levels of detection effectiveness in the monitored area. The module has the following functionalities [14]:

- movement detection,
- changing the background of the field of vision,
- information on loss of quality,
- counting objects in the field of vision,
- identification of unattended objects left in a predefined area,
- detection of crossing virtual lines with identification of movement direction,

- detection of an object stopping in a predefined area,
- detection of an object loitering in a predefined area,
- detection of an object entering or exiting,
- automatic number plate recognition.

ITS can automatically react to the events identified by video cameras, for instance by sending text messages, emails, sounding and alarm, sending a signal to devices coupled to the camera, such as a boom gate or door, or by directing the camera lens at the point of interest.

4 Characteristics of the Lodz ITS

On December 1, 2015, Lodz launched an ITS called the Area Traffic Management System, equipped with cameras with the parameters specified in the table below. The ITS architecture consists of a network of logical, physical, and communications connections between ITS elements to create a scalable solution which is easy to maintain and manage [15].

The cost investment till December 2015 was about 80 mil PLN. It took 16 months, since the first phase: planning, through project making, ending with the ITS system launch in city. Required technical documentation was prepared for ITS. Intersections were equipped with special communication devices, which connected them to the system, and optical network data transmission and Traffic Management Centre were constructed. This project was around public transport communication. Second thing important was improvement of the road traffic. The priority to public transport modes move effectiveness ahead. In conclusion travel time were shortened and the comfort and safety is enhanced. System of Area Traffic Management is a main component of Lodz ITS using shortened polish name SOZR. It is based on two GIS applications [18] (Table 2).

Table 2. Components of Lodz ITS [18]

ANPR – traffic intensity	Vehicle localization subsystem
SCATS – system SOSR (Area Traffic Control System)	Public transportation subsystem
CCTV	Tunnel subsystem
VMS – driver information	Parking subsystems
Passenger information board subsystem	Street lighting subsystem
Chosen eDIOM layers	Winter maintenance subsystem

VMS tables of Road Traffic information for Drivers gives information on boards that are standing (Fig. 2) in a different locations in the city. They present information about real time of traffic between them around the city by using appropriate roads (Fig. 3 and Table 3).

Data transmission and processing in the system is enabled by a telecommunications network developed for the needs of the ITS with the objective of connecting the different parts of the system [14].

Fig. 2. Map of VMS tables localisation around the city [18]

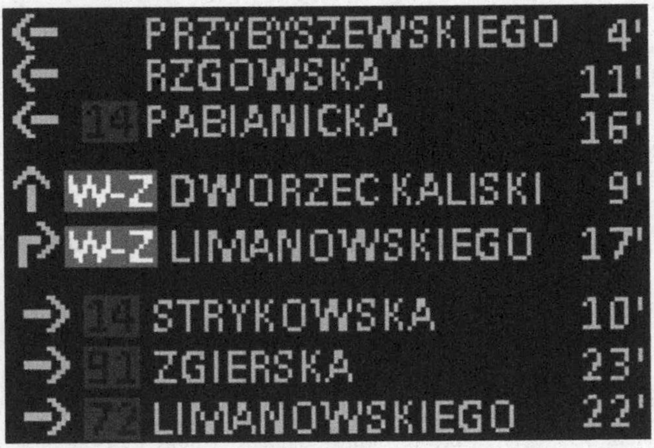

```
←   PRZYBYSZEWSKIEGO    4'
←   RZGOWSKA           11'
←  14 PABIANICKA        16'

↑ W-Z DWORZEC KALISKI    9'
↱ W-Z LIMANOWSKIEGO     17'

→ 14 STRYKOWSKA         10'
→ 91 ZGIERSKA           23'
→ 72 LIMANOWSKIEGO      22'
```

Fig. 3. VMS table of Road Traffic information for Drivers [16]

The monitoring software enables continuous recording of video material from the cameras at the maximum resolution available for those devices, while the operators enjoy uninterrupted access to real-time streams. The Lodz ITS is entirely based on IP technology and client–server architecture, which forms an open and extensive system providing access to video materials from different locations, also by means of mobile devices in the form of frames from cameras transmitted to the passenger information system at 5 fps. The ITS camera system is also equipped with the "magnifying glass" function whereby any fragment of the image can be enlarged on a predefined scale by indicating the area of interest with the mouse cursor [14].

Table 3. Parameters of cameras in the Lodz ITS [own study]

Stationary cameras	Rotating cameras
• Number of pixels: 1920 (H) × 1080 (V) • Image processing speed at least 30 fps at full resolution • Image compression, e.g. MPEG4 • Integrated lens with a focal length range of 3.6 mm to 16 mm with motozoom and autofocus functionalities • Support for multiple video streams • Automatic and manual aperture and shutter speed control • Automatic and manual daylight/nocturnal modes • Automatic and manual white balance adjustment • Audio input/output • Analog video output • Movement detection • Private zone configuration • Digital alarm input/output • More than one power supply option • Network interface standard: 100BASE-TX • "Vandal-resistant" casing with protection class IP66 • Integrated infrared illuminator • SD or micro SD slot • Working temperature range: at least -40 to +50 °C • Compatibility with ONVIF standard	• Number of pixels: at least 1920 (H) × 1080 (V) • Image processing speed 30 fps at full resolution • Image compression: H.264 and MJPEG • Supports multiple video streams • Adjustable resolution • Daylight and nocturnal modes (ICR filter) • Broad a dynamic range: at least 100 dB • Integrated lens with a focal length range of at least f=4.7–94 mm, • Motozoom and autofocus functions • Automatic and manual aperture and shutter speed control • Automatic and manual white balance adjustment • Movement detection • Network interface standard: 100BASE-TX • Audio input/output • Alarm input/output • Private zone configuration, support for 3D masking • At least 100 presets and 10 patrol tracks can be defined • Horizontal rotation angle: 360° • Vertical rotation angle: 186° • Rotation rate: 0.05°–360°/s • Automatic image flipping (E-flip) • Power supply 24 V AC or 24 V DC or Hi PoE • Support for ONVIF standard • Working temperature range: at least −25 °C to +50 °C

One can review all the events recorded in the system, replay the stored files, as well as delete and exports video files. It is also possible to save screenshots to the desktop [14]. It should also be noted that, according to the initial requirements, the video material from cameras in Lodz is to be stored for 14 days [14].

The camera system used by the Lodz ITS was designed to enable cooperation with external systems by access control, alarm systems, fire protection systems, as well as flexible and scalable architecture for fast integration of new security and safety devices technologies.

Thanks to its open nature, the ITS is compatible with a wide range of commercially available solutions for image analysis based on access to images from many cameras as well as export of video material correlated with search results. Thus, the camera system may be readily expanded, and it would also be possible to create several independent systems based on the Lodz ITS infrastructure. This could be accomplished by using existing servers or adding new ones, acquiring new software licenses, creating new viewing stations and camera points, or perhaps combining them into large systems integrated as part of one overarching monitoring system.

5 The Expanded Potential of the ITS in Lodz Following the Installation of Additional Cameras

The objective of the camera subsystem in the Lodz ITS, which is an open vision monitoring system, is to enable fast adaptation to new requirements, which are certain to emerge in the future. It can also be easily expanded by adding more cameras, thus extending the area covered by the system. In turn, automatic number plate recognition (ANPR) cameras may improve the accuracy of the tasks performed by the system and its operators with the aim of optimizing traffic in the city [14].

Systems of cameras coupled to ITS have the following automated, algorithm-based functionalities [14]:

- intruder detection,
- traffic direction control,
- detection of crossing virtual lines,
- detection of camera sabotage and altered field of vision,
- crowd detection, detection of unattended items,
- queue length detection,
- detection of loitering,
- detection of objects stopping,
- counting of objects,
- tracking of object movement,
- calculation of object movement speed.

Upon the implementation of the ITS, the monitoring system was equipped with fast-rotating and fixed focal length dual-stream CCTV cameras. The cameras had to combine all the functionalities needed for ANPR in one device in order to enable successful number plate recognition. Within the basic functionality of the ITS, ANPR make it possible to implement average speed measurements over road segments designated by the municipal transport system. Such solutions are implemented using special algorithms for number plate recognition from recorded images. Average speed measurement has been developed in such a way as to acquire, store, process, and distribute data within system solutions coupled to information for drivers. The system displays information about traffic speed on variable message signs (VMS). These data are transmitted to the users through the traffic Management System using not only VMS, but also a mobile information system for drivers [14].

At some intersections, it would be necessary to automatically detect road accidents and monitor running a red light. While the system is not equipped with that functionality yet, it is technically feasible. The detection of other traffic events is enabled by the implementation of an ANPR system and other available image analysis systems.

ANPR cameras transmit the following information:

- location,
- date and time of vehicle identification,
- license plate number,
- country identifier of the vehicle,
- black-and-white image of the license plate,
- data on reading reliability.

ANPR devices may operate during the day and at night, under poor lighting conditions, and even in complete darkness. They may gather data from all lanes of a road. License plate recognition can be done for vehicles moving at speeds of up to 150 km/h. The reliability of license plate recognition and identification to is at least 95% accurate. ANPR devices coupled to IP cameras may not only acquire images of license plates, but also of entire vehicles (Fig. 4).

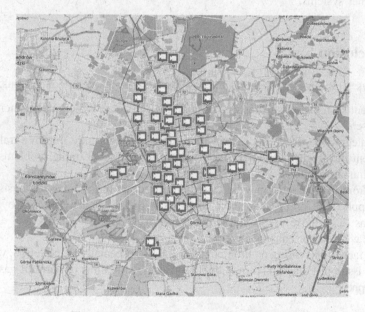

Fig. 4. Map of CCTV cameras in Lodz [16]

Images from CCTV cameras are viewed in the Traffic Management Center, and conveyed to websites concerned with passenger information.

In 2016, another camera-based city monitoring system in Lodz was expanded with a system for identification of hazardous situations involving human gatherings identified through image analysis. The system covers, amongst others, the main pedestrian area in the city (Piotrkowska Street) as well as other places in Lodz attracting large numbers of visitors [17]. If the system detects a large group of people, whether during the day or by night, it alerts the operator by displaying a message on the monitor. This information may be conveyed to the emergency services, such as police, municipal police, and medical services, which may be deployed to the site if necessary. Integration of ITS with the city monitoring system would also help to manage traffic in the city.

ITS may also be used to integrate all new vision monitoring systems in the city into one management system, enabling the connection of many locations from many owners; by using a common interface those owners can share cameras and other detectors available in the system.

Extension of a monitoring system through ITS enables cooperation with the following independent vision monitoring systems:

- the police,
- municipal police,
- managers of city, regional, or even national systems,
- systems and sports facilities,
- systems in train stations,
- and other systems.

6 Conclusion

The analyses presented in this chapter have led to the following conclusions:

- Different systems of cameras can be combined to create synergies for their owners and users.
- The integration of ITS and urban monitoring systems may improve traffic management in the city, for example by identifying potential traffic problems such as large human gatherings.
- Data from ITS cameras may be stored and used for future image analysis.
- The application of ITS and their coupling with the capacity offered by external systems (e.g., those used by emergency services, sports stadiums, and exhibition halls), will lead to enhanced safety.
- In the face of increasing environmental risks, and especially air pollution, which is a major issue in Poland, ITS may help reduce the emissions of harmful gases to the atmosphere.

Acknowledgement. Authors would like to thank to Lodz Board of Roads and Transport and special for Traffic Management Centre of City of Lodz for help and support.

References

1. Iwan, S., Małecki, K.: Analiza wybranych krajowych architektur ITS i wskazania dla architektury dla Polski. Drogi lądowe, powietrzne i wodne **5**, 33–43 (2011)
2. Marczak, M., Kozłowski, R.: Budowa inteligentnych systemów transportowych jako szansa dla zrównoważonego rozwoju regionów. Ekonomia i Zarządzanie **6**(2), 34–42 (2014). Białystok
3. Kozłowski, R.: Wpływ infrastruktury transportu na rozwój klastrów na przykładzie regionu łódzkiego w. In: Sosnowski, J. (ed.) Klastry logistyczne na tle procesów rozwoju regionu, pp. 148–159. Wydawnictwo Uniwersytetu Łódzkiego, Łódź (2011)
4. ITS Polska. Inteligentne Systemy Transportowe. http://www.itspolska.pl/?page=11. Accessed 18 Jan 2017
5. Kabaciński, W., Żal, M.: Sieci telekomunikacyjne. Wydawnictwa Komunikacji i Łączności, Warszawa 2008, p. 231 (2012)
6. ITS Polska, Inteligentne Systemy Transportowe. http://www.itspolska.pl/?page=11. Accessed 18 Jan 2017
7. Kożlak, A.: Inteligentne systemy transportowe jako instrument poprawy efektywności transportu, pp. 1–2. Uniwersytet Gdański Wydział Ekonomiczny, Sopot (2008)
8. Łakota, K.: Źródła i bazy danych związane z nadzorem w its systemach. Вісник Національного транспортного університету **20**, 156 (2010)
9. Litwin, M., Oskarbski, J., Jamroz, K.: Inteligentne Systemy Transportu – Zaawansowane Systemy Zarządzania Ruchem za. http://www.itspolska.pl/?page=11. Accessed 18 Jan 2017
10. Oskarbski, J., Jamroz, K.: Zarządzanie bezpieczeństwem ruchu drogowego w systemie Tristar. Konferencja Gambit 2006, Gdańsk maj 2006. In: Kożlak, A. (ed.) Inteligentne systemy transportowe jako instrument poprawy efektywności transportu, p. 4. Uniwersytet Gdański, Wydział Ekonomiczny, Sopot (2006)
11. Gadawa, M.: Kamery ITS nie nagrywają obrazu z ulic? Nie wszystkie. http://www.gazetawroclawska.pl/artykul/3897623,wroclaw-kamery-its-nie-nagrywaja-obrazu-z-ulic-nie-wszystkie-sprawdz-lista,1,1,1,id,t,so,nk,sa.html. Accessed 21 Apr 2017
12. Modelewski, K.: Zoneminder – darmowy system monitoringu w ITS. http://przeglad-its.pl/2011/03/11/zoneminder-darmowy-system-monitoringu-w-its/. Accessed 21 Apr 2017
13. Sumiła, M., Kasprzyk, Z.: Koncepcja wykorzystania inteligentnych kamer IP do wspomagania nadzoru wizyjnego ITS. Logistyka 4 (2012)
14. Materiały Centrum Sterowania Ruchem Zarządu Dróg i Transportu w Łodzi
15. http://www.itspolska.pl/?page=11. Accessed 21 Apr 2017
16. Portal Informacji pasażerskiej. www.its.lodz.pl. Accessed 23 Apr 2017
17. Urazińska, A.: Już 185 kamer monitoruje Łódź. Będzie jeszcze więcej. http://lodz.wyborcza.pl/lodz/1,35153,19713763,juz-185-kamer-monitoruje-lodz-bedzie-jeszcze-wiecej.html?disableRedirects=true. Accessed 26 Feb 2017
18. Kozlowski, R., Palczewska, A., Jablonski, J.: The scope and capabilities of ITS – the case of Lodz. In: Mikulski, J. (ed.) Challenge of Transport Telematics. CCIS 640. Springer, Heidelberg (2016). doi:10.1007/978-3-319-49646-7_26

Intelligent Container in Water – Land Transport. MBSE Approach for System Design

Wojciech Ślączka[1], Krzysztof Pietrusewicz[2(✉)], and Marcin Marcinek[2]

[1] Faculty of Navigation, Maritime Risk Analysis Center, Maritime University of Szczecin, Waly Chrobrego 1-2, 70-500 Szczecin, Poland
w.slaczka@am.szczecin.pl
[2] Faculty of Electrical Engineering, West Pomeranian University of Technology, Sikorskiego 37, 70-313 Szczecin, Poland
{krzysztof.pietrusewicz,marcin.marcinek}@zut.edu.pl

Abstract. The article described the concept of an intelligent system of spreading and scanning information about shipping cargo in containers. The system (in accordance with assumption of Industry 4.0) described in the research contains four system modules enabling to support monitoring and crisis management during the transport of dangerous materials: 1. Monitoring system of the containers' movement classified as shipping dangerous materials with the following subsystems: (a) monitoring subsystem of the containers' movement in inland shipping; (b) monitoring subsystem of the containers' movement on roads in urbanized city network; (c) monitoring subsystem in the railway; (d) monitoring subsystem of the containers' movement in the docks; (e) data acquisition subsystem concerning containers in maritime transport. 2. Monitoring/scanning system of containers' content. 3. Integration/harmonization/transfer system of data among IMDG, ADR, ADN systems. 4. Spreading of encoded information system about shipping load in the container. 5. Interface for services: the police, the fire brigade, the emergency medical service in the range of equipment and programming.

Keywords: Telemetry · Internet of things · Industry 4.0 · Intelligent broad-cast · Dangerous goods · Intelligent monitoring

1 Introduction

Transportation of dangerous goods over the land and water needs a special supervision. Such supervision must be conducted according to proper regulations, different in land and on the water. Such transportation is usually conducted with the use of special packaging systems matched to chemical and physical properties of the goods to be transported. Then such packages are transported among the others with the use of containers.

© Springer International Publishing AG 2017
J. Mikulski (Ed.): TST 2017, CCIS 715, pp. 344–359, 2017.
DOI: 10.1007/978-3-319-66251-0_28

1.1   Transportation of Dangerous Goods

Dangerous goods are classified according to their influence on the natural environment of physical and chemical properties. Example division in sea transport is introduced by the International Maritime Organization (IMO).

Main hazards resulting from dangerous goods transport are listed as follows [1]:

- fire, which could lead to uncontrolled response of the material,
- penetration of water, which could cause uncontrolled reaction materials with water,
- reaction of hazardous materials between themselves,
- uncontrolled leak causing risks arising from the nature of the material.

For the transportation of these dangerous materials different material codes are used for identification during transportation on land and water. Problem of monitoring and ICT integration of information with crisis services is such complex that to solve it efficiently, Model-Based Systems Engineering should be introduced, as proposed in this paper.

1.2   Model-Based Systems Engineering

Nowadays modeling is a very important phase in complex mechatronic control system design process. It requires complete change of approach to research, development and the way for new product introduction to the market.

For many years a common linear approach called *waterfall* was used for conducting research and development mechatronic projects. It was proven that such a scheme (Fig. 1) is effective, but only for low risk well defined projects. In such cases life cycle of the prototype is quite linear. There is no need for process robustification and the number of decision loops is low as well.

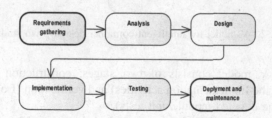

Fig. 1. Waterfall linear way of doing projects [own study]

In *waterfall* approach construction design (after its verification and validation, usually on a construction prototype) is an initial step for development of measurements and data acquisition subsystems, electronic and power electronic devices integration. From requirements, through analysis, system design and implementation, at the very last stage of the project testing is conducted. After successful testing the system is deployed to the final product and maintained as planned within project.

From the practical point of view it seems that prototype construction is finalized as a separate product and then control system as a completely new one. *Waterfall* approach is not optimal (effective) for development of complex, multi-variant mechatronic systems.

Work [2] proposes significant extension of the cascaded model (*waterfall*, Fig. 1). Completely new approach, shown in Fig. 2 (very innovative in 2004 [2], being one of the pillars for Industry 4.0 approach in 2017 [3]) introduced new way for development of complex mechatronic systems.

V-model from Fig. 2 ensures effective validation for research hypotheses at each stage of the project, from business model (usually defined and documented according to [4]), through requirements, system design including its use cases, up to implementation and coding. V-model approach describes general methodology for design of mechatronics systems. Nowadays V-model is used within procedures for ensuring functional and operational safety of machines [5–9]. Good practices in automotive [10] are based on the V-model concept as well.

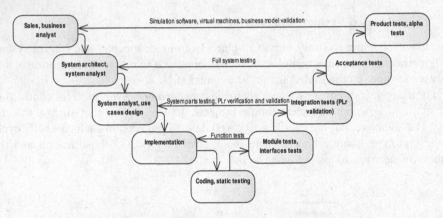

Fig. 2. V-model for intelligent container design [own study]

The so-called V model consists of two stages: construction (from the Sales requirements up to the Implementation activities) and verification (from the upper level alpha tests up to the detailed module/unit tests).

On each stage of the product development there is a closed loop created with the selected activities, which is essentially the path for communicating the expectations to the developers of the system.

The achievement of the required functional safety Performance Level (PLr) can be verified as a part of the system/integration test activities. The attempt to this without the defined use cases and the discussed/accepted system architecture can be difficult or even impossible.

Within the scope of this paper design assumptions for intelligent container in water-land transport are presented. They were prepared according to design methodology for mechatronic systems [2], with the use of dedicated software tools [11].

1.3   Intelligent Container Concept

IntelCon technology will enable continuous monitoring of the container transporting dangerous goods. It will provide in the form of encoded broadcast legal and operational information on its cargo.

Among the others, IntelCon will improve loading and unloading processes, especially scheduling of the containers that cannot be transported in close distance to each other. IntelCon technology will shorten heavily operational times on a container yards. To achieve these targets we propose modeling of the intelligent container system with IntelCon technology that consists of the functional modules that support monitoring and crisis management during transportation of dangerous goods with the use of containers.

2   Proposed Intelligent Container Solution

General concept of intelligent self-broadcasting and intelligent container is presented in this section by the means of the functional description of its subsystems. Also some initial technology discussion is given as follows.

2.1   Functional Description of System Modules

Proposed here solution of intelligent container is a functional combination of different subsystems, creating complex behaviour of mechatronic system. According to integration of IT, mechanical, electrical legal components it is classified as highly risk for implementation.

Intelligent container system consists of the following functionalities:

- Monitoring system of the containers' movement (inland, railway, in cities, in docks, with data acquisition),
- Scanning of container's content,
- Transform data between IMDG, ADR, ADN file formats,
- Coding/encryption/data-information conversion on dangerous goods for intelligent safe broad-cast purposes,
- Data exchange with crisis management services thanks to standardized multipurpose data interface.

2.2   More on Technologies

Assuming that proposed system will be implemented, proposal of a hardware setup is required. Authors propose and discus two hardware configurations. First proposal is organized as follows. System is consisted of two main layers: a mobile and a stationary, they are depicted in Fig. 3. Stationary layer could be considered as a cloud within all data processing, access rights verification and low level hardware drivers is done. There is not direct communication between public services (PS) and the cargo container (CC). With this approach all PSs have access to the last transmitted data from each cargo container.

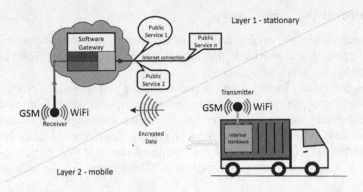

Fig. 3. The first configuration of the proposed system [own study]

Messages from gateway to the end-nodes (in our case public services) should be transmitted through the internet connection. Infrastructure used to build this layer is typical also to implement algorithms high level programing languages would be used, thus a detailed description is omitted.

Structure of second proposal is depicted in Fig. 4.

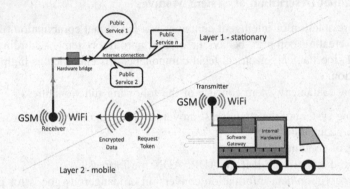

Fig. 4. Direct connection between the cargo container and end nodes (public services) [own study]

The gateway is located inside CC, and this is the main difference between these two topologies. Such approach implicates totally different communication scheme. In this scenario, a direct communication between PS and CC is obtained. Hence, each data exchange must be initialized by one of the end-nodes. One of the PS transmits "Request Token" and the specific CC respond sending encrypted data. In this topology, there is no need to maintain additional equipment or software (gateway) which could be considered as an advantage. However, while absence of the communication, none of actual or historical data are present about CC. There is no such possibility, that some of data – the most critical – would be stored in the external cloud (which is in favor for the first scenario). Additional issue will occur when more than one CC will be present in the network, it forces to use additional algorithms for CC addressing and management.

In authors opinion, a standardized communication frame should be proposed. Such frame could be organized, as it is presented in Fig. 5. Due to possibility of lack of the communication infrastructure, proposed frame should has as "light" as it possible. To be able to transmit all necessary data to an adequate public services. In our proposal some of data are represented in form of the standardized codes. Here we propose to use the simplest way, which are the decimal numbers. For instance, there are three possible physical states of cargo: a solid, a liquid, a gas, thus we have proposed to assignee numbers to them: 1, 2, 3, respectively. In much the same approach would be used for others fields listed in Fig. 5, like the type of hazard or an explosive nature of cargo (the bit – possible two values True or False). Some of data will be presented in public form, like Sender Company or Receiver Company. Nature of the transmitted data is rather sensitive thus every single frame should be encrypted.

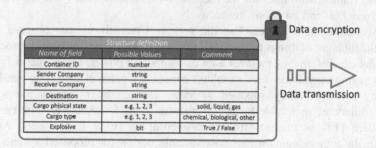

Fig. 5. Proposed frame structure used in our system (example) [own study]

Hardware used to construct both topologies would be similar. Therefore a detailed description of each part will be done for first configuration. The merits of this paper corresponds to the mobile layer. The hardware scheme of the possible topology is depicted in Fig. 6. It consists of three sub layers: a communication (sublayer 1), data processing and acquisition (sublayer 2) and the sensors (sublayer 3). Bellow a brief descriptions for each of them is presented.

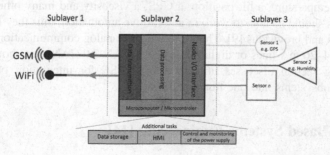

Fig. 6. Proposed scheme of the cargo container hardware – movable layer [own study]

Initially, the communication layer is described. Currently, we are facing some kind of race to deliver, fastest, robust and efficient communication technologies. Some of them are dedicated for commercial solutions other for industrial. As an example of commercial protocols: Wi-Fi, Bluetooth, GSM and the industrial Zigbee should be mentioned here. From mentioned communication schemes authors propose to use at least two different types, it should be Wi-Fi protocol and some type of the GSM (generation 3G or 4G - LTE). GSM communication should be used as a main, it allows to transmit data in the biggest area (range covers almost whole area of Poland) and it has the highest reliability. However, necessity of the additional costs related with data transitions are the biggest drawback. Thus in urbanized area, if it is possible, system should be switched to the open/public free Wi-Fi connection.

Communication speed for proposed aren't issue. 3G technology is able to transmit data with rate of 5 Mbps (Megabits per second) and 4G peak upload rate is equal to 500 Mbps. Transmission rates for the modern Wi–Fi standards are also high. N and G versions have 600 and 54 Mbps, respectively.

The second layer, is a "heart" of the system. Currently we can observe, a rapid grow of the inherent of things philosophy. Such systems allows to: collect, process and transmit data, therefore they requires a lot of computing power. Simultaneously energy efficiency is important issue, they shouldn't consume a much electrical power, because often they are the battery supplied systems. Hence, the microcomputers or the microcontrollers are preferable for use. For instance platforms like Raspberry PI [12], Intel Galileo [13] or microcontrollers delivered by Texas Instruments, Atmel or Microchip. In authors opinion the most suitable option for proposed system are the microcomputers. First of all, they are products that are common, their cost is not high and have a good technical support. Operation with usage of an operating system (e.g. Linux) is another advantage of this family. Secondly, they have a high rating of a computing power, described as MIPS (Million Instructions Per Second; for RPi it is 2500 MIPS) and relatively low electrical power consumption. For example at maximum load Raspberry PI consume less than 15 W. are supplied from low level voltage source (5–15 V) [12], thus no additional power inverters are necessary and system could be supplied from batteries.

Additionally proposed control units are able to communicate with the sensors layer. This layer is going to be used for monitoring the cargo. Many different sensors like: a humidity, a temperature, a tilt position, a GPS, a viscosity and many others could be used. Sensors should be able to work in automotive automation, thus they should fulfil proper norms and provisions [9]. Therefore the safe analog communications interfaces like: current 4–20 mA lines or digital buses like: CAN 2.0/CANopen should be used.

Following section discusses utilization of MBSE for integration of intelligent container's subsystems into one functional unit.

3   Model-Based Systems Engineering [14]

Effective work on intelligent container concept from Sect. 2 is possible and supported thanks to integration of software tools for modeling, conducting dynamic computer simulation and automatic code generation for different control platforms.

There are six areas of interest in case of systems modeling:

- Modeling requirements for system to be designed. It is a combination of selected business model (business or stakeholder needs), with legal requirements. Within the scope of the requirements model system engineers have to combine functional requirement with business target for solution [4];
- Modeling interfaces for information exchange between main system components and its environment (including cooperating systems). Main interface for described here intelligent container can be listed as follows: interface for power source (charging), interface for data exchange with crisis services and interfaces for IT systems with container route monitoring functionality over the land and water;
- Modeling software and hardware architecture of the selected solution. Thanks to architecture modeling it is clear and easy to show and explain its constraints resulting from selected business model and requirements (legal, functional);
- Modeling use cases of top-level and more detailed. Use cases define interaction between different users and system (intelligent container). Usually they are combined with activity based scenarios describing these interactions;
- Modeling architecture and specific properties of the controlled system. Such modeling enables parametrization of computer simulation during virtual verification and validation [14];
- Modeling architecture and software functionality of control system.

Model-Based Systems Engineering together with proper methodology can be treated as a multiphase complex modeling process supporting mechatronic systems integration. MBSE combines requirements, constraints and stakeholder expectations with proposed system solution.

The importance of modeling for integration of intelligent container subsystems will be explained in the next section of this paper.

3.1 Modeling Issues

The *model* is an abstraction of the system (sometimes including environment), that describes the system from the chosen perspective, with the selected details level. There is always notation connected with the model. It can be textual or graphical, depending on the tools used for modeling.

When modeling systems, one need to decide in the beginning, if it is necessary from the business target point of view to build a model or not. There is a strong need for model creation, when:

- the original system is too huge (for doing real prototypes),
- the original system is too complex (i.e. have many complex subsystems),
- the original system combines many domains (electrical, mechanical, hydraulic, IT),
- the original system is too miniature (for doing high expensive real prototype),
- the experiments on the original system are costly, by the means of time and money,
- the original system is not existing.

In all these cases modeling of the system give big time and project money savings, even if in the beginning some level of the investment is needed, mainly in software and human resources on quite high level of technical knowledge and system integration experience.

From the practical point of view there is no need for model-based design if:

– the domain of the problem is well known (nothing new to develop),
– it is easy to create the product that solves the problem stated (with the use of business requirements),
– the project team is small, and there is a small number of end users of the product,
– a minimal maintenance of the final product is needed,
– the possibility of the further product development is close to zero.

In these cases using Model-Based System Engineering is purposeless.

Taking intelligent container in water-land transportation into account one can see, that in the case of proposed system there is a strong need of introduction of model based development, since:

– the domain of the problem is still full of innovation possibilities, technical and technological from one side and legal and organizational from the others,
– solving new business and legal targets in the field is not easy, also a versioning of the possible problem solutions needs to be introduced, i.e. taking into account regulations in different countries,
– there is no need for project teams to be big, but they will be surely distributed across different places around the world, and sectors (business and universities),
– the number of end users of the product is big,
– a continuous maintenance of the product must be provided,
– the possibility of new products development is close to one, while new materials and fatigue requirements came into action.

Modeling is connected with the selected level of abstraction, from the code visualization (i.e. code coloring frameworks), through the graphical code representation, up to the business processes and relations explanation.

The most important is system modeling level, while code visualization is something that can help efficiently implement control functions.

The system modeling, usually supported with the graphical languages/notations, starts with the requirements, and leads to implementation models, that can be used for automatic generation of the control code structure.

While Model-Based Systems Engineering is a combination of language, methodology and supporting software tools, we need to understand some methodologies behind it. It is discussed in the following subsection of the paper.

3.2   MBSE Workflows by INCOSE and NoMagic

Nowadays there are many software tools for systems modeling and engineering available on the market. The most popular are listed as follows:

– Cameo Systems Modeler (formerly known as: MagicDraw with SysML plugin), Web: http://www.nomagic.com/products/cameo-systems-modeler.html,
– Innoslate, Web: https://www.innoslate.com/,

- Enterprise Architect with SysML plugin, Web: http://www.sparxsystems.com/pro ducts/mdg_sysml.html,
- Modelio, Web: http://www.modeliosoft.com/en/products/solutions/system-architect-solution-overview.html,
- Papyrus 4 SysML, Web: http://www.eclipse.org/modeling/mdt/papyrus/,
- IBM Rhapsody, Web: http://www.ibm.com/software/rational/products/rhapsody/sy architect/,
- ARTiSAN Studio, Web: http://www.atego.com/products/artisan-studio/.

These tools are the most popular thanks to its advanced functionality and good support for modeling projects usually created in distributed teams.

Modeling software architecture, conducting simulation research work on new products including control systems is possible thanks to the functionality of the following programs:

- Matlab/Simulink, from Mathworks,
- MapleSim, prepared by Maplesoft,
- LMS Imagine.Lab Amesim, from Siemens PLM,
- Scilab, open source project, initially started by INRIA, France,
- HOPSAN, from Linkoping University of Technology, Sweden,
- OpenModelica, open-source Modelica-based modeling and simulation environment, supported mainly by the Open Source Modelica Consortium (OSMC),
- Dymola.

Nowadays most of the software Integrated Development Environments for industrial control systems compliant with IEC61131-3 standard [15, 16] enables possibility of algorithms validation before compilation of the production code with the use of offline or real-time simulation [14].

MBSE cannot exist without proper work methodology. Without workflow definition MBSE is nothing but the nice set of tools supporting selected engineering aspects. INCOSE OOSEM and Magic Grid from NoMagic company are among the others the most popular, well defined and proven with many industrial case studies.

INCOSE Object-Oriented Systems Engineering Method (OOSEM) is based on the functional decomposition approach. It means modeling order from most general to the most detailed areas of interest.

SysML [11] language is preferred way for modeling, managing, tracing and documenting projects compliant to INCOSE methodology.

INCOSE workflow is quite linear and combines requirements model, use cases definition, architecture design and implementation within single project toolchain, as shown in Fig. 7. INCOSE methodology workflow is based on a hierarchical realization of the following steps, from modeling and project setup, through analysis, definition n of requirements and logical architecture, up to final verification and validation: 1. Setup project iteration; 2. Analyze needs; 3. Define system requirements; 4. Define logical architecture; 5. Synthesize allocated architecture; 6. Capture domain & assumptions; 7. Optimize & evaluate alternatives; 8. Reuse opportunity analysis; 9. Trace requirements and allocations; 10. Validate & verify system.

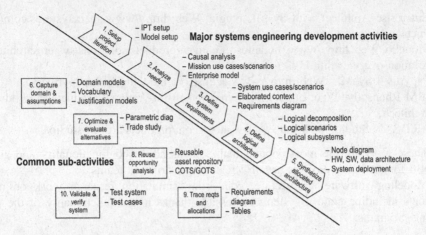

Fig. 7. INCOSE OOSEM in systems engineering [own study]

INCOSE OOSEM ensures rapid risks and possible problems identification at the early project stages. Steps 1–5 (Fig. 7) define the most important modeling parts, necessary to fully describe the project and architecture of implemented solution. These steps are usually most time and resources consuming among the whole project.

Steps 6–10 (Fig. 7) consist of the practices and activities conducted during modeling and system development.

INCOSE OOSEM workflow is presented in Fig. 8.

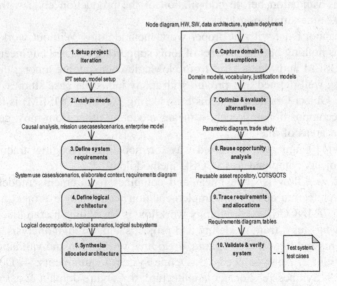

Fig. 8. INCOSE OOSEM – activity diagram [own study]

In case of complex mechatronic systems combining many domains, control distributed hardware with real-time multi-tasking software, much nicer and easier to understand workflow was proposed by the group of engineers from NoMagic company. Magic Grid workflow is shown in Fig. 9. It will be used here to discuss development and integration phases for proposed intelligent container in water-land transport. It is pointed out which phases of Magic Grid are the most suitable for creation of functional prototype.

Parametrization and measurements of effectiveness (C4, P4, Fig. 9) creates the main difference between industrial research (phases 1–10, Fig. 9) on prototypes and development work on final products (transfer of scientific results to the business reality, i.e. market).

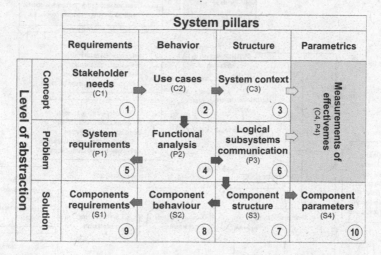

Fig. 9. Magic Grid for systems engineering [own study] (Color figure online)

Magic Grid for systems modeling is defined with three layers, called "Level of abstraction" (Concept, Problem, Solution) and four system pillars (Requirements, Behaviour, Structure, Parametrics). Thanks to the detail level and completeness of Magic Grid every project team member is able to precisely share activities in the whole process. From the other hand there is no need to think and consider functionality of the whole system. It heavily supports modularity of the resulting solutions.

We used different transitions coloring on Fig. 9:

- Green (C1-C2, S3-S4) – concept works and activities, solution proposals discussed with stakeholders;
- Orange (C2-C3, P2-P1, S2-S1) – workflow phases being a result of the decisions made, selected solutions, summarizing project stages (C3), resulting from the system certification needs or future versions of the solution (P1, S1);
- Blue (C2-P2, P2-P3, P3-S3, S3-S2) – Magic Grid workflow activities being a proposals for solving problems pointed by the stakeholders in the field of necessary

functions (P2), communications between subsystems (P3), system components structure (S3) and their behaviour (S2);

– Yellow (C3-C4, P3-P4) – for the final products (as compared with functional prototypes) it is necessary to conduct "Measurement of effectiveness" for the system solution (C3) and for proposed logic interactions for subsystems (P3). These factors are to be analyzed when developing new products.

Magic Grid workflow from Fig. 9 is depicted in the Fig. 10 in the form of activity diagram describing flow of decisions during adoption of this methodology for intelligent container development.

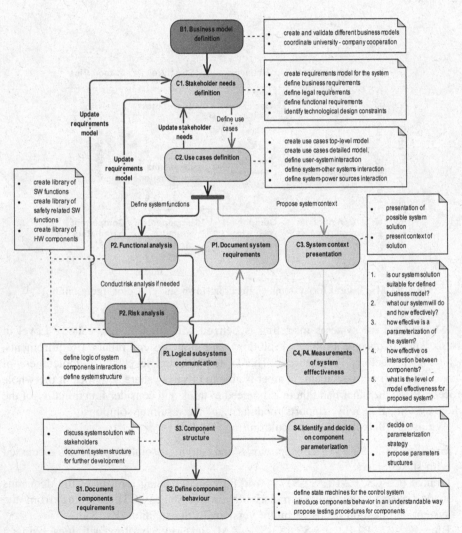

Fig. 10. Magic Grid for integration of intelligent container modules [own study]

3.3   Magic Grid Workflow for Intelligent Container

At the very beginning results from selected business models must be summarized within set of the stakeholder needs (C1). MBSE usually results with modular architectures. That supports utilization of different business models for selected subsystems. Expectations for the product/prototype and the whole project are modeled with the use of SysML Requirements Diagram [17–19].

Definition of the use cases (C2) conducted on the basis of client needs and the business target is one of the most important step in the system modeling process.

Within this phase (no matter how it will be implemented) system engineers define interaction scenarios for the users and other systems. Top level and detailed use cases are defined within step C2.

In the presented here project (and typically in most of the projects for mechatronic systems) definition of the use cases model influences and updates stakeholder needs (C2-C1). When requirements (C1) and use cases (C2) are defined the system realization context (C3) can be concluded.

It is another crucial decisive moment in product lifecycle. It is the acceptance gateway for the further development.

After system context acceptance functional analysis takes place (P2). Use case definition and documentation lets the project team analyze, identify and summarize set of functions that needs to be implemented (P2). While intelligent container with described subsystems will create different hazards during operation conducting risk analysis is necessary. Risk analysis is crucial when working on the products. Functional and risk analysis (P2) must be documented precisely. Functional analysis including risk procedures and guidelines are presented in book [20] and proper standards [8, 21]. Outcomes from these publications are applicable to the most of the mechatronic devices available on the market. Risk analysis is heavily supported in software tools for systems modeling.

Functional and risk analysis (P2) will update the stakeholder needs surely. After realizing the complexity of the system (stakeholders usually are not technology oriented) final clients can exclude some of their requirements (C1) or create completely new ones. It can even influence defined use case model (C2).

Functional analysis (P2), use cases (C2), model of business requirements (C1) are resulting with the system requirements (P1). Model for P1 consists of: functional requirements, control hardware constraints, measurement types and their parameters (resolution, compliance with proper standards). After finalizing phase P2, the very next step according to Magic Grid (Fig. 10) is definition of logic for communication between container subsystems (P3).

Proper documentation of the interfaces between components (P3) and optional parametrization (P4) closes the stage of formulating problems and tasks to be solved during implementation of the final system.

According to Magic Grid (Fig. 9) workflow Component Structure (S3) is the first step at the Solution level of abstraction. Definition of the component behaviour (S2) and summarizing requirements for components (S1) fulfills set of Magic Grid methodology based activities for development of the intelligent container.

SysML language will be of use during adoption of the Magic Grid for modeling intelligent container in water-land transport.

4   Conclusion

As it is shown in the paper, proposed system and workflow for its implementation and integration enables the following:

- Clear vision between selected business model for intelligent container, its business value proposition and final integrated solution,
- Robust integration process thanks to selected MBSE methodology, based on combination of SysML language with Magic Grid approach for systems engineering,
- Precise risk identification for high priority business and legal requirements,
- Powerful project management and progress analysis at each stage of possible implementation.

Proposed intelligent container thanks to its data processing functionalities is great alternative for actual state of the art in the field of transport systems.

Initial analysis of nowadays available technologies that could be applied to final implementation proves an effective integration is challenging but possible. Also the most risky part is satisfying different legal requirements in EU countries.

Presented here MBSE methodology thanks to its agile and robust character ensures completeness for successful implementation of selected business model at every stage of the system V-model.

References

1. Grzybowski, L., et al.: Kontenery w transporcie morskim (1997)
2. VDI - Association of German Engineers, VDI 2206 – Design methodology for mechatronic systems, Design, no. June. p. 118 (2004)
3. Davies, R.: Industry 4.0. Digitalisation for productivity and growth (2015)
4. Osterwalder, A., Pigneur, Y.: Business Model Generation: A Handbook for Visionaries, Game Changers, and Challengers (2010)
5. EN ISO 13849-1:2006. Safety of machinery – Safety-related parts of control systems - Part 1: General principles for design (2008)
6. EN ISO 13849-2:2012. Safety of machinery – Safety-related parts of control systems - Part 2: Validation. (2012)
7. SIS-ISO/TR 23849:2010. Guidance on the application of ISO 13849-1 and IEC 62061 in the design of safety-related control systems for machinery (2010)
8. IEC 62061:2005. Safety of machinery – Functional safety of safety-related electrical, electronic and programmable electronic control systems (2005)
9. EN 61508. Functional safety of electrical/electronic/programmable electronic safetyrelated systems, Parts 1–7
10. MAAB. Control algorithm modeling guidelines using Matlab, Simulink, and Stateflow (2012)

11. Omg. OMG Systems Modeling Language (OMG SysML) v.1.4, Source, no. June. p. 260 (2010)
12. Raspberry Pi website. https://www.raspberrypi.org/. Accessed 15 Feb 2017
13. Intel Galileo Documentation Website. https://software.intel.com/enus/iot/hardware/galileo/documentation. Accessed 15 Feb 2017
14. Roger, A.: Managing Model-Based Design (2015)
15. IEC 61131-3:2013. Programmable controllers. Part 3: Programming languages (2013)
16. John, K.H., Tiegelkamp, M.: IEC 61131-3: Programming Industrial Automation Systems. Springer, Berlin, Heidelberg (2010)
17. Roques, P.: Modeling Requirements with SysML. How modeling can be useful to better define and trace requirements, 2015. http://remagazine.ireb.org/issues/2015-2-bridging-the-impossible/modeling-requirements-with-sysml/. Accessed 15 Feb 2017
18. Holt, J., Perry, S.A., Brownsword, M.: Model-Based Requirements Engineering Model-Based Requirements Engineering
19. Friedenthal, S., Moore, A., Steiner, R.: A Practical Guide To SysML: The Systems Modeling Language (2015)
20. Barg, J., Eisenhut-Fuchsberger, F.: 10 Steps to Performance Level: Handbook for the implementation of functional safety according to ISO 13849 (2012)
21. ISO 12100:2010, Safety of machinery – General principles for design - Risk assessment and risk reduction (2013)

Maritime Systems for Automatic Exchange of Information and Vessel Traffic Monitoring

Ryszard Wawruch[(✉)]

Gdynia Maritime University, Morska 81-87, 81-225 Gdynia, Poland
wawruch@am.gdynia.pl

Abstract. The paper describes briefly the legal basis and the rules for the implementation, operation and utilization of different types of automatic ships monitoring and data exchange systems. It presents a comparative analysis of their advantages and disadvantages and the author's view on the possibility of further development.

Keywords: Ships monitoring · AIS · LRIT

1 Introduction

The process of establishing and development of automatic systems for data and information exchange between maritime floating units and for ships traffic monitoring began in 2002. It was initiated after the fitting of sea going vessels in the years 2002–2010 with ship's devices of the Automatic Identification System (AIS) and the Long Range Identification and Tracking (LRIT). First of all, AIS was introduced in order to reduce the impact of radar detection and tracking limitations on safety of navigation and to facilitate and simplify the establishment of voice radio communication in the VHF band between vessels in the potential collision situations. Reception of information transmitted by ship's AIS by properly equipped VHF shore stations enabled establishing of the coastal vessels traffic monitoring system working within the range of these stations and systems to automatic data exchange in the VHF band between ships and Vessel Traffic Services (VTS) or Vessel Traffic Management Services (VTMS). Later International Maritime Organisation (IMO) introduced LRIT being a global satellite ships' monitoring system delivering information on vessels to authorised receivers (respectively flag state, coastal state, port state and Search and Rescue (SAR) and marine environment protection services) only. At the same time there were established commercial costal and global monitoring systems providing information to all interested recipients and utilizing messages transmitted by ship's AIS in the VHF band and received by AIS receivers installed on shore and the Low Earth Orbit (LEO) satellites. Currently operate and are developed described in this paper so-called governmental (available for authorized users only) and commercial worldwide and coastal monitoring systems using different radio communication links.

J. Mikulski (Ed.): TST 2017, CCIS 715, pp. 360–374, 2017.
DOI: 10.1007/978-3-319-66251-0_29

2 Automatic Identification System (AIS)

2.1 The Reason and Method of Introduction

AIS is a digital information exchange system introduced on the basis of the requirements of International Convention for the Safety of Life at Sea (SOLAS), regulation V/19 "Carriage requirements for shipborne navigational systems and equipment", as mandatory on all seagoing passenger ships irrespective of their size, and cargo ships of 300 gross tonnage and upwards engaged on international voyages and of 500 gross tonnage and upwards engaged on home trades [1]. Some national and regional regulations extend the obligation to install AIS on other ships. For example, European Union (EU) regulations extend the obligation to install AIS on all seagoing cargo ships of 300 gross tonnage and upwards, and on fishing vessels with the length of not less than 15 m.

Described system was introduced primarily in order to:

- reduce the impact of limitations of radar tracking aids (Automatic Radar Plotting Aid – ARPA and Automatic Tracking Aid – ATA) on safety of navigation;
- make it easier to establish voice communications between vessels in potential collision situations; and
- automate the submission of vessel's reports in the Ship Reporting Systems (SRS) and the process of data exchange in relation: ship – VTS centre – ship.

Ship's AIS broadcasts in VHF band in predefined reporting time intervals between 2 s and 6 min, depending on type of the AIS onboard equipment and ship's speed and stability of course, and on the request, in so called autonomous and pulling modes, four groups of information:

- static including ship's identification (name, call sign, IMO number assigned to the ship's hull and MMSI (Maritime Mobile Service Identity) number), type, main dimensions, etc.;
- dynamic: values of current ship's geographical position, speed and direction of movement indicated by connected to the AIS onboard GNSS (Global Navigation Satellite System) receiver, speed and distance measuring device (log) and gyrocompass, angular rate of turn and manually introduced ship's navigational status indicating the preference level of the vessel in accordance with the rules of the International Regulations for Prevention Collision at Sea (COLREG);
- voyage related: current maximum ship's draught, number of persons onboard information about carried dangerous, hazardous and polluting goods, name of the destination place and Estimated Time of Arrival (ETA) there, the route plan; and
- short safety related message.

2.2 AIS as a Tool for the Exchange of Information Between Ships for Anti-collision Purposes

ARPA and ATA calculate positions, values of the true and relative motion vectors and the approaching parameters (Closest Point of Approach (CPA) and Time to the Closest

Point of Approach (TCPA)) of detected target on the basis of consecutive radar measurements of distances and bearings of this target and indications of the log and gyrocompass installed on the own ship and connected to radar. Due to the low accuracy of radar measurements and own ship's yawing, tracking devices average results of their calculations and present mean values of the true course, true speed and CPA and TCPA of detected and tracked targets. This averaging causes a delay in the detection of manoeuvres conducted by observed and tracked targets. AIS transmits the current values of parameters presented by connected onboard navigational instruments and, due to that, indicates manoeuvres immediately. It enables presentation of current values of opposite ship's position, CPA, TCPA and motion vector with the same accuracy as they are available onboard that vessel, higher than the accuracy of their values calculated by radar tracking devices. This advantages of AIS may be observed in Figs. 1, 2 and 3 showing values of opposite ship's CPA and direction of movement (course) presented by ARPA and AIS installed on the own ship. More information on the accuracy of AIS and ARPA indications for the purpose of collision avoidance can be found for example in [2–4].

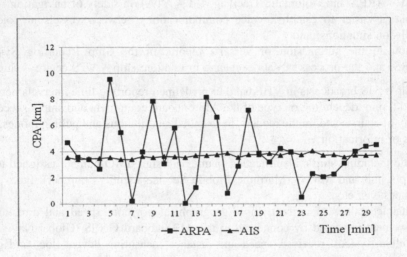

Fig. 1. Information on CPA of opposite vessel presented by onboard AIS and ARPA. Both ships (own and observed) were proceeding with steady courses and speeds, sea state 5 [3]

Additionally AIS presents, not available for person keeping radar observation only, navigational status of opposite vessel informing on responsibilities between ships in meeting situation and its identification enabling calling it by the radiotelephone to establish voice communication and agree manoeuvres.

The above-mentioned advantages of described equipment make its use increases the safety of navigation. Due to that it is also installed on vessels where it is not required by international, regional or national regulations. Less expensive types of device, so-called AIS class B or AIS receivers only are installed on these crafts often.

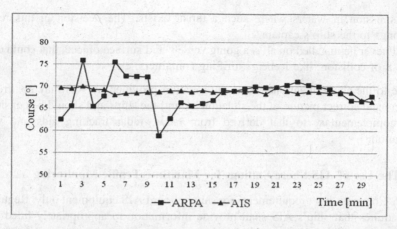

Fig. 2. Information on the course of opposite vessel presented by onboard AIS and ARPA. Both ships (own and observed) were proceeding with steady courses and speeds, sea state 4 [4]

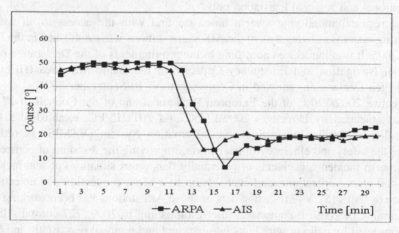

Fig. 3. Information on the course of opposite vessel performing turn presented by onboard AIS and ARPA, sea state 3 [3]

As anti-collision tool AIS also has drawbacks. It:

- Requires manual input part of the broadcasted information (navigational status, voyage related data and short safety related message).
- Transmits and receives information introduced in automatic or manual manner without checking its quality, reality and accuracy. Ship's captain and officers of watch are responsible for all information entered into the AIS manually and received from the onboard sensors and should validate them.
- Broadcasts unencrypted messages on VHF channels available for all and anyone can buy AIS receiver to receive messages transmitted by ships. This fact may pose a threat to the ship's security, especially in waters exposed to pirate attacks or terrorists. For this reason, international regulations allow the cessation of ship's AIS

transmission in waters where such a threat exists. The decision in this respect belongs to the ship's captain.
– Will never be installed on all sea going vessels and surface objects that could create a risk of collision, like icebergs, drifting containers, etc.

Due to the above mentioned facts, information provided by AIS may not create a complete and correct picture of the situation around the ships and should be considered to be complementary to that derived from radar, radar tracking aids and visual observation.

2.3   The Use of AIS Shore Stations for Maritime Traffic Monitoring

SOLAS Convention set requirements for ship onboard AIS equipment only. Regulation V/19 requires that ship's AIS shall provide information to appropriately fitted shore stations, other ships and aircraft but it does not contain any requirements or recommendations to coastal states regarding construction of AIS shore stations and establishing maritime traffic monitoring by means of these stations. These requirements stem from national and regional legislation only.

First regional monitoring system based on and with the access to all national monitoring systems utilizing nets of the AIS shore stations was established in the Baltic Sea in 2005. It was introduced according to the requirements of the Declaration on the Safety of Navigation and Emergency Capacity in the Baltic Sea Area (HELCOM Copenhagen Declaration) adopted in Copenhagen on 10 September 2001.

Directive 2002/59/EC of the European Parliament and of the Council of 27 June 2002, as amended by Directives 2009/17/EC and 2011/15/EC, established a Community Vessel Traffic Monitoring and Information System (VTMIS) in order to enhance the safety and efficiency of navigation, improving the response of competitive authorities to incidents, accidents or potentially dangerous situations at sea, including search and rescue operations, and contribution to a better prevention and detection of pollution by ships [5]. VTMIS using nets of shore AIS stations has been working since the beginning of 2009. It comprised, after establishing in 2009, 727 coastal AIS stations connected by radio or wire links into national and regional (e.g. Baltic and North Sea) subnets and European VTMIS net using SafeSeaNet system. VTMIS is operated by European Maritime Safety Agency (EMSA). Coastal AIS stations, radio and wire links connecting these stations with the operational centres and SafeSeaNet create the European Shore-based Traffic Monitoring Infrastructure Database (STMID). Over 2000 users sent reports to and made requests of the system on average, per year. According to information published by EMSA more than 8 millions queries were made in 2011. SafeSeaNet allowed the real time tracking of the approximately 17000 ships transiting in UE waters on a daily basis, through the exchange of more than 5 million ship position reports per day [6].

AIS monitoring system allows:

– presentation on an Electronic Navigational Chart (ENC) current positions of all ships equipped with activated AIS located within the range of the shore AIS stations;

- presentation data transmitted by ship's AIS indicated on the ENC or by giving its identification;
- selection of information presented graphically on ENC, using different criteria: draughts of the vessels, their types, flags, destinations, etc.; and
- presentation statistic data about ships' routes and number of the ships' passages within a certain period of time, in a user-defined areas, for defined by him groups of vessels.

Figure 4 shows, for example, marked by black triangles, the current positions and directions of movement of vessels equipped with activated AIS in maritime waters of the European Union and Mediterranean Sea at the time of registration by the VTMIS centre in Lisbon in 2009 [7].

Fig. 4. Position and direction of movement of vessels equipped with activated AIS in the European waters and Mediterranean Sea current at the time of registration in 2009 [7]

Consecutive Fig. 5 presents AIS data displayed on the ENC and in the digital form on operator's console in the centre of VTS "Zatoka Gdańska" in Gdynia. Shown in this figure incomplete digital data of ship "North Express" indicates the extent to which the possibility of a thorough and proper assessment of the situation in the monitored area depends on the data to be entered manually into the AIS onboard ships.

Figure 6 shows, as an example of the statistical analysis possible to perform on the basis of data collected by a network of AIS shore stations, routes of ships in the Baltic Proper in summer 2013.

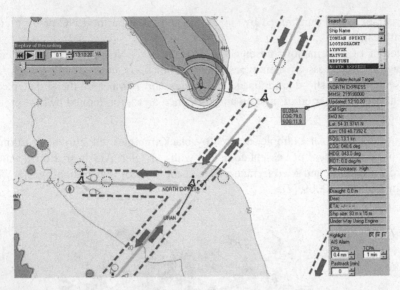

Fig. 5. AIS data displayed on the ENC and in digital form on operator's console in the centre of VTS "Zatoka Gdańska" in Gdynia [own study]

Fig. 6. Ships' routes in the Baltic Proper in summer 2013 received on the basis of data collected by a network of AIS shore stations [own study]

In addition to the ships traffic monitoring systems utilizing networks of AIS shore stations established by the maritime administrations of individual coastal states and regional organisations (e.g. Helsinki Commission and the European Union), operate commercial systems offering similar functionality to any buyer. Utilizing the fact that the ship's AIS transmits unencrypted messages on non-proprietary, publicly available frequencies, commercial companies had created a networks of agents fitted with AIS receivers connected through the Internet to the company server. Data received from ships AIS messages is updated at regular intervals (usually every 5 or 6 min) and available on companies' websites. It may be provided to the buyer in any electronic data format including ASCII, Excel, KML, MS Access, NMEA, Oracle, OTH-Gold and XML or as a movie to display on ENC in PC. Hard copy reports may be supplied on demand [8]. By placing the cursor over the symbol of specific vessel on ENC users can display additional data transmitted in AIS messages like ship's position, course, speed and port of destination. Companies offer interactive options allowing the user to zoom and filter presented data by flags, types of ships, their gross tonnage, etc. Subscribers of Internet Ships Register have a link to existing ship reference databases. Users have to log into get access to the services and pay subscription charges (usually yearly) which cost depends also on the number of users licenses. Specific companies allow access to regional specific AIS information for free too [9]. Information about these services can be found on the website (www: aishub.net, fleetmon.com, marine-traffic.com, myshiptracking.com, orbcomm.com, shipfinder.com, SkyWave.com, vesselfinder.com, vesseltracker.com, etc.). Examples of information presented graphically on the electronic chart on the PC show Figs. 7 and 8.

The shore AIS stations and their nets create an effective tool for monitoring and control of vessels fitted with activated onboard AIS in the coastal waters but have drawbacks resulting from limitations of the AIS technique and equipment. These are:

- dependence of quality of information presented in monitoring and control centres on the correctness of the data introduced by ship's crew into onboard AIS manually;
- costly land infrastructure; and
- limiting the monitored area to the coverage range of the coastal radio station working in VHF band.

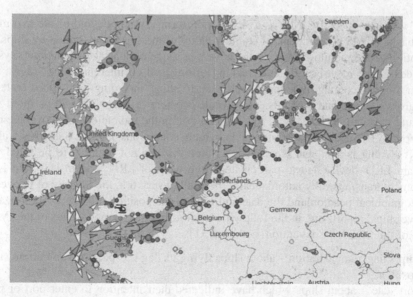

Fig. 7. Ships positions received from the commercial service Vesselfinder [10]

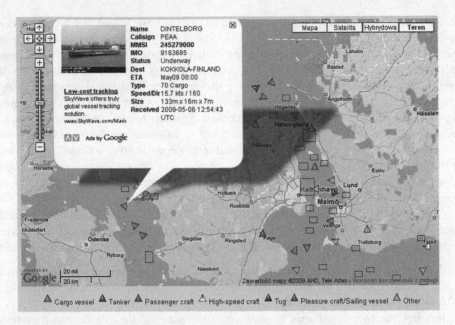

Fig. 8. Ships positions received from the commercial service SkyWave [11]

3 Long Range Identification and Tracking (LRIT)

LRIT was introduced by IMO as a vessel monitoring system devoid of the two major drawbacks of AIS system: lack of confidentiality of messages transmitted by ships and limitation of the monitored area to the coverage range of coastal AIS stations. System provides global identification and tracking of engaged on international voyages: passenger ships, including high-speed crafts, cargo ships, including high-speed crafts, of 300 gross tonnage and upwards, and mobile offshore drilling units. Ships fitted with AIS and operated exclusively in the sea area within the radiotelephone coverage of at least one VHF coastal radio station in which continuous Digital Selective Calling (DSC) alerting is available (so called sea area A1) are not required to be fitted with onboard LRIT device and to participate in the work of LRIT system [1]. Ship LRIT equipment transmits fully automatically, normally every 6 h, the identity of the vessel, its geographical position and the date and time of the position provided. Authorized to receive ships' reports are, in the following situations, only member states of the IMO – sides of the SOLAS Convention, acting as [1]:

- ship's flag administration – about ships flying its flag irrespectively of current ship's position;
- port state – about ships which have indicated their intention to enter port or place under the jurisdiction of that government, provided that they are not within the inner water of another state or within the territorial sea of the government whose flag the ship is entitled to fly; and

- coastal state – about ships entitled to fly the flag of other governments, not intending to enter port or place under the jurisdiction of that state, navigating within a distance not exceeding 1852 km of its coast provided such ships are outside inner waters of another states and territorial sea of the government whose flag the ship is entitled to fly.

LRIT data may additionally receive following services established by particular governments:

- search and rescue (SAR) - in relation to conducted by them search and rescue operations of persons in distress at sea; and
- responsible for marine environment protection - in relation to actions provided by them in accordance with international law.

Particular IMO member states are responsible for further distribution of LRIT data to recipient authorised by them. Ships' reports are transmitted by commercial satellite communication systems selected by the flag state administration. The costs of the system bear the recipients of ships' reports, who have to pay for ordered and received reports. Ship reports are available free of charge for SAR services only.

Ship's LRIT device, as well as the ship's AIS may be switched off if the transmission of reports poses a threat to the security of the ship. The decision belongs to the ship's captain, who shall communicate it to the administration of the flag state of the ship. Flag state administration may at any time decide that information from LRIT system about ships entitled to fly its flag shall not be available for all, for particular states or when the ships are inside areas defined by it. Information about the decision, its amendment, suspension or annulment should be communicated to the IMO, which should notify it to all member states participating in the system.

The frequency of transmission of ship's reports can be changed automatically by authorized recipients. The time interval between successive reports can vary from 15 min to 24 h.

The main problems in the establishing of the system were formal and legal issues, not technical. Due to them, as shown in Fig. 9, system has a complicated architecture comprises different types of data centres, different application and communication services providers, international data exchange, international data distribution plan server and system coordinator supervising its work. According to the information provided by the International Mobile Satellite Organization (IMSO) designated by the IMO as LRIT coordinator, on 2 December 2016, 56 SOLAS contracting governments out of 163 had yet to participate in the described system.

Table 1 presents current composition of the LRIT data centres. Permanent International LRIT Data Exchange (IDE) is maintained and operated by EMSA in Lisbon. EMSA also maintains a business continuity facility site in Madrid for the immediate restoration of service in the event of its interruption at the primary site. The United State Coast Guard maintains the disaster recovery site for permanent IDE in Kearneysville in West Virginia (USA). LRIT Data Distribution Plan (DDP) server is located in the IMO building in London and the United Nations International Computing Centre

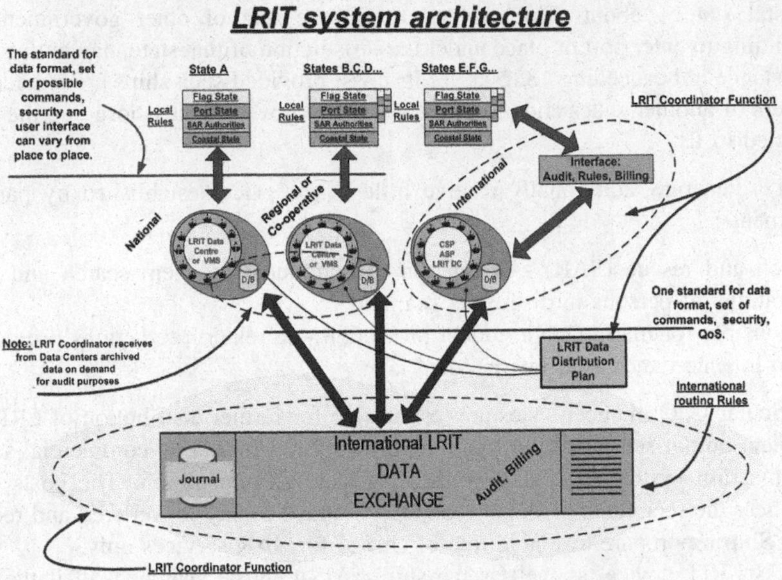

LRIT system architecture

Fig. 9. LRIT system architecture [12]

Table 1. Composition of LRIT data centres on 2 December 2016 [13]

Type of data centre	National with one flag administration	National with more than one flag administration	Regional	Cooperative
Number of data centres	42	9	1	3
Participation by flag administrations				
SOLAS contracting governments	42	27	3	35
Non-metropolitan territories	–	5	–	5
Special administrative regions	–	2	–	2

(UNICC) in Geneva is used for the disaster recovery arrangements. The performance of all data centres and of the IDE is subject to annual review and audit conducted by the LRIT Coordinator [13].

Figure 10 shows marked by yellow triangles positions of vessels transmitting LRIT reports to the LRIT cooperative data centre maintained by EMSA, current at the time of registration in 2009 [7].

Fig. 10. Positions of vessels transmitting LRIT reports to the LRIT cooperative data centre maintained by EMSA, current at the time of registration in 2009 [7] (Color figure online)

Ships had to be fitted with the LRIT devices till 1 July 2010 (vessel sailing inside the coverage of an INMARSAT geostationary satellite in which continuous alerting is available, till the end of 2008). This means that the LRIT system should be fully operational for many years but according to the already mentioned information presented by the IMSO, in the operation of the system is not involved yet third countries - parties to the SOLAS Convention. There are still some problems with the financial settlement of the system and concerns about the need for annual audits of data centres and IDE and fees for those audits.

4 Satellite AIS Monitoring

AIS messages transmitted in the VHF band are detected by LEO satellites. Reacting to this possibility, World Radio Conference in 2007 (WRC-07) added a secondary mobile-satellite service working on AIS VHF frequencies to allow satellites receiving AIS messages on secondary basis. As indicated in the Report ITU-R M.2084 "Satellite detection of automatic identification system messages" there are some limitations in detection those messages by satellites resulting from the principle of operation of the AIS working on the frequencies not exclusively reserved for its work [14].

AIS equipment uses the Self-Organised Time Division Multiple Access (SOTDMA) as the basic method to find access to the VHF Data Link (VDL). Access is in this method synchronised in one minute cycles, so called frames. Each frame is divided into 2250 slots. One AIS message may occupy from one to five consecutive slots, but standard position report needs one slot only. Each AIS class A plans its own transmission basing on information about the slots already occupied by other AIS stations received during the current frame and pre-announces slots for its own planned transmission during the next frame. The other AIS class A stations are able to take this slot use into account when planning in the same manner their own transmissions. If AIS class A is not available to find the required number of free consecutive slots, rejects transmissions of AIS equipments currently located in the largest distances, limiting the range of communication equal, under normal propagation conditions and for heights of ships' antennas equal to 25–30 m, to a maximum of 90 km. The AIS class B using CSTDMA algorithm transmits its message in found free slots without prior notice about it. The broadcast mode allows the system to be locally overloaded by 400 to 500% and still provides nearly 100% throughputs for ships closer than approximately 15 km to each other [14].

Satellite receives all messages broadcasted on frequencies AIS1 and AIS2 by transmitters that are within its field of view. Low is an orbit around Earth with an altitude between 160 and 2000 km (orbital period from of about 88 to 127 min). For example, LEO satellites used by commercial company exactEarth circle at an altitude of about 650 km above the Earth. They have a field of view with a radius of about 2500 km. That means, in the field of view (coverage pattern) of one satellite can be found groups of ships transmitting in the same time slots as they are in mutual distances greater than the range of communication between them in the VHF band. Transmissions from these vessels may overlap, causing problems with detection of individual AIS messages in signal received by satellite. Due to that in areas with heavy traffic of ships may be impossible to read AIS messages directly through their AIS receiver located in the satellite. In these cases is used Spectrum Decollision Processing (SDP). Satellite receives all radio signals from AIS VHF channels and retransmits them to the Earth station where AIS messages are separated from the noise and interference, and assigned to individual ships.

In coastal waters may appear a problem with the work of different land users at the AIS frequencies. VHF channels 87B and 88B known as AIS1 and AIS2 are not exclusively dedicated for the transmission of AIS messages (like VHF channels 16, 75 and 76 which are dedicated to maritime use only and restricted from terrestrial use on a global basis). On land, they may be utilized by other users. LEO satellite, due to the large antenna footprint receives ships' AIS messages together with transmissions of those land users. It creates additional problems with the separation of AIS messages transmitted by ship's AIS.

The last issue is the question of the length of AIS message transmitted inside one slot. It is adapted to the terrestrial communication range in VHF band equal for ship's transceivers approximately to 90 km. AIS transmission packet includes time buffer for correction of time delay connected with distance between two AIS devices, sufficient to ensure effective exchange of AIS messages at distances up to about 185 km but too short for the satellite detection range.

In order to solve the above mentioned problems ITU-R Study Group 8 proposed to [14]:

– introduce new AIS Message No 27 for so called long-range application with a similar content to AIS position reports (messages No 1, 2 and 3) but with compressed total number of bits to allow for longer propagation delay associated with the satellite detection; and
– establish VHF channel exclusively dedicated for maritime satellite AIS service and restricted from terrestrial use on a global basis.

Commercial companies, without waiting for introduction of the above-mentioned regulations, expanded possibilities of offered monitoring services working on the basis of AIS shore receivers by adding data on vessels acquired through the LEO satellites or built separate satellite ships monitoring systems, offering services not only to the industry but also to governments and government agencies. For example, the exact-View utilises currently nine LEO satellites, including the recently launched EV-9, the first equatorial satellite in the constellation which greatly improves global revisit times. Launches also begin in 2017 for the new exactView RT powered by Harris, providing an additional 65 satellites to the system [15].

EMSA uses data from the AIS satellite systems supplied by private companies chosen by the public contract in the scope of EU programme of the Integrated Maritime Data Environment (IMDatE) developed in 2013 in order to integrate and correlate data from existing information systems (SafeSeaNet, CleanSeaNet and LRIT) and from external source including AIS satellite service, and deliver on this basis a number of Integrated Maritime Services (IMS) to EU member states and EU organisations like the European Agency for the Management of Operational Cooperation at the External Borders of the Member States of the European Union (Frontex), the European Fisheries Control Agency (EFCA) and EU Naval Forces in Somalia (EUNAVFOR). At the same time EMSA commenced together with the European Space Agency (ESA) satellite AIS (SAT-AIS) programme [6].

5 Conclusion

Currently, it is used only one system of automatic exchange of information between ships and between ships and shore stations, used simultaneously as a sensor for global and coastal maritime traffic monitoring. Ship's devices of this system broadcast on unrestricted frequencies unencrypted data, introduced partly manually, without verification of its correctness. System was introduced in order to reduce the impact of radar detection and tracking limitations on safety of navigation and to facilitate and simplify the establishment of the voice radio communication in the VHF band between ships in the potential collision situations. Receiving through coastal radio stations and LEO satellites, collecting and processing of data transmitted by ship's AIS devices enables establishing of ships monitoring systems. The introduction of AIS satellite monitoring raises doubts as to the further usefulness of LRIT system. LRIT system was established by IMO in order to allow member states to track commercial vessels fitted with onboard devices of this system all over the world but in accordance with international law and rules established by this organisation. The system has a complicated structure, is expensive to maintain and allows states and government authorities to obtain information about the current positions of ships only. Utilising micro- and nanosatellites commercial satellite AIS systems are the source of much wider, relatively cheap information available without any formal and legal restrictions every person, firm and institution ready to pay for the received data.

References

1. SOLAS: Consolidated edition 2014. Consolidated text of the International Convention for the Safety of Life at Sea, 1974 and its Protocol of 1988: articles, annexes and certificates, incorporating all amendments in effect from 1 July 2014. IMO, London (2014)
2. Wawruch, R.: Study reliability of the information about the CPA and TCPA indicated by the ship's AIS. Int. J. Mar. Navig. Saf. Sea Transp. "TransNav" **10**(3), 417–424 (2016)
3. Wesołowski, J.: Analiza porównawcza dokładności danych o parametrach ruchu względnego i rzeczywistego statku obcego prezentowanych przez ARPA i AIS. Engineering thesis. Gdynia Maritime University, Gdynia (2016)
4. Wilczyński, M.: Analiza porównawcza dokładności śledzenia systemów AIS i ARPA. Engineering thesis. Gdynia Maritime University, Gdynia (2015)
5. Directive 2002/59/EC of the European Parliament and of the Council of 27 June 2002 establishing a Community vessel traffic monitoring and information system and repealing Council Directive 93/75/EEC. Official Journal of the European Communities 5 August 2002
6. European Maritime Safety Agency: Annual Report 2013. EMSA, Lisbon (2013)
7. European Maritime Safety Agency: Annual Report 2009. EMSA, Lisbon (2009)
8. Wawruch, R.: Comparative assessment of the satellite and shore based ships monitoring systems. In: Annual of Navigation No. 15, pp. 109–116 (2009)
9. Report to the European Commission containing a preliminary discussion paper on co-ordinated AIS shore-based implementation through development of an AIS master plan for Europe, EMSA, Lisbon (2007)
10. www.vesselfinder.com. Accessed 10 Feb 2017
11. www.SkyWave.com. Accessed 10 Feb 2017
12. Resolution MSC.210(81) adopted on 19 May 2006. Performance standards and functional requirements for the long-range identification and tracking of ships. IMO, London (2006)
13. Updates to the LRIT system. Performance to the LRIT system and recommendations by the LRIT Coordinator submitted by the International Mobile Satellite Organization (IMSO). NCSR 4/4. IMO, London (2016)
14. ITU matters, including radiocommunication ITU-R Study Group 8 matters. Liaison statement from Working Party 5B to IMO. Preliminary draft new report ITU-R M. [SAT-AIS]. Improved satellite detection of AIS. Note by the Secretariat. NAV 54/INF.2. IMO, London (2008)
15. http://exactearth.com/technology/exactview-constellation. Accessed 10 Feb 2017

Traffic Safety Analysis of Intersection Based on Data from Red Light Enforcement System

Artur Ryguła[1]($\boxtimes$), Krzysztof Brzozowski[1], and Dawid Brudny[2]

[1] University of Bielsko-Biala, Willowa 2, 43-309 Bielsko-Biała, Poland
{arygula, kbrzozowski}@ath.bielsko.pl
[2] APM PRO sp. z o.o., ul. Barska 70, 43-300 Bielsko-Biała, Poland
dawid.brudny@apm.pl

Abstract. The paper presents selected analysis of traffic safety of the intersection equipped with a red light enforcement system. Violations recorded by the system are a substantial breach of the traffic participants' safety and its effects, in the form of an incident or accident, have unacceptable social and economic dimension. Authors presented the scale of existing traffic violations and described relations between event occurrence and factors such as daytime, traffic speed and the time between the beginning of red signal and moment of passing the stop line. In addition, the paper proposes a measure to evaluate the traffic risk level as a function of time of passing the stop line and the speed of vehicle enters the intersection. This measure was mainly related to the indicator of traffic safety level of conflicting movements at signalized intersection (intergreen time) and the maximum allowed speed at the intersection entry.

Keywords: Road traffic safety · Red-light enforcement · Risk level

1 Introduction

Vehicle ignoring traffic signal and entering the intersection during a red light often leads to serious accidents, including fatalities. According to the Insurance Institute for Highway Safety [1] in the US for more than half of death cases of such accidents are road users, such as pedestrians and cyclists. Several studies clearly show that running a red light is unfortunately a common offense in many countries of the world. The impact on the driver's decision is directed by a number of factors, among which is driver's reaction time, vehicle speed, traffic intensity as well as the road gradient and weather conditions [2]. Polish National Program of Road Safety 2013–2020 lists running at red light as one of the most common offences made by drivers [3]. Thus, as one way of reducing this phenomenon is appropriate educational activities within the framework of social campaigns (e.g. campaign "Three Colours" conducted by the Warsaw Municipal Road Administration Authority [4], where the red light violation in 20% is the cause of accidents). On the other hand, as a preventive solution, a dedicated systems for red light enforcement are used. Due to the scale of the phenomenon and its effects, it is not surprising that in Implementation Programme for the years 2015–2016 of National Program of Road Safety in Poland in the framework of ongoing tasks assumed inter alia the installation of red light enforcement systems [5]. In this paper authors analysed

© Springer International Publishing AG 2017
J. Mikulski (Ed.): TST 2017, CCIS 715, pp. 375–384, 2017.
DOI: 10.1007/978-3-319-66251-0_30

the level of risk, which is associated with those type of offenses on the basis of the data recorded by the traffic light enforcement system.

2   Characteristics of the Research Object

The object of research is the intersection with three inlet which the traffic organization is shown in Fig. 1. Analyzed intersection is a connection of the national and the municipal road with the speed limit at the main communication route (inlet A-B) of 70 km/h. For C inlet speed limit is 50 km/h. At presented intersection operates acyclic accommodative traffic lights working on the basis of group control.

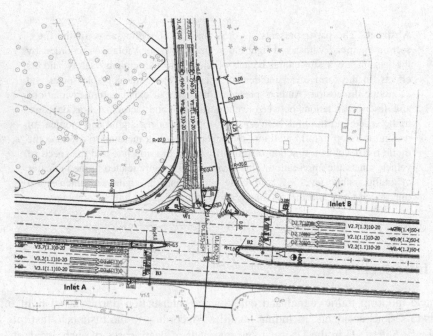

Fig. 1. Traffic organization at intersection equipped with traffic light enforcement system [6]

At the analysed intersection, two main inlets are equipped with traffic light enforcement system. The system consists of cameras observing the current traffic signal and cameras which are recording the vehicle stream. In total, inlet A and B is equipped in a set of three detection cameras and three overview cameras. For each recorded traffic light violation system provides information about the date and time of the offense, vehicle speed, set of photos and video and the time elapsed from the start of emitting a red signal to time when vehicle runes through the stop line.

3   Data Analysis

The frequency of recorded violation was analysed for the two intersection inlets from September 1, 2015 to October 7, 2015. For whole period system registered more than 6,000 events in total, of which more than 82% was recorded on weekdays. Summary of statistical measures describing the speed of vehicles running at red light and the time elapsed from the start of emitting a red signal until running the stop line for weekday is presented in Table 1 and for the public holidays and weekends in Table 2.

Table 1. Statistical measures characterizing the events that occurred on weekdays [own study]

	Average	Standard deviation	Median
Vehicle speed	68 km/h	33 km/h	60 km/h
The time elapsed from emitting a red signal	0.85 s	0.87 s	0.63 s

Table 2. Statistical measures characterizing the events that occurred on weekends [own study]

	Average	Standard deviation	Median
Vehicle speed	73 km/h	35 km/h	60 km/h
The time elapsed from emitting a red signal	0.89 s	0.78 s	0.63 s

Analysis of the indicators presented in Tables 1 and 2 leads to the conclusion that on weekends events were characterized by a higher, in relation to the weekdays, average speeds of vehicles and an extended average time elapsed since the emitting red light until crossing the stop line. This observation may be a results of less traffic intensity. It is also worth noting that the median values determined for both indicators are the same for events recorded on weekdays and weekends. In the case of a set of offenses at analysed intersection there was no evident correlation between the parameters characterizing the event (e.g. speed, time, etc.).

For a more detailed analysis, authors prepared histograms. Figures 2, 3, 4, 5, 6 and 7 respectively presented data including the number of events at a given speed for

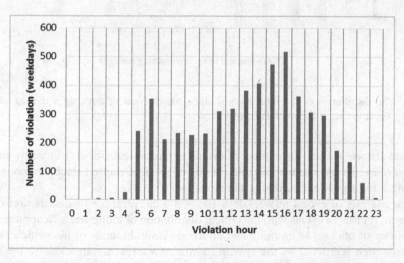

Fig. 2. The distribution of the number of events recorded in hours of the day in the weekdays [own study]

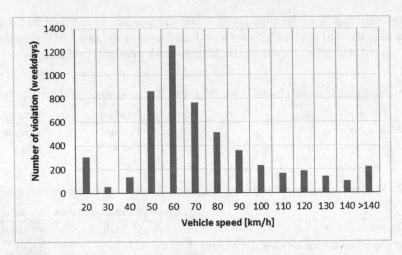

Fig. 3. The distribution of the speed of vehicle which committed the offense in the weekdays [own study]

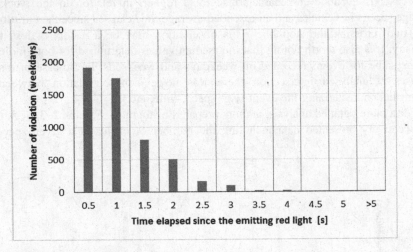

Fig. 4. The distribution of the time elapsed since the moment of emitting a red signal until crossing the stop line in the weekdays [own study]

assumed discrete classes, the number of events in the particular hours of the day and the number of events for assumed discrete classes of the time elapsed since the moment of emitting a red signal until crossing the stop line.

The analysis of events in weekdays showed a convergence of distribution of the number of registered events with the typical distribution of traffic and a clear increase in the number of offenses in morning and afternoon peak. In terms of the vehicles speed that have been registered by the system, dominate values that are close to the speed limit at the inlet (70 km/h). Importantly, the vast majority of offenses has been made till the 2 s after the start of emitting a red signal.

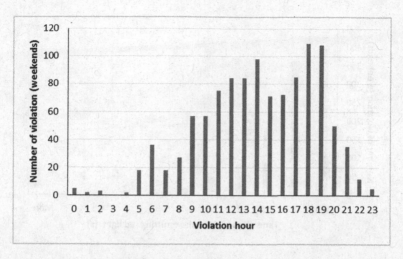

Fig. 5. The distribution of the number of events recorded in hours of the day in the weekends [own study]

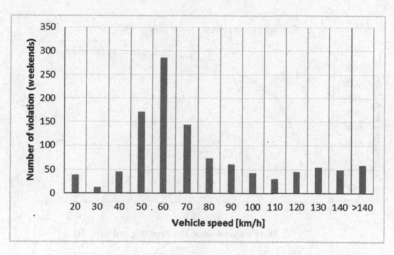

Fig. 6. The distribution of the speed of vehicle which committed the offense in weekends [own study]

For non-working days (Saturday, Sunday) the graph of total number of events also shows a similarity to the typical distribution of traffic over the weekend with the increased number of events in the corresponding peak during the afternoon. Moreover, within the scope of the velocity distribution, it was observed that dominate speeds which are close to the speed limit at the inlet. Just as in the weekdays, the time elapsed from the start of emitting a red signal until crossing the stop line in most cases was less than two seconds. Distribution function of this parameter for weekdays and weekend are shown in Fig. 8.

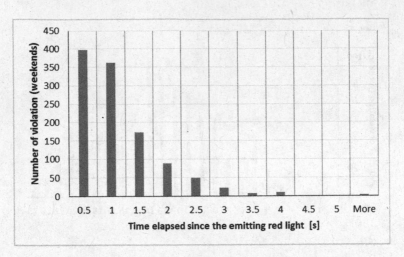

Fig. 7. The distribution of the time elapsed since the moment of emitting a red signal until crossing the stop line in the weekends [own study]

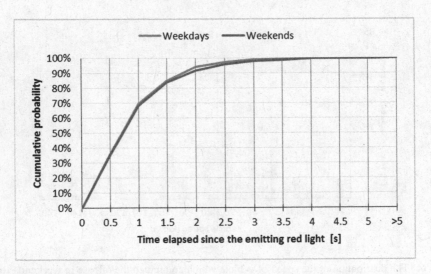

Fig. 8. Distribution of entry time after start of emitting a red signal [own study]

Analysis of the cumulative distribution function of entry times after start of emitting a red signal leads to a conclusion that there is a lack of clear differences in the case of working days and weekends. Only relatively minor differences exist in the offenses which entry time is between 1.5–2.5 s (20% of all events). In both cases, nearly 70% of offenses, time of running after emitting a red signal does not exceed 1 s.

4   Definition of Risk Measure

In order to define a measure of safety threats caused by the vehicle running at a red light, in accordance with regulation [7], authors calculate the clearance t_e and entry time t_d for the particular traffic flows. For the purposes of analysis conflict streams are chosen which are shown in Fig. 9.

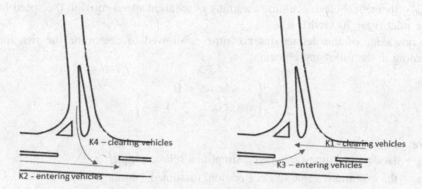

Fig. 9. Analysed conflict streams [own study]

The clearance time of the stream (vehicle entering at a red light) is defined using the following equation:

$$t_e = \frac{s_e + l_p}{v_e} \qquad (1)$$

where:
   S_e – length of clearance route (in the case of analyzed intersection 35 m),
   l_p – the value extending the clearance (assumed 10 m),
   v_e – clearing speed – the actual speed of the vehicle running at a red light.
   While entry time of stream (vehicle entering from the collision stream) was calculated using the formula:

$$t_d = \sqrt{\frac{2(s_d + 1.5)}{a_d}} \qquad (2)$$

where:
   s_d – entering distance (in the case of analyzed intersection 28 m),
   a_d – acceleration rate of entering vehicle (assumed 3.5 m/s^2).

Assuming the theoretical clearance time (clearance speed equal to the speed limit at the inlet) and taking into account the entry time of vehicles gaining right of way and the intergreen time (parameter defined according with the regulation [7] at the stage of the intersection traffic organisation project) defined a measure named as reserve time z_t:

$$z_t = t_d + (t_m - t_{\dot{z}}) - (t_{e0} + t_r)[\text{s}] \tag{3}$$

where:

t_r – the time elapsed from the start of emitting a red signal [s] - defined on the basis of red light enforcement system measurements,

t_m – the intergreen time (in this case 5 s),

$t_{\dot{z}}$ – the duration of the yellow light signal (3 s),

t_{e0} – theoretical clearance time assuming evacuation speed equal to the speed limit at the inlet ($v_d = 70$ km/h).

Knowledge of the defined reserve time z_t allowed to determine the risk index according to the following equation:

$$W_k = \begin{cases} 1 \ \text{when } z_t < 0 \\ \max\left\{0; 1 - \frac{z_t}{t_e}; 1 - \frac{z_t}{t_h}\right\} \end{cases} \tag{4}$$

where:

t_h – deceleration time (assumed a simplified relationship $t_h = \frac{v_e}{a_e}$),

a_e – the maximum value of deceleration, (assumed 10 m/s^2).

Defined in formula (4) the risk index W_k assumes values between 0 and 1, where the lowest value corresponds to the situation for which there is little probability of collision, while the value 1 corresponds to a situation in which the probability of a collision reaches a maximum.

When reserve time $z_t > 0$, value of $1 - \frac{z_t}{t_e}$ quantifies risk index for passing through the conflict point without braking, while the value $1 - \frac{z_t}{t_h}$ quantifies the risks associated with the situation by passing conflict point while braking. For the risk assessment, authors assumed that the most unfavorable situation occurs, i.e. in case $z_t > 0$ the indicator takes the largest of the calculated values. In addition, as a part of analysis, they defined the risk levels on the basis of W_k factor (Table 3).

Table 3. Definition of risk levels [own study]

Risk levels	Value of W_k
1	<0; 0.2)
2	<0.2; 0.4)
3	<0.4; 0.6)
4	<0.6; 0.8)
5	<0.8; 1>

The defined risk levels mean that the risk level 1 occurs, when there is no significant increase in the probability of a collision, while the risk level of 5 signify an maximal increase in the probability of road accident. The dependence of the defined risk level for the offenses recorded at the analyzed intersection, in terms of time elapsed from the start of emitting a red signal until crossing the stop line and the vehicle speed is shown in Fig. 10.

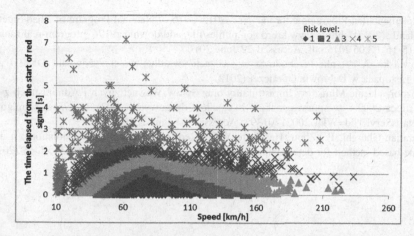

Fig. 10. The dependence of the defined risk level in terms of time elapsed from the start of emitting a red signal until crossing the stop line and the vehicle speed [own study]

Dependence presented in Fig. 10 allows to identify a particular areas of risk as a function of driving speed and the time elapsed since the start of the red light. In general, for violations with entry speed v_e within a range $\langle \frac{1}{2} v_{d;} 2v_d \rangle$, where v_d is the speed limit at the inlet, and in less than one second from the start red light, risk level in widely area remains basic (value 1). Moreover, for relatively low speeds (below $\frac{1}{2} v_{d;}$), risk level for offense with the same time lapse since emitting a red signal varies faster with changing speed than the risk level for offenses with speed of entry of above $2v_d$.

5 Conclusion

The proposed procedure allows to estimate the level of risk associated with the entry of the vehicle at the intersection during the red signal emission. In contrast to the proposal presented in [8] the issue is not reduced to description of the probability of collisions in the zero-one diagram, but introduces risk level gradation. This approach allows an assessment of the impact of individual parameters on risk level at the stage of defining traffic organisation and therefore the conscious risk management at the intersection design stage.

References

1. Red light running (2016). http://www.iihs.org/iihs/topics/t/red-light-running/qanda. Accessed 22 June 2016
2. Jahangiria, A., Rakhaa, H., Dingus, T.A.: Red-light running violation prediction using observational and simulator datam. Accid. Anal. Prev. **96**, 316–328 (2016)
3. Polish National Program of Road Safety 2013–2020 (2013). http://www.krbrd.gov.pl/files/file/Programy/KRBRD-Program-P1a-20140422-S4-K1-PL.pdf. Accessed 22 June 2016
4. Campaign "Three Colours" (2016). http://brb.waw.pl. Accessed 22 June 2016

5. Implementation Programme for the years 2015–2016 of National Program of Road Safety in Poland (2015). http://www.krbrd.gov.pl/files/file_add/download/176_program-realizacyjny-2015-16_03.06.2015.pdf. Accessed 22 June 2016
6. Kowalski, A., et al.: Przebudowa sygnalizacji świetlnej na skrzyżowaniu DK 94 z ul. 11 Listopada w Dąbrowie Górniczej (2012)
7. Rozporządzenie Ministrów Infrastruktury oraz Spraw Wewnętrznych i Administracji z dnia 31 lipca 2002 r. w sprawie znaków i sygnałów drogowych (2002). http://isap.sejm.gov.pl/DetailsServlet?id=WDU20021701393. Accessed 22 June 2016
8. Baratian-Ghorghi, F., Zhou, H., Zech, W.C.: Red-light running traffic violations: a novel time-based method for determining a fine structure. Transp. Res. Part A **93**, 55–65 (2016)

Risk Analysis in Air Transport Telematics Systems Based on Aircraft's Airbus A320 Accident

Michał Kozłowski$^{(\boxtimes)}$ and Ewa Dudek

Faculty of Transport, Warsaw University of Technology,
Koszykowa 75, 00-662 Warsaw, Poland
{m.kozlowski,edudek}@wt.pw.edu.pl

Abstract. This article is a continuation of the Authors' study on the ways to ensure safety of aeronautical operations in the aspect of aeronautical data and information quality assurance in the entire process (considered as the supply chain) of those data and information creation, collection, processing, publication as well as final operational utilization, extending the discussed context to the risk of proper information utilization. In its content the accident of Airbus A320 aircraft (LH2904) that occurred on September 14th, 1993 at Warsaw airport [EPWA] during landing was described. Particular attention was paid to the telematic systems (especially those responsible for information and automatic control), that took part in the unsuccessful deceleration process of the aircraft. Then the brief characteristics of operational risk management in aviation was presented and hazard and operability study of the aircraft's telematic systems, based on the described case, and using structured examination method (HAZOP), was conducted.

Keywords: Hazard and risk analysis · Telematics · Air transport

1 Introduction

Air transport is a domain that develops very dynamically. The resulting therefrom action, aimed at increasing airspace capacity, however cannot take place at the expense of air traffic's safety, but only with the improvement of information transmission systems. In Air Traffic Management (ATM) the growing dependence on telematics systems (particularly those related to data sharing and transmission of information) can be observed. Requirements, procedures and regulations being in force in the civil aviation, discussed in more details in other Authors' works [1–3], aim at ensuring the adequate level of flight operations' safety. As the efficiency and effectiveness of air transport operation is strongly dependent on the supply of proper quality information in previous works (e.g. [2, 3]) the Authors proposed a comprehensive and systematic approach to quality assurance at all stages of the aeronautical data and information chain. They also drew attention to the fact that a data error can occur at any stage of the aeronautical data and aeronautical information supply chain also or better said most of all in its final operational utilization while its cause, source, place, and manner of materialization can be extremely different.

© Springer International Publishing AG 2017
J. Mikulski (Ed.): TST 2017, CCIS 715, pp. 385–395, 2017.
DOI: 10.1007/978-3-319-66251-0_31

This article is a continuation of the Authors' study on the ways to ensure safety of aeronautical operations in the aspect of aeronautical data and information quality assurance in the entire process of those data and information creation, collection, processing, publication as well as final operational utilization, extending the discussed context to the risk of proper information utilization. Its first paragraph describes the accident of Airbus A320 aircraft (LH2904) that occurred on September 14th, 1993 at Warsaw airport during landing. Special attention is paid to the telemetric systems (especially those responsible for information and automatic control), that took part in the unsuccessful deceleration process of the aircraft. This particular case is later used for risk assessment presentation as well as hazard and operability study of the aircraft's telematic systems that is conducted in the following part of the paper.

2 Description of Aircraft's Airbus A320 Accident

On September 14th, 1993 flight LH 2904 of German Lufthansa, from Frankfurt [FRA], to Warsaw [WAW], operated by aircraft Airbus A320-200 (registration D-AIPN) overran the runway at Warsaw Airport (EPWA) causing a serious aeronautical accident.

At 15:30:46 [UTC] LH 2904 crew reported to the Warsaw Approach Control Unit (Warsaw APP) stabilized status and received clearance to continue precision approach procedure on runway 11 (RWY 11) together with meteorological information: wind 160°, 2–5 km/h, wind shear reported by just landed crew. During LH 2904 approach to RWY 11 atmospheric front passed from west to east over Warsaw Airport. At 15:33:20 [UTC] LH 2904 passed ILS middle marker (MM) and continued approach with landing configuration (landing gear down, full flaps [35°], manual control of thrust and of aircraft flight control surfaces) until touchdown.

According to Flight Manual instructions pilot flying (PF) used increased approach speed - about 20 kts (37 km/h) bigger than usual (which was the correct procedure for the reported weather conditions) and with this speed softly first time touched runway surface ca. 770 m from runway threshold (THR11) only on right main gear leg (see Figs. 1 and 3).

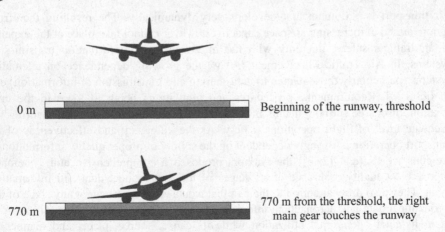

0 m   Beginning of the runway, threshold

770 m   770 m from the threshold, the right main gear touches the runway

Fig. 1. Airbus A320 LH2904 touches the runway ca. 770 m from the threshold [own study]

Unfortunately due to dynamically changing weather conditions the received weather report was already out of date. At the moment of touchdown, the assumed crosswind turned out to be a tailwind of approximately 20 kts (37 km/h). In this situation the aircraft hit the ground with increased speed of approximately 170 kts (310 km/h) far beyond the normal touchdown point (see Fig. 3). The effect of this fact, in conjunction with a lack of compression of the left landing gear leg resulted in delayed deployment of spoilers and thrust reversers.

1 525 m both wheels are on the ground, only 1 275 m of the runway left to stop the aircraft

Fig. 2. Both wheels of Airbus A320 LH2904 on the ground ca. 1 525 m from the threshold [own study]

Which means that only when the left landing gear also touched the runway did the telematic aircraft systems allow the ground spoilers and engine thrust reversers to operate. Delay was about 9 s, at the distance ca. 1525 m from THR 11 (see Figs. 2 and 3).

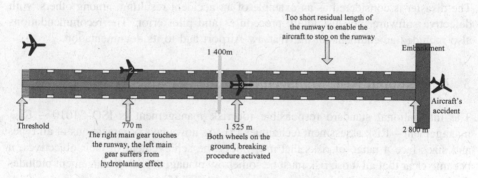

Fig. 3. Airbus A320 LH2904 accident's scheme [own study]

Thus the braking commenced with delay and in condition of heavy rain and strong tailwind (storm front passed through aerodrome area at that time) the aircraft did not manage to stop on the runway. Rollout of the aircraft progressed in conditions of heavy rain and with a layer of water on the runway. A320-200 was decelerated according to possibilities in actual conditions, but on the distance of last 180 m of runway deceleration decreased only by about 30%. Residual length of the runway (left from the moment when braking systems had begun to work) was too short to enable the aircraft to stop on the runway. Seeing the approaching end of runway, and the obstacle behind it (ILS embankment), the pilot managed only to deviate the aircraft to the right. The A320-200 rolled over the end of RWY11 with the ground speed (GS) = 72 kts and having passed next 90 m collided by its left wing with the embankment, slipped over it destroying ILS LLZ aerial located on the embankment, and stopped right behind the embankment (at 15:43 UTC). In effect of this movement the landing gear of the aircraft and the left engine were also destroyed.

From the 64 passengers on board and 6 crew members one crew member and one passenger lost their lives as an effect of the crush. The aircraft itself sustained damage caused by fire.

Taking into account the analysed issue it is worth mentioning that according to the Main Commission Aircraft Accident Investigation Final Report [4] the main cause of the accident were flight meteorological conditions (wind component changes from: 150°–22 km/h, to 210°–36 km/h within 15 min) and incorrect automation flight management system (FMS[1]). Simply said the aircraft's software did not activate the most effective braking system as the conditions for its activation (according to the software's knowledge) were not fulfilled. This protection was used to ensure that the braking systems (thrust reverse and the spoilers) do not start their operation before the aircraft is certainly on the ground. To sum up – the aircraft's automatic systems' logic prevented the activation of ground spoilers and thrust reversers until 1 525 m from the normal touchdown point. Moreover, in case of the described accident there was no way for the pilots to override the decision of the telematic systems implemented in the cockpit.

As a result of this accident's investigation the aircraft's manufacturer has amended the technical sensors (but not software) and has advised to change procedures for pilots. The disaster is considered as an example of an accident resulting, among others, with defective software, not just faulty procedures and pilot error. The recommendations also included amendments at the Warsaw Airport and to its documentation.

3   Operational Risk Management and Assessment

The international standard responsible for risk management is ISO 31010 – Risk management – Risk assessment techniques [5]. As nowadays organizations of all types and sizes face a range of risks that may affect the achievement of their objectives, it became clear that all those risk must be somehow managed. Risk management includes the application of logical and systematic methods for [5]:

- communicating and consulting throughout this process;
- establishing the context for identifying, analysing, evaluating, treating risk associated with any activity, process, function or product;
- monitoring and reviewing risks;
- reporting and recording the results appropriately.

---

[1] FMS – Flight Management System – a specialized computer system that automates a wide variety of in-flight tasks, reducing the workload on the flight crew. Its primary function is in-flight management of the flight plan. Using various sensors to determine the aircraft's position, the FMS can guide the aircraft along the flight plan. The FMS sends the flight plan for display to the Electronic Flight Instrument System (EFIS), Navigation Display (ND), or Multifunction Display (MFD).

Therefore it can be said that risk management process contains the following elements [5] (see Fig. 4):

– communication and consultation;
– establishing the context;
– risk assessment (comprising risk identification, risk analysis and risk evaluation);
– risk treatment;
– monitoring and review.

Fig. 4. Elements of the risk management process (own study based on [5])

Attention should be paid to the fact that risk management process is cyclic, which means that after the completion of phase 5 (monitoring and review), possible risks should be re-determined, or previously determined risks should be checked if they are still identified adequately (step 1) and then all steps of the process should be completed again.

One of the most significant parts of the risk management procedure is risk assessment. It can be said that it is an overall process of risk identification, risk analysis and risk evaluation (see Fig. 5), however it is not an independent activity and should be fully integrated with other risk management parts.

Attention should be paid to the fact that risk management concept is in literature defined in various ways. One of the most popular definitions determines risk as the likelihood of adverse effects occurrence [6]. However it can also be understood as a possibility of negative phenomena occurrence [7]. No matter which definition is taken into account and where does it come from one thing is sure, that the total risk can be divided into detected and remaining risk. The last one consisting of acceptable and residual risk – Fig. 6.

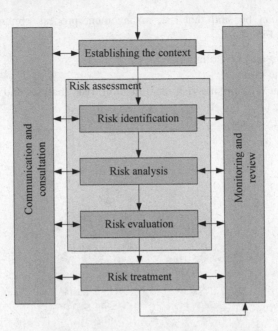

Fig. 5. Risk assessment process in an overall process of risk management process (own study based on [5])

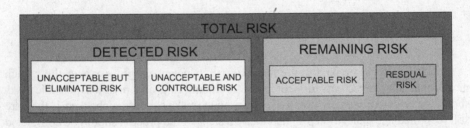

Fig. 6. Risk's types (own study based on [7])

Referring to documents related to aviation another risk division (however not excluding the first one) can be presented. It is a triangle in inverted position [8], representing the entire safety management process (see Fig. 7).

It can be noted that most safety risks, as a consequence of hazards, are assigned to the intolerable risk part, while less of them can be attributed into the tolerable region and the smallest number of safety risks fall straight into the acceptable risk level. While analysing a specific case it is of course expected to obtain the result in the acceptable region.

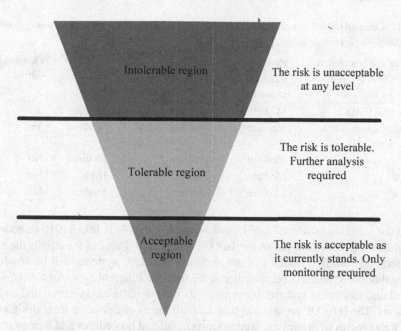

Intolerable region — The risk is unacceptable at any level

Tolerable region — The risk is tolerable. Further analysis required

Acceptable region — The risk is acceptable as it currently stands. Only monitoring required

Fig. 7. Risk's types in safety management process (own study based on [8])

In order to conduct the risk assessment process correctly it is important to properly select the techniques used. Suitable techniques should exhibit the following characteristics [5]:

– it should be justifiable and appropriate to the situation or organization under consideration;
– it should provide results in a form which enhances understanding of the nature of the risk and how it can be treated;
– it should be capable of use in a manner that is traceable, repeatable and verifiable.

Factors influencing the choice of risk assessment techniques in accordance with [5] refer to:

– complexity of the problem and the methods needed to analyse it;
– the nature and degree of uncertainty of the risk assessment based on the amount of information available and what is required to satisfy objective;
– the extent of resources required in terms of time and level of expertise, data needs or cost;
– whether the method can provide a quantitative output.

Examples of a few selected risk assessment methods (from the group of function analysis) are indicated in the Table 1. Relevance of each method is rated as high, medium or low in terms of attributes described above.

Table 1. Comparison of risk assessment techniques from the functional analysis group (own study based on [2])

Name of the selected risk assessment technique	Risk assessment process			Quantitative output
	Resources and capability	Nature and degree of uncertainty	Complexity	
FMEA and FMECA	Medium	Medium	Medium	Yes
Reliability-centred maintenance	Medium	Medium	Medium	Yes
Sneak analysis	Medium	Medium	Medium	No
HAZOP	Medium	High	High	No
HACCP	Medium	Medium	Medium	No

All the methods listed in Table 1 are discussed in details in ISO 31010 standard [5]. In the following part of this paper, for the case described in Part 2 of the article, the hazard and operability study (HAZOP) of the aircraft's telematic systems will be conducted. This method was selected because it can be used in different types of systems (mechanical and electronic systems, procedures, as well as software systems) and complex operations. The HAZOP process can deal with all forms of deviation from design intent due to deficiencies in the design, component(s), planned procedures and human actions.

4   Hazard and Operability Study of Aircraft's Telematics Systems

As we can read in [5] "HAZard and OPerability study (HAZOP) is a technique to identify risk to people, equipment, environment and/or organizational objectives. The HAZOP process is a qualitative technique based on use of guide words which question how the design intention or operating conditions might not be achieved at each step in the design, process, procedure or system. It is generally carried out by a multi-disciplinary team during a set of meetings. The study team is also expected, where possible, to provide a solution for treating the risk".

The first important task of this team is to define the objectives and scope of the study and to establish a set of key- or guidewords for the study: *No* or *not*, *More* or *higher*, *Less* or *lower*, *As well as*, *Part of*, *Reverse* or *opposite*, *Other than*, *Compatibility*, which are applied to parameters such as [5]:

- physical properties of a material or process;
- physical conditions such as temperature, speed;
- a specified intention of a component of a system or design (e.g. information transfer);
- operational aspects.

Using the guidewords and splitting the system, process or procedure into smaller elements or sub-systems or sub-processes the study team conducts a qualitative and quantitative research and review of relations between the important parameters and requirements (e.g. technical, operational, procedural), using the auxiliary keywords, e.g.: *Other*, *Fluctuation*, *Early*, *Late*.

The main objective and the result of the HAZOP study is a confrontation of functional and operational requirements with the requirements concerning safety. This involves the systematic cross-checking with less probable circumstances of the deviations and inconsistencies. Experts analyzing the system ask themselves the following exemplary questions:

- *What deviations can arise?*
- *How can they affect the safety and operability?*
- *What actions are necessary to prevent this?*

As a result, HAZOP method is more effective than checklists, by identifying and describing the outcomming data:

- *Deviation* - what is the deviation expressed by the combination of relationship keywords.
- *Potential cause of deviation (reason)* - a description of the conditions which must be fulfilled so that a deviation occurs.
- *Effects* - description of the consequences of deviation.
- *Preventive* - how to protect the system or mitigate the risk?
- *Action* - what actions should be taken to remove or mitigate the effects of the cause?

For HAZOP analysis of an air accident described above, Authors determined the parameters and guiding words, as shown in Table 2.

Table 2. HAZOP analysis for flight LH2904 accident [own study]

Parameter	Requirements	Deviation	Reason	Effects
Speed of approach	Nominal, as in the aircraft operating manual	• Too fast; or • Too slow	• Speedometer error; • FMS hardware or software error	• Increases – RWY overrunning; or • Decreases – RWY undershooting
Course of approach	Nominal, as in the standard approach procedure	Incorrect	• Compass error; • Wind direction and/or wind speed sensors or indicator error; • ILS/LOC receiver or Indicator error; • FMS hardware or software error	Missed approach point (MAP or MAPt) – missed approach procedure
Angle of glide slope	Nominal, as in the standard approach procedure	• Larger; or • Smaller	• Speedometer error; • ILS/GP receiver or indicator error; • FMS hardware or software error	• Larger – overrunning the runway; or • Smaller – undershooting the runway; or • Missed approach procedure

(*continued*)

Table 2. (*continued*)

Parameter	Requirements	Deviation	Reason	Effects
Flaps extended	Nominal, as in the aircraft operating manual in given conditions	• Too little; or • Too much	• Mechanical fault; • Wind direction and/or wind speed sensors or indicator error; • Aircraft flap position indicators error; • FMS hardware or software error	• Too little – overrunning the runway; or • Too much – RWY undershooting
Touchdown speed	Nominal, as in the aircraft operating manual	• Too fast; or • Too slow	• Speedometer error; • FMS hardware or software error	• Increases – overrunning the runway; or • Decreases – RWY undershooting
Touchdown distance from the runway threshold	Nominal – in touchdown zone length, as in the standard landing procedure	• Too far; or • Too close	• Wind direction and/or wind speed sensors or indicator error; • ILS/GPreceiver or indicator error; • FMS hardware or software error	• Too far – RWY overrunning; • Too close – extended length of the rolling-out
Engine reverse turn on	Nominal, as in the aircraft operating manual in given conditions	• Too early; or • Too late; • Not turned on	• Mechanical fault; • Speedometer error; • FMS hardware or software error	• Too early – loss of steering and damage of aircraft or extended length of the rolling-out; or • Too late or not turned on – RWY overrunning
Gear brakes turn on	Nominal, as in the aircraft operating manual in given conditions	• Too early; or • Too late; • Not turned on	• Mechanical fault; • Speedometer error; • FMS hardware or software error	• Too early – damage of gear and/or wheels or extended length of the rolling-out; or • Too late or not turned on – RWY overrunning

5   Conclusion

Term telematics is defined as telecommunication, information and IT solutions as well as automatic control solutions, tailored to the needs of the supported physical systems and highly integrated with them. Transport telematics is a wide range of applications, in which the most important role play intelligent transport systems used not only in road transport but also in air applications. Transport telematics systems use a variety of devices and applications: radio systems, geographical databases, satellite navigation systems, radars, etc. Flight Management System described in this paper is an example of such telematic solutions. It is used to automate a wide variety of tasks, and to reduce the workload of the flight crew. However, its incorrect software, erroneous application or not proper information utilization may lead to an aircraft accident or serious incident, as it was in the presented case where the most effective braking systems were not activated as the conditions for their activation (according to the software's knowledge) were not fulfilled.

The results of the hazard and operability study conducted in this paper indicate that the crucial meaning on the occurrence of the Airbus A320 aircraft accident, which at the same time mean the most common cause, concern the FMS system. Not only data and information quality must be monitored in the entire data chain but also the applied telematic systems at all stages of their usage.

The general Authors' conclusion is the necessity to conduct research and operational risk analysis at each FMS process stage: software planning, development, verification and configuration management process, as it is described in [9]. The Authors' recommendation is to perform a FMS HAZOP analysis with the significant participation of pilots and air traffic controllers.

References

1. Dudek, E., Kozłowski, M.: Koncepcja procesu oceny jakości danych lotniczych, Prace Naukowe Politechniki Warszawskiej – Transport, z. 113, str. 131-140, Warszawa (2016)
2. Dudek, E., Kozłowski, M.: Koncepcja zarządzania jakością danych lotniczych, Prace Naukowe Politechniki Warszawskiej – Transport, z. 113, str. 141–150, Warszawa (2016)
3. Dudek, E., Kozłowski, M.: The concept of a method ensuring aeronautical data quality, J. KONBiN 1(37), 319–340 (2016). Warszawa (2016)
4. http://www.rvs.uni-bielefeld.de/publications/Incidents/DOCS/ComAndRep/Warsaw/warsaw-report.html. Accessed 2 Jan 2017
5. ISO 31010, Risk management – Risk assessment techniques
6. Molak, V.: Podstawy analizy ryzyka i zarządzania ryzykiem. CRC Lewis Publishers, Floryda (1996)
7. Klich, E.: Bezpieczeństwo lotów, Wydawnictwo Naukowe Instytutu Technologii Eksploatacji. PIB, Radom (2011)
8. ICAO Doc. 9859 – Safety Management Manual (SMM), International Civil Aviation Organization, 3rd edn. (2013)
9. RTCA/DO-178B – Software Considerations in Airborne Systems and Equipment Certification, Radio Technical Commission for Aeronautics (RTCA) and European Organisation for Civil Aviation Equipment (EUROCAE)

Aerodrome Traffic Support with the Use of Infrastructure-to-Vehicle Communication

Jacek Skorupski[✉]

Faculty of Transport, Warsaw University of Technology,
ul. Koszykowa 75, 00-662 Warsaw, Poland
jsk@wt.pw.edu.pl

Abstract. Increasing air traffic imposes the need to seek for solutions improving airport capacity. The standard approach is to expand the infrastructure. However, this is costly and time-consuming. This is why solutions increasing the capacity merely by changing the organization of aerodrome traffic are sought. The purpose of this paper is to present a new concept involving the optimization of braking on the runway. Standard braking profiles can be inefficient because of many possible disturbances and uncertainties. By applying the concept of infrastructure-to-vehicle communication it is possible to modify the standard braking profile so as to reach the desired speed in the vicinity of the runway exit and at the same time to not extend the runway occupancy time. Preliminary version of braking profile adjustment algorithm has been developed and implemented into the ACPENSIM simulator, built as hierarchical, coloured Petri net. Results for the simulation of the scenario where an aircraft touches down in a different place than it was planned and has a different mass, show the effectiveness of the algorithm. Modified braking profile allowed for achieving the appropriate final velocity with almost unchanged runway occupancy time, which determines the capacity of the airport.

Keywords: Aerodrome traffic management · Petri nets · Infrastructure-to-vehicle communication · Airport capacity

1 Introduction

Air traffic is steadily increasing in recent years. This makes it necessary to take measures aimed at increasing airports capacities. Organizational solutions, much cheaper than the investment in the infrastructure, are getting increasing importance. Some of them require close cooperation of ATM (Air Traffic Management) with the aircraft crews in managing the aerodrome traffic.

The airside capacity of an airport depends directly on the runway system capacity (runway's layout). This, in turn, depends on many factors, but the most important is the runway occupancy time (ROT). This time is the basis for determining the theoretical maximum runway throughput. A significant part of the runway occupancy time for landing operations falls on the landing roll, which is under consideration in this paper. The relations are [16]:

© Springer International Publishing AG 2017
J. Mikulski (Ed.): TST 2017, CCIS 715, pp. 396–410, 2017.
DOI: 10.1007/978-3-319-66251-0_32

$$T_{max} = \frac{3600}{\overline{ROT}} \tag{1}$$

where:

T_{max} — maximum runway throughput with continuous takeoff and landing operations, expressed in the number of operations per hour

$\overline{ROT}$ — the average runway occupancy time in seconds. This time depends on many factors, such as aircraft type, touchdown speed, touchdown point, selected runway exit, runway surface condition and many others. One of the key factors is the average aircraft speed between the touchdown point and the runway exit point. This speed depends on the braking profile (BP), including the use of wheel brakes and the thrust reverser. In practice, we usually use the concept of the so-called practical runway capacity, which takes into account many random factors affecting airport traffic operations and the delays resulting from this. Regardless of the selected definition of capacity, the general relationship between runway capacity and runway occupancy time is the same [12].

Airport capacity management is one of the most frequently found topics in the literature on air transport management. It is discussed, for example, in [1, 6, 8, 9, 11, 20, 21]. These papers focus on various factors affecting airport capacity, e.g. aircraft speeds, traffic mix, separation minima, passenger operations, taxi-out times, etc. Relatively little is devoted to issues regarding runway occupancy time [22], and particularly the influence of what technique is used when the aircraft braking process is being carried out.

Chen et al. [3] pointed to the extremely important issue of uncertainty regarding the aerodynamic forces and moments which arise, for example, from a lack of information regarding the plane's mass at touchdown, which may significantly change the parameters of the braking process. They propose the use of artificial neural networks in order to find a good approximation of these forces. In the current paper we also assume that these forces are unknown, but it is possible to compare current position of the aircraft on the runway with the planned one. This allows finding the difference from the assumed parameters.

In this paper we propose a model and a decision support system in which the pilot plays an active role in a task which is closely related to ATM. This task is runway capacity management. The pilot can select the braking force in such a way as to keep the ROT as small as possible or so that occupancy time is compatible with the intentions of the air traffic controller. Unfortunately, there are many factors that cause that it is extremely difficult without a proper support of the braking process. As a part of the previous research we have developed the so-called standard braking profiles that allow achieving the variety of goals, including minimizing the ROT [19]. This paper proposes an extension, involving the use of the infrastructure-to-vehicle (I2V) communication concept. It can be used to update the information about the location of the aircraft on the runway. This enables adjustment of the standard BP in order to take into account the possible disturbances of the braking process. Location of the radio transmitters, so-called position markers, and their use in the BP correction algorithm was proposed. This algorithm is currently under development. Preliminary results obtained using the ACPENSIM simulator with correction algorithm implemented show that the proposed system has a great potential for implementation.

The remaining part of this paper has the following structure. Section 2 discusses applications of infrastructure-to-vehicle communication in air traffic. Section 3 contains the description of the landing roll model used in the ACPENSIM simulator. In Sect. 4 some guidelines for creating standard braking profiles are given, together with the recommended BPs with different evaluation criteria. In Sect. 5 the concept of the system for BP adjustment is explained, together with the general algorithm used to modify the profiles. Section 6 contains a case study, which presents how the system works for the scenario involving the touchdown in a different location than expected and at the same time for different aircraft mass. This example was·developed for the Warsaw Chopin Airport. In Sect. 7 a summary, final conclusions and further work plans are presented.

2   I2V Communication in Air Traffic

In air traffic we can clearly distinguish several subsystems that make up a federative structure. They are characterized by a high degree of autonomy with its own internal organization, but at the same time cooperating for achieving a main goal. In this paper we deal with a specific usage of I2V communication, so as examples of such subsystems we can distinguish:

- an aircraft performing the flight according to a given trajectory and
- radio navigation aids which assist the aircraft crews in determining their position.

Currently used radio navigation aids belong to different technology generations, utilize different physical phenomena, have different accuracy, but the general idea of their work consists in:

- indicating the angle between the aircraft heading and the direction to radio navigation beacon which has a known location on the ground, or
- indicating the distance from the navigation aid.

An example of a system implementing the former of these functions is the VHF Omnidirectional Range beacon (VOR), generally used to support aircraft that are in the cruise phase of flight, landing approach or climbing after takeoff. The principle of VOR operation is the continuous emission of fixed and variable radio signal. The time difference between receiving both signals allows determining the angle between the heading of the aircraft and the beacon. An example of a device utilizing a latter function is a Distance Measuring Equipment (DME). It works on the principle of question-and-answer initiated by the on-board device. Ground-based equipment receives question, identifies it and sends back a response. The device installed on board the aircraft receives the response and determines the distance from the DME beacon using the duration of the transmission.

A special case of I2V communication used in aviation is a function of precise location of the aircraft - directly above a navigational aid. It is important as it minimizes the error during determining the position of the aircraft. The error is all the greater the greater is the distance from the navigation aid. This kind of navigational support is implemented by devices typically used in cruise navigation, such as VOR, but also by devices supporting landing, such as Instrumental Landing System (ILS). It contain elements (so-called markers) that emit a narrow radio beam directed vertically upwards. Passing aircraft can accurately locate the position in relation to the runway threshold which they intend to use for landing. Typically, there are three ILS markers:

- outer, located approximately 7 km from the runway threshold,
- middle, located about 1 km from the runway threshold,
- inner, located approximately 60 m from the runway threshold.

Air traffic navigation systems operating in I2V communication technology consist of an appropriate ground component (I) and onboard component (V). Their mutual co-operation allows supporting the pilot in controlling the aircraft. It is also the basis for the construction of systems performing automatic execution of certain functions. The classic example is a flight using the autopilot.

3 Model of a Landing Roll Process

The phase of flight which occurs immediately after touchdown is a phase of braking (landing roll). The aircraft moving with a touchdown speed reduces it to make it possible to turn into the runway exit. In civil aviation, the aircraft brakes by using the wheel brakes, aerodynamic brakes or thrust reverses.

Braking with the use of wheel brakes can be done in two ways [7]:

- by manual braking, where the braking force is controlled by the pilot who uses the brake pedal; this type of braking is used during emergency situations and on short runways when it is needed to stop the airplane as soon as possible
- by automatic braking (autobrake), which starts to operate after touchdown when an aircraft's force pressure on the runway is large enough to make the braking effective; the braking force is determined based on the range selected by the pilot as well as by the aircraft speed.

For aerodynamic braking, the aircraft uses [7]:

- aerodynamic brakes; these are usually tilting surfaces on top of the aircraft's wings which increase the aerodynamic drag without changing the aircraft lift force,
- flaps, which serve mainly to increase the aircraft lift force but at the same time they also increase the aerodynamic drag; they are usually in the extended position after touchdown, thus they increase the efficiency of aerodynamic braking,
- spoilers, whose main task is to decrease the aircraft lift force as soon as possible so that the automatic braking process becomes possible; spoilers also increase the aircraft aerodynamic drag.

The main purpose of this paper is to show the possibility of supporting the efficient use of available braking resources during the aircraft landing roll. The landing process model which was used is relatively simple but reliable. This was intended because the simulator is based on a numerical prediction of further aircraft movement, which should be performed in real time. With the excessive complexity of the mathematical model, one may find that the time required for the calculation is too large, which would put the concept into question. However, the measurements that were carried out initially confirm the model's adequacy.

3.1   Mathematical Model

The model which was created here will be used to support the braking process. The following assumptions were adopted during development of the model:

- the area of analysis is the aircraft landing roll, beginning from the touchdown point and ending at the point where the aircraft turns into a taxiway, usually via a rapid-exit taxiway
- the designated runway exit is determined in advance, for example by the air traffic controller
- the final speed of the landing roll is the input variable; this speed may depend on the exit taxiway configuration
- other available data are: aircraft lifting surface, aircraft weight and the wheels' braking drag coefficient, which depends on the runway surface condition
- the optimization criterion is one of the following: minimization of ROT, noise, tire wear or maximization of passenger comfort
- the output variables (decision variables) are: the moments in time when the aircraft begins and ends braking with the use of the thrust reverser and with wheel brakes; moreover, the rotation speed of the low-pressure rotor, which unambiguously defines the thrust reverser braking power as well as the wheels' braking power.

In the model, the time was adopted as a discrete value and the discretization step Δt was adopted to be equal to 0.1 s. During the landing roll the aircraft moves by decelerated movement [5]. The details of the model can be found in [19]. Generally we use the formula for aircraft speed at time t_i

$$V_i = \frac{\Delta t}{m} \left(P_i + C_x \cdot S \cdot \frac{\rho \cdot (V_{i-1})^2}{2} + \mu_d \cdot m \cdot g \cdot k + T_p + W \right) + V_{i-1} \qquad (2)$$

where
m - mass of the landing aircraft,
P_i - reversed thrust at time t_i produced by the engine with the thrust reverser deployed (obtained from [2]),
C_x - drag coefficient determined for the lift force coefficient $C_z = 0$,
S - aircraft lifting surface,
ρ - mass density of air,
V - aircraft speed,
μ_d - drag coefficient during braking with the use of wheel brakes,
g - gravitational acceleration,

k - coefficient that shows the use of the wheel braking force, related to the force used for pressing on the brake pedal, $k \in [0, 1]$,

T_p - rolling friction force which occurs during the landing roll,

W - wind force component which causes an additional increase or decrease of the braking force depending on which side the wind blows.

The calculation tool (ACPENSIM simulator) was created by using the presented mathematical model. The tool allows to estimate the landing roll distance, runway occupancy time (ROT) and deceleration affecting the passenger. The simulator is described in more detail in Sect. 3.3.

3.2   Petri Net for Braking Process Simulation

Petri nets [10, 13, 15] can be an excellent tool for simulation analysis of BPs during the landing roll. Originally, they were developed for modelling computer systems working synchronously. However, their high versatility has resulted in their having many other applications in recent years, including modelling and support of air traffic management processes [4, 14, 17, 18, 23].

In this paper, a hierarchical Petri net was implemented. Its essence is decomposition of a model for multiple subnets (pages) that are combined through "substitution transitions" and "fused places". These indicate nodes that simultaneously belong to several sub-models, thereby connecting them and allowing to transfer tokens between subpages.

3.3   Braking Process Simulator – ACPENSIM

The braking process simulator supporting management of the runway capacity was developed using the CPN Tools 4.0 software package. This is an advanced and increasingly popular package for creating models in the form of Petri nets, in their simulation and analysis in state space. In this section, the basic idea of this simulator will briefly be presented.

The ACPENSIM simulator consists of five subpages:

– *PrflMv* – this page models the aircraft's movement from touchdown to the runway exit from the BP point of view. It uses movement parameters (speed, acceleration) as assumed in the BP and takes into consideration the system time flow. Additionally, it stores information about the movement parameters in consecutive time moments.

– *RlMv* – this page models the real aircraft's movement. It uses actual movement parameters (speed, acceleration) which in reality can differ from those assumed in the braking profile.

– *Check roll* – this page determines the distance that is necessary to achieve the appropriate speed to exit the runway if the plane was moving without using wheel brakes and the thrust reverser.

- *APCS* – this page models the process of receiving information from the position markers. The planned and the actual locations are compared and updated here. Depending on the results of the comparison, the elements of the BP adjustment algorithm are initiated and partially executed.
- *ParChg* – this page models the process of braking parameters change. This includes lengthening or shortening various phases of braking and also changing the braking force.

The place *Start* on the page *ParChg* (Fig. 1) initiates the simulation and allows determination of the distance that the modelled aircraft has moved. The transition *Touchdown* is responsible for operations related to aircraft touchdown. In particular, the weight of the arc connecting transition *Touchdown* with the place *Roll* determines the time that the plane moves with aerodynamic braking only. The place *Roll* is used to connect pages in the hierarchy and is generally responsible for counting the time remaining until the end of the current phase of the landing roll. The transition *Change parameters* is responsible for changing the phase of braking.

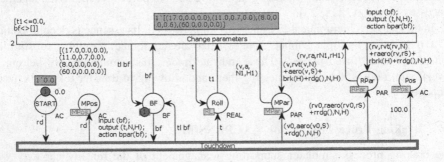

Fig. 1. The *ParChg* subpage of the ACPENSIM simulator [own study]

The place *Time* on the page *PrflMv* (Fig. 2) is responsible for the system time flow. The transition *Move* synchronizes the calculation of the movement parameters along with the time flow (places *Pos* and *Par1*). The place *Profile* is responsible for storing the BP data. Figure 2 also shows references to functions *aero*, *rvt*, *brk*, and *rdg* which define the parameters of braking, respectively, aerodynamically with the thrust reverser, wheel brakes and by rolling drag. These functions are consistent with the model described in Sect. 3.1; their software code is not presented.

The subpage *Check roll* (Fig. 3) represents a simulation model of the aircraft's movement in the case when only aerodynamic braking takes place. Its main tasks include determining the speed during motion (place *Roll speed*) and comparing it to the preset target speed (transition *Stop roll*). If these are equal, then the distance necessary to achieve the desired speed is determined and returned to the *PrflMv* page through the place *Distance to Stop*.

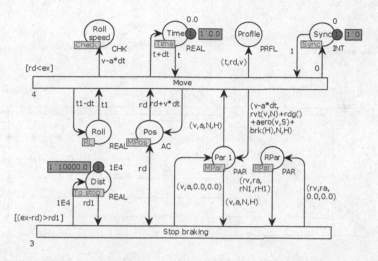

Fig. 2. The *PrflMv* subpage of the ACPENSIM simulator [own study]

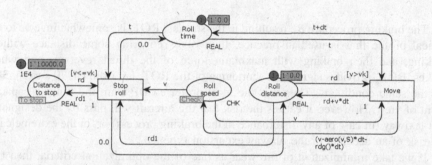

Fig. 3. The *Check roll* subpage of the ACPENSIM simulator [own study]

4 Standard Braking Profiles

The mathematical model of the braking process and ACPENSIM simulation tool based on coloured Petri nets allows to determine the BP that meets the control objectives. As will be shown in this section, these objectives may be different. The detailed description of standard BP's can be found in [19].

A typical braking procedure B_0 begins just after aircraft touchdown with an almost maximal braking force both with the use of wheel brakes and the thrust reverser. Then, after substantial speed reduction, the aircraft rolls freely, braking only by aerodynamic drag and rolling friction drag. This kind of braking profile results with a long ROT. In case of using exit S from runway 33 at Warsaw Chopin Airport (Fig. 4), and the maximum speed at this point equal to 17 m/s the total ROT while using this BP is about 70 s.

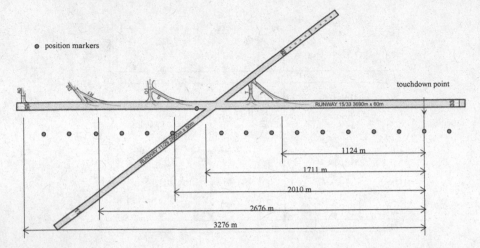

Fig. 4. The layout of Warsaw Chopin Airport and proposed locations of position markers [own study]

The braking procedure B_s, resulting in the shortest ROT, is somewhat inverse to the typical profile that is used in practice. It involves covering some distance without braking, and then braking with maximum force of the thrust reversers and wheel brakes. BP B_s provides significant shortening of the ROT, i.e. from ca. 70 s to ca. 38 s for the same conditions. Thus it is the most preferred BP from the runway capacity point of view. However, it carries the risk that the aircraft will pass by the designated exit taxiway (in case of any disturbance in the braking process) or, in the extreme case, the risk of an accident of the runway excursion type.

If we take minimization of tire wear as one of the optimization criteria, then two cases are possible. The first occurs when the use of wheel brakes is not necessary at all. In such a case it is possible to obtain ROT that is equal to 43.2 s, which is very close to the minimum of 37.6 s. However, this kind of BP is difficult to recommend for general use because it requires to brake with the thrust reverser at very low speeds when its efficiency is low. The second case which may occur with such an optimization criterion is when the slowdown to the assumed runway exit is impossible without using wheel brakes. The wheel brakes should then be activated as late as possible, at the lowest possible speed.

Braking by thrust reversers causes emission of a large amount of noise. At many airports, environmental (noise) constraints preclude such procedures. This applies mainly to the night period, when it is prohibited to exceed certain levels of noise. However, restrictions on the total emission of noise may also apply at other times of the day. Accordingly, three braking strategies that reduce the level of noise can be proposed. The first of them, B_{n1}, does not provide for the use of a thrust reverser, but only of braking by using wheel brakes. The second, B_{n2}, involves the need for the reverser. Such a situation occurs when, for example, the runway is covered by a mixture of water and snow; then it is necessary to take into account the reduced friction coefficient. In this BP we propose to use the smallest possible reverse thrust. The third strategy, B_{n3}, consists in very short use of the thrust reverser.

All of the previously presented BPs assumed that the braking equipment (reverser, brakes) may be used with high intensity. Such action results in the existence of large gradients of deceleration, which is uncomfortable for passengers. It is also possible to determine a BP (designated as B_c) in which the basic criterion is passenger comfort. This will be achieved through such use of the braking devices that deceleration will be less than 2 m/s². The ROT of profile B_c differs only slightly from profile B_s, which maximizes the runway capacity. However, in the case of landing during lower traffic, when minimizing ROT is not that important, this way of braking should be considered because the passengers' negative perception is relatively small for this BP.

It is also possible to determine heuristic BPs which largely meet all of these criteria. This kind of BP will be considered in our case study in Sect. 6. Their general characteristics are as follows:

- after touchdown the aircraft moves without braking for some time; deceleration takes place due to aerodynamic drag and rolling friction drag
- then thrust reverser braking is applied with the low-pressure rotor at moderate rotational speed (60–70% of the maximum)
- applying the wheel brakes follows with medium intensity (60–70% of the maximum); at the same time the thrust reverser is stowed
- then the wheel brakes are disengaged and the plane rolls to the exit of the runway; this piece of the BP aims to create a safety buffer that can be applied if the actual braking parameters are worse than anticipated.

5 Braking Profile Adjustment System

Standard braking profiles can provide a significant support for the achievement of the high runway capacity or other goals mentioned in Sect. 4. This paper presents the next stage of the research aiming at creation of automatic braking system which could be installed as a pilot assistance unit. Its main task is to present the concept of a BP adjustment system which will adapt a standard BP to the current traffic conditions.

The main purpose of the proposed system, and the concept of its further use, is to aid one specific landing. Many factors influence the course of the braking process. They were briefly discussed in Sect. 3. It is important that most of these factors are not known precisely at the time when standard BP is being determined. In this context, we can distinguish three groups of factors.

- Possible to determine from the aircraft level, thus possible to automatic inclusion in the braking assistance unit. Touchdown speed is an example.
- Possible to determine by ground services. Examples of such parameters are the runway coefficient of friction or wind speed and direction. They can be measured immediately before landing and transmitted to the aircraft crew which could manually enter them into the braking assistance unit.

– Impossible to be determined accurately. This group is the most numerous. It includes factors such as: the landing aircraft mass, air density, aircraft lifting surface. They are all known approximately, but due to measurement inaccuracies, local conditions or inaccuracies in operation of devices they may vary slightly from the nominal values. On the other hand, it includes parameters such as available landing roll distance resulting directly from the actual touchdown point.

It should be noted that the factors determined from the outside of the aircraft may be inaccurate due to the high dynamic of changes. For example, both the strength and direction of wind can vary, which may reduce the usefulness of the information communicated to the aircraft crew.

Inaccuracy of the execution of the braking procedure complements all of this information uncertainty. Even assuming that the individual parts of the braking operation will be performed automatically, it may turn out that the thrust reverser or wheel brakes can be deployed or stowed with some delay. The same applies to applying different rotational speed values of the low-pressure rotor or the force used to press the brake pedal.

All these circumstances indicate that simulation is an adequate approach to the implementation of the BP adjustment system. Instead of treating data by methods appropriate to the uncertainty (e.g., using probability or fuzzy theory), the idea is to repeat calculations of the braking profile in real time, wherein each successive calculation is performed taking into account new information about the traffic parameters.

Such new information can come from many sources. However, the analysis of methods used in I2V communication so far indicates that the best solution will be installation of special radio transmitters along the runway. The principle of their operation will be similar to ILS markers, with the difference that the radio beam is emitted horizontally. They will be positioned at known locations, and will transmit simple information about the position on the runway to the aircraft. On this basis, the BP adjustment system will be able to compare the actual position with the expected by standard BP and to modify it adequately. The general algorithm of the initial version of the BP adjustment system is as follows:

– Consider the first position measurement as a verification of touchdown point. Compare the real position with the nominal one, described by BP used.
  • In case of the touchdown before the nominal point - extend the phase of movement without braking. Go to step 3.
  • In case of the touchdown beyond the nominal point - shorten the phase of movement without braking. Go to step 3.
  • In case when the difference is greater than the length of the section where no braking was planned - go to step 2.
– Calculate the braking time according to the B_0 profile (See Sect. 4). Then calculate the time according to the standard BP, but for the next runway exit. Choose the option that minimizes the runway occupancy time.
– Compare the actual location with the nominal one and specify Δs_1. Modify the braking parameters by the earlier or later beginning of braking with both types of devices (thrust reverser and wheel brakes).

- Compare the actual location with the nominal one and specify Δs_2. If $|\Delta s_2| > |\Delta s_1|$ modify the braking parameters by a proportional increase/decrease of rotational speed of the low pressure rotor and the wheel brakes force.
- If the runway exit point was reached - STOP. Otherwise go to step 3.

The above algorithm has been implemented in the ACPENSIM simulator as a preliminary version of the braking profile adjustment system. An example of its operation for the case of disturbances scenario is given in Sect. 6.

6 Case Study: BP Adjustments for Disturbances Scenario

As a case study, adjustments for the BP compromise due to different criteria will be presented. This BP was developed for the Warsaw Chopin Airport. It involves using a taxiway S (Fig. 4). We will analyze a scenario in which the aircraft touches down 100 m beyond the nominal point, and at the same time the actual weight of the aircraft is about 10% lower than the nominal, assumed in the scenario. The analyzed profile consists of four steps:

- after touchdown the aircraft moves without braking for 17 s,
- the thrust reverser braking is applied with the low-pressure rotor at 70% of the maximum rotational speed for 11 s,
- the wheel brakes are applied with 60% of the maximum force for 8 s, at the same time the thrust reverser is stowed,
- the wheel brakes are disengaged and the plane rolls to the exit of the runway for 7 s.

The concept of the BP adjustment system requires the provision of I2V information from transmitters (position markers) located along the runway. The density of these transmitters will be of key importance for the speed and efficiency of the system. For this example, we assume that transmitters are placed every 200 m. The location information is transmitted at the time the aircraft passes the marker. These are the only moments when adjustment of braking parameters is carried out. The proposed locations of the position markers for the runway 33 at Warsaw Chopin Airport are shown in Fig. 4.

The standard BP adopted for the experiment is illustrated in Fig. 5.

According to the analyzed disturbances scenario, the touchdown takes place 100 m beyond the nominal point so the first measurement of the position is possible after 100 m of movement in accordance with the standard BP. The correction of the actual position and the possible change in braking parameters, according to the general algorithm described in Sect. 5, takes place in the subsequent measurement points spaced at 200 m. The resulting braking profile which is the result of the application of BP adjustment system is shown in Fig. 6.

A step change in distance approximately 1.5 s after touchdown can be clearly seen on the graph of actual braking profile. This is the moment when the aircraft passes the first position marker. At this point, the actual position is updated and corrected by 100 m. The change is due to the displaced touchdown point. In the following moments, we can see smaller adjustments of the position. They are caused by the difference of the

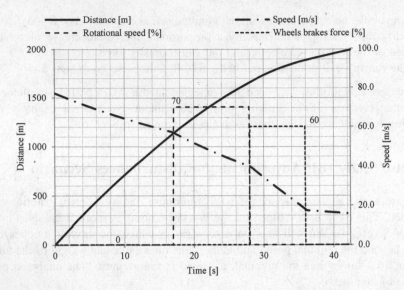

Fig. 5. Braking profile B_W used in a case study for Warsaw Chopin Airport [own study]

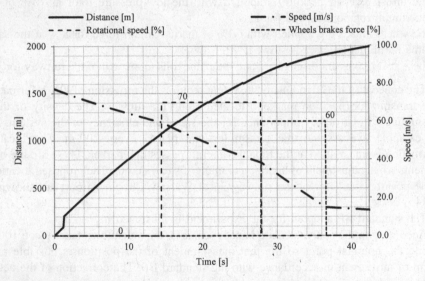

Fig. 6. Braking profile B_WA modified by the BP adjustment system [own study]

actual weight of the aircraft. Particularly well they are visible in the vicinity of 25th and 31st second after the beginning of landing roll. We can also observe a change in the use of braking devices resulting from the operation of BP adjustment algorithm. The phase when only aerodynamic braking is used was shortened, and the phase with a thrust reverser use was extended. This is correct, since in the case when we have less place to brake we need more intensive braking. The phase with the use of wheel brakes was also slightly longer.

7   Conclusion

The mathematical model and the computer calculation tool based on coloured Petri nets (ACPENSIM simulator) allow to determine the BP using the thrust reverser and the wheel brakes, which allows for optimization of the ROT or another criterion (minimization of noise, minimization of tire wear, maximization of passenger comfort). It is also possible to find solutions combining several criteria.

In the paper [19] we showed the possibility of effective control of the braking process so as to obtain the expected operational results. Among them, an increase of the runway capacity seems to be the most important, and therefore an increase of the entire airport capacity. The model can be the basis for creation of a decision support system for the pilots, it will also be of benefit for aerodrome traffic management. The target solution is a system that automatically, continuously and smoothly chooses the rotational speed of the low-pressure rotor and the strength of wheel brakes' application. This paper shows the assumptions and preliminary results achieved by the developed system.

The paper presents a general concept of the standard braking profile adjustment system. It is based on position markers installed along the runway. Their locations have been proposed. The algorithm is relatively simple, but it generates satisfactory solutions for many typical traffic situations. The case study analyzed the situation when touchdown occurs farther than planned, and at the same time the aircraft is slightly lighter. Thus we are faced with the need to plan for a more intensive braking in the first phase of landing roll (because the available distance is shorter), then a less intense (because the aircraft is lighter). By using the concept of I2V communication it is possible to adjust a braking profile so that the ROT practically does not change (the difference is less than 1%). This is possible thanks to comparing the planned and the actual position. At the same time, by using a modified BP, somewhat lower velocity was obtained, which shows that there is still potential to improve the BP adjustment algorithm.

As a part of further work it is planned to improve the BP adjustment algorithm. We need to find an answer to the question "which operating parameters are inconsistent with the assumptions?" in the real time and to take account of this in the algorithm. Additionally it is planned to include a total change of the sequence of using braking devices in the algorithm. At the same time it is planned to study for proper selection of the location of position markers.

References

1. Balakrishna, P., Ganesan, R., Sherry, L.: Accuracy of reinforcement learning algorithms for predicting aircraft taxi-out times: a case-study of Tampa Bay departures. Transp. Res. Part C Emerg. Technol. **18**(6), 950–962 (2010)
2. CFM. Normal Operating Considerations, Flight Ops Support, Snecma & General Elecric., 14 February 2014. www.smartcockpit.com
3. Chen, B., et al.: Robust adaptive neural network control of aircraft braking system. In: 2012 10th IEEE International Conference on Industrial Informatics (INDIN), pp. 740–745 (2012)

4. Davidrajuh, R., Lin, B.: Exploring airport traffic capability using Petri net based model. Expert Syst. Appl. **38**(9), 10923–10931 (2011)
5. Dommasch, D.O., Sherby, S.S., Connolly, T.F.: Airplane Aerodynamics. Pitman Publishing Corporation, New York (1961)
6. Farhadi, F., Ghoniem, A., Al-Salem, M.: Runway capacity management—an empirical study with application to Doha International Airport. Transp. Res. Part E Logist. Transp. Rev. **68**, 53–63 (2014)
7. Flight Safety Foundation. Approach-and-Landing Accident Reduction Briefing Note 8.4: Braking devices (2009)
8. Gelhausen, M.C., Berster, P., Wilken, D.: Do airport capacity constraints have a serious impact on the future development of air traffic? J. Air Transp. Manag. **28**, 3–13 (2013)
9. Irvine, D., Budd, L.C.S., Pitfield, D.E.: A Monte-Carlo approach to estimating the effects of selected airport capacity options in London. J. Air Transp. Manag. **42**, 1–9 (2015)
10. Jensen, K.: Coloured Petri Nets. Basic Concepts, Analysis Methods and Practical Use. Springer, Berlin (1997)
11. Kalakou, S., Psaraki-Kalouptsidi, V., Moura, F.: Future airport terminals: new technologies promise capacity gains. J. Air Transp. Manag. **42**, 203–212 (2015)
12. Horonjeff, R., McKelvey, F.X., Sproule, W.J., Young, S.B.: Planning and Design of Airports. McGraw-Hill, New York (2010)
13. Marsan, M.A., et al.: Modelling with generalized stochastic Petri Nets. Dipartamento d'Informatica, Universita degli Studi di Torino, Torino (1999)
14. Oberheid, H., Söffker, D.: Cooperative arrival management in air traffic control—a coloured Petri net model of sequence planning. In: van Hee, K., Valk, R. (eds.) Applications and Theory of Petri Nets, vol. 5062, pp. 348–367. Springer, Berlin (2008)
15. Reisig, W.: Understanding Petri Nets. Modeling Techniques, Analysis Methods, Case Studies. Springer, Berlin (2013)
16. Sherry, L.: Runway Capacity Workbook. George Mason University, Center for Air Transportation Systems Research, Fairfax (2009)
17. Skorupski, J.: Method of analysis of the relation between serious incident and accident in air traffic. In: Soares, G. (ed.) Advances in Safety, Reliability and Risk Management, vol. 2, no. 11, pp. 2393–2401. CRC Press/Taylor & Francis, London (2011)
18. Skorupski, J.: The risk of an air accident as a result of a serious incident of the hybrid type. Reliab. Eng. Syst. Saf. **140**(140), 37–52 (2015)
19. Skorupski J., Wierzbicki, H.: Airport capacity increase via the use of braking profiles. Transp. Res. Part C-Emerg. Technol., 1–18 (2016). doi:10.1016/j.trc.2016.05.016
20. Sölveling, G., Clarke, J.-P.: Scheduling of airport runway operations using stochastic branch and bound methods. Adv. Comput. Commun. Impact Transp. Sci. Technol. **45**, 119–137 (2014)
21. Stelmach, A., Malarski, M., Skorupski, J.: Model of airport terminal area capacity investigation. In: Safety and Reliability for Managing Risk, Proceedings of the European Safety and Reliability Conference ESREL 2006, pp. 1863–1868 (2006)
22. Trani, A.A., et al.: Flight Simulations of High Speed Runway Exits. Department of Civil Engineering, Transportation Systems Laboratory, Virginia Tech University, Blacksburg (1996)
23. Werther, N., Moehlenbrink, C., Rudolph, M.: Colored Petri net based formal airport control model for simulation and analysis of airport control processes. In: Proceedings of the 1st International Conference on Digital Human Modeling (ICDHM 2007), pp. 1027–1036 (2007)

Application Model Fuzzy-Probabilistic in Work Designation of Routes in Transport Internal

Katarzyna Topolska[✉]

WSB University in Wrocław, Fabryczna 29/31, 53-609 Wrocław, Poland
katarzyna.topolska@wsb.wroclaw.pl

Abstract. The article presents a model combining the theory of fuzzy sets of Probability to the task of routing internal transport. This model apart from the literature of neural networks can be used to plan routes in intelligent transport systems in stock. The article described a model realizing the optimal setting of internal transport using the operator fuzzy Hamacher. By optimizing internal transport as well as internal roads, we can count on a significant reduction in manufacturing costs. Transport plants plays an important role in the mechanism of production. Depending on the type and nature of the enterprise, transport costs may be high on the hierarchy of the costs incurred by the company. A well-designed internal transport is not only lower costs but also the time and the quality of production.

Keywords: Fuzzy model · Probabilistic model · Transport of systems

1 Introduction

In recent years knowledge management systems were developed in different directions, to understand and improve business processes, to generalize them in process models, as well as company specific fittings. Lots of concepts were acquired for different situation-depended problems. Many of them focused in user-friendly and intuitively application with approaching in human mind. Especially in business administration and knowledge management great store is put in verbal information and sketchy worded statements, assumptions, purposive ideas or restrictions. Many decision-making-processes are marked by not being formed as mathematical models or give concrete numerical values. They are deduced by fuzzy assumptions. Although these fuzzy assumptions are not precise, it is possible to link significant and additional information with specifically captured company situation.

At this point, the Fuzzy-Logic starts. With the Fuzzy-Set-Theory, or Fuzzy-Logic, an appropriate tool is developed that get over the difference between the precision of model and process engineering and considering qualified information and tolerate uncertainties at modeling process [1–4].

In the following chapters I will arrange the Fuzzy-Logic in business processing context. I will describe the functionality of Fuzzy-Logic and display an example of transport forecasting.

© Springer International Publishing AG 2017
J. Mikulski (Ed.): TST 2017, CCIS 715, pp. 411–423, 2017.
DOI: 10.1007/978-3-319-66251-0_33

2 Fuzzy-Logic in General

Fuzzy-Logic is used in many different application areas and systems. It is useful when fuzzy information like speech should process in semantic meaning or even complex systems should be described by models in short time and cost-efficiently.

Except of pure mathematical application or direct using of the Fuzzy-Set-Theory, like used in social science, there are two main directions of methodical approach, the *knowledge-based* and the *algorithmic* applications [5].

The knowledge-based applications are based on the human experience. In case of difficult or not been solved mathematically problems or in case of absence of efficient algorithm to create mathematical models, the human problem solving behavior is simulate in a data processing system. To get this, knowledge has to be captured and complex prepared [6].

The algorithmic applications are based on existing mathematical models or methods to solve the problem. The problems in these models are the exact values. The Results are often insufficient described problems. The Fuzzy-Logic tries to extend the missing parts in these methods as well as to improve them [7].

The most successful applications of Fuzzy-Logic come from the field of automation and control. In addition, it is also used in other areas, e.g. in databases, in medicine and in expert systems. They are particularly suitable for control problems, in which not only the correct action, but also its dosing must be derived (e.g. temperature control, vehicle control). Fuzzy-Logic-Systems are also suitable for problem solving in cases where a large number of input parameters move in continuous value ranges (measured data in technical or natural processes). In addition to the control and regulation technology, special mention should also be made of applications in the business context. Following are some examples of possible applications for Fuzzy-Logic [12]:

- Personal deployment planning: controlling of field worker
- Waste incineration plant: controlling firing and steam intense (optimization of the incineration process, reducing pollutant emission)
- Controlling business processes: controlling warehouse stock by business rules (good relation of security of supply and storage costs)
- Subway, elevator, air-condition, heating-systems, video cameras: automation of different devices
- Artificial neural networks
- Fire detection: decreasing false alarm rate by interpreting signal characteristics.

Today the knowledge management area is highly diverse and has a complex structure. But three main types of knowledge management systems (Fig. 1) have emerged:

- Supporting systems for knowledge processing
- Enterprise-Knowledge-Management-Systems
- Intelligent Technologies.

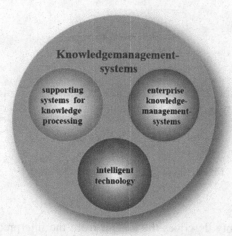

Fig. 1. Main types of knowledge management [own study]

Supporting systems for knowledge processing are special systems to improve and create new kinds of knowledges, created for engineers, scientists and other knowledge worker.

Enterprise-Knowledge-Management-Systems are general, integrated, related to entire company systems that collects, saves, transmit and use of digital values and knowledges.

The *intelligent technologies* compare many different systems and have many different targets. These technics are data mining, expert-systems, neural networks, genetic algorithm, intelligent agents and the Fuzzy-Logic.

The *Fuzzy-Logic* is used in all main types of knowledge management, but especially in the intelligent technologies. In the following chapter I will focus on these intelligent technics and especially on the application of transports forecasts [10].

3 Fuzzy- Logic

Before explaining the basics of Fuzzy-Logic, I would like to explain what the term Fuzzy-Logic actually means. The word "fuzzy" is similar to "vague", "unsafe" or "blurred". In the case of this expression, it is noticeable that fuzzy and logic are initially mutually exclusive. In the Fuzzy-Logic, the knowledge seems to be predominant in linguistically formulated rules. The linguistic formulations usually have uncertainties that can be divided into the following three categories of uncertainties (Fig. 2) [13].

Stochastic Uncertainty
Stochastic uncertainty means, that the occurrence of an event cannot predicted with certainty. For example the rolling of the event "6" that can be achieved with a probability of 1/6 [13].

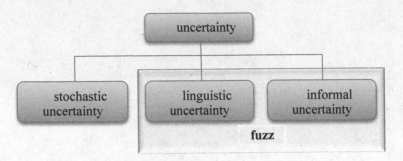

Fig. 2. Types of uncertainty [own study]

Linguistic Uncertainty
The linguistic uncertainty describes the arising from the interpretation of a statement such as "warm" or "cold". Thus, a person's feeling of "cold" can already be a temperature of 15 °C or even a temperature of 10 °C. People are capable of condensing linguistic uncertainty with context into less uncertain terms. Computer cannot [13].

Informal Uncertainty
This type of uncertainty uses the fact, that certain terms are based on very subjective definitions, for which there are in some cases no objective evaluation variables or cannot be obtained. This refers to concepts such as "trustworthiness", "creditworthiness" or "sympathy". Fuzzy-Logic is a rule-based technique that can be described as artificial intelligence. It represents this uncertainty by creating rules that use approximate or subjective values. It describes a particular phenomenon or process linguistically and then presents it as a small number of flexible rules [13].

Development
The origins of the Fuzzy-Logic are already in the old Greek philosophy. Plato, for example, suspects that there must be another third value-difference between "true" and "false" in the classical set-theory. His disciple Aristoteles did not follow this teaching, whit which he influenced mathematical thinking for the last two millennia. Modern philosophers resumed the views of Plato. Thus B. Russel wrote in 1923:

"The law of excluded middle is true, when precise symbols are employed, but it is not true, when symbols are vague, as, in fact, all symbols are."

This inaccuracy represented B. Russel as a set of values rather than exact values. At the same time, J. Lukasiewicz was the first to introduce a systematic alternative to divalent logic. He argued that there are sentences to which the values "true" and "false" cannot be assigned. From this, Lukasiewicz concluded the existence of a third value, which he put between "true" and "false" and gave him the designation "possible". Later he developed four- and five-digit logics and also mentioned that an infinitely valuable logic with values between [0, 1] could exist. In 1937 M. Black took the idea of B. Russel and introduced the numerical representation of uncertain symbols with the name *consistency-profile*. The mathematical concept of uncertain sets (Fuzzy-Sets-Theory) led to L. Zadeh in 1965. With this concept, he combines the ideas of M. Black and J. Lukasiewicz. Zadeh looked for a mathematical treatment of vague, fuzzy concepts to simplify the modeling of complex systems. The bases for the description of vague terms are fuzzy sets, of which theory the Fuzzy-Logic is based [13].

Concept

Zadeh's motivation was to deal with the uncertainty of human thought in terms of data presentation and decision-making. The classical set-theory says, that an element x from a basic set X ($x \in X$) either belongs uniquely to a set A or is not unique to this set A. For many problems, such a sharp classification is not an appropriate representation of knowledge. Because even in our language of speech and even our way of thinking we often use formulations of uncertain amounts, such as "fairly", "something", "a little", etc. In reality, gradually graded affinities can also be observed. Let us take the temperature sense for the sake of illustration. For example, a temperature of 25 °C can be felt as warm for one person, and the value of 20 °C for another. Thus, it appears to make little sense to start the warm temperature from 25 °C. Values of 24 °C would no longer be associated with the quantity of the warm temperature range [13].

In language as well as our way of thinking, we often use formulations of uncertain quantities such as "fairly", "something", "a little". This type of formulations covers the classical set theory inadequately.

The Fuzzy-Logic allows the representation of such gradual affinities and speaks in this context of Fuzzy-Sets. Compared to the classical concept of quantity with sharp limits, a Fuzzy-Set has fuzzy limits. The Fuzzy-Set-Theory is not based on the two-valued statements "true" (1) or "false" (0), but on the values between "true" or "wrong", i.e. between 1 and 0. This allows each element to belong to a certain set with a certain probability. Each of these elements of a "fuzzy" set is characterized by the membership value (probability of belonging). The description is made by the introduction of Fuzzy-Sets and the characterization of simple relations with fuzzy-conditional statements of the form:

If statement A,

then statement B.

Characterization of Fuzzy-Sets
Fuzzy-Sets, if x is a set of objects, then

$$\tilde{A} := \{(x, \mu(x)p(x)), x \in X\} \tag{1}$$

$\tilde{A}$ is a Fuzzy-Set of X. Probability of an event x is p. A Fuzzy-Set $\tilde{A}$ is described by the fact that the membership of an element x to $\tilde{A}$ can be specified by a real number which is normalized to the value range [0, 1]. Formally, a Fuzzy-Set $\tilde{A}$ can be described by a real-valued function $\mu_{\tilde{A}}$:

$$\mu_{\tilde{A}} \text{ and } p : X \rightarrow [0, 1] \tag{2}$$

This membership function assigns to each element of a definition set X a number from the real-valued interval [0, 1] of the target set, which indicates the membership degree $\mu_{\tilde{A}}(x)$ of each element x to the defined Fuzzy-Set $\tilde{A}$.

A value of $\mu_{\tilde{A}}(x) = 0$ implies that x does not belong to the Fuzzy-Set $\tilde{A}$, whereas a value of

$$\mu_{\tilde{A}}(x) = 1 \text{ indicates a full affiliation.}$$

Values in the interval $0 < \mu_{\tilde{A}} < 1$ therefore show a partial affiliation from x to $\tilde{A}$.

The following Fig. 3 shows the representation of the vague concept "warm temperature", where the basic set of possible temperatures is continuous.

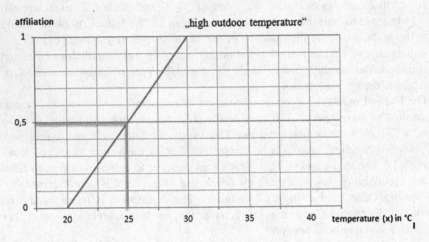

Fig. 3. Representation of "high temperature" by blurred quantity [own study]

In addition to the graphical representation, there are further descriptions for Fuzzy-Sets:

Parametric representation in the form of analytical functions describing the course of the characteristic curve as a whole or piecewise, e.g.:

$$\mu(x) \begin{cases} 0 & \text{for} & x < 20 \\ \frac{x-20}{10} & \text{for} & 20 \leq x \leq 30 \\ 1 & \text{for} & x > 30 \end{cases}$$

Discrete representation by specifying *discrete pairs* of values (element, membership value), e.g.:

$$\mu(x) = \{(25, 0.5), (24, 0.4), (26.5, 0.7), (28, 0.9)\}$$

Fuzzy-Sets can be interpreted semantically in two different ways:

- *Physical Interpretation*: Fuzzy-Set describe a vague object.
- Example: gray-value-image of radiography.
- *Epistemic Interpretation:* Fuzzy-Set describe a vague observation of a sharp object.

Example

Usually, trapeze functions are used to describe Fuzzy-Sets. This can be used to express terms such as "approximately between" the set m_1 and the set m_2. But also triangles or bell curves are used. These best represent terms such as "close to m_1", or individual points for discrete Fuzzy-Sets. These forms, such as trapezes, triangles, and individual points, are often used to describe Fuzzy-Sets because of their simplicity and mathematical properties (Fig. 4).

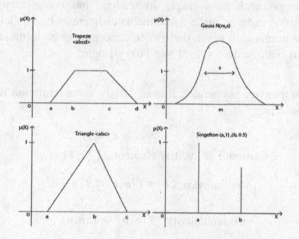

Fig. 4. Description of fuzzy-sets [own study]

Height of Fuzzy-Sets

The height H of the Fuzzy-Set is the highest degree of membership which all elements of the Fuzzy-Set have. For a Fuzzy-Set μ over a basic set $\tilde{A}$, its height is defined as:

$$\mathbf{H}(\tilde{\mathbf{A}}) = \mathbf{max}\{\mu(\mathbf{x}) : \mathbf{x} \in \mathbf{X}\} \tag{3}$$

The two-valued logic therefore operates with sets whose maximum is 1.

α-section of a Fuzzy-Set

The α-section $\tilde{A}_\alpha$ of a Fuzzy-Set $\tilde{A}$ is the set of all elements whose membership to $\tilde{A}$ is at least α.

$$\tilde{\mathbf{A}}_\alpha = \{\mathbf{x} | \mathbf{x} \in \mathbf{X}, \mu_{\tilde{\mathbf{A}}}(\mathbf{x})\mathbf{p}(\mathbf{x}) \geq \alpha\} \tag{4}$$

With this given α and a Fuzzy-Set Ã, their so-called α-sections can be determined. However, a Fuzzy-Set can also be approximated solely from the knowledge of the α-sections, so that it is sufficient to store individual α-sections (usually of 0.1, 0.2 ... 1.0) in the computer system. From these, the original function can be calculated, which is facilitated by the use of triangular or trapezoidal amounts.

The Sect. 1

$$\tilde{A}_1(\mathbf{x}) = \left\{ \mathbf{x} | \mathbf{x} \in \mathbf{X}, \mu_{\tilde{A}}(\mathbf{x})\mathbf{p}(\mathbf{x}) = 1 \right\} \tag{5}$$

of Ã is named core of Ã.

Operations on Fuzzy-Sets

Now that we have elaborated some of the basic concepts, we will look at the well-known concepts such as average, unification, and complementarity. This is therefore necessary in order to make a reasonable estimate of Fuzzy-Sets. To this end, we introduce two norms with which these three concepts can be formulated so broadly that they apply to both the two-valued and Fuzzy-Logic:

t-Norm (AND)

In order to create an average operator, it must satisfy some minimum requirements of mathematics. These are:

$$\text{Existence of neutral elements: } T(a, 1) = a \tag{6}$$

$$\text{Monotony: } a \leq b \Rightarrow T(a, c) \leq T(b, c) \tag{7}$$

$$\text{Commutativity: } T(a, b) = T(b, a) \tag{8}$$

$$\text{Associativity: } T(a, T(b, c)) = T(T(a, b), c) \tag{9}$$

The most common average operator, which just meets these requirements is a function T and is called t-norm:

$$T : [0, 1]^2 \rightarrow [0, 1] \tag{10}$$

This is often referred to as the AND operator of Fuzzy-Logic. Where T is monotonic and T(a, 0) = 0.

The first point states that the section of a Fuzzy-Set with an ordinary set results in the degree of membership remaining. The second point ensures that a degree of membership to the average cannot be reduced when a Fuzzy-Set with a larger Fuzzy-Set is cut. The third and forth point should be self-evident to average operators.

t-Conorm (OR)

An association operator must also meet certain minimum requirements:

$$\text{Existence of neutral elements: } \perp(a, \mathbf{0}) = a \tag{11}$$

$$\text{Monotony: } a \leq b \Rightarrow \perp(a, c) \leq \perp(b, c) \tag{12}$$

$$\text{Commutativity: } \perp(a, b) = \perp(b, a) \tag{13}$$

$$\text{Associativity: } \perp(a, \perp(b, c)) = \perp(\perp(a, b), c) \tag{14}$$

The most common association operator, which just meets these requirements, is a function $\perp$

$$T\perp : [0, 1]^2 \rightarrow [0, 1] \tag{15}$$

and is named t-Conorm. This is often referred to as the OR operator of the Fuzzy-Logic.

Negation (NOT)

Each constructed negation function is a special case of the generalized negation. This is a function n: $[0, 1] \rightarrow [0, 1]$ with $\alpha \rightarrow 1 - \alpha$. By this definition of the negation one obtains the t-conorm from the t-norm and vice versa, so that the dual principle is preserved as in the classical logic. That means, the t-conorm can be determined from the negation and the t-norm, and the t-norm from negation and t-conorm. The simplest functions which correspond to these general operators and their requirements are the "min" and "max" functions, which correspond to the average and association formation. The following graphics are intended to illustrate this, and these terms can be interpreted as the colloquial AND (ref. Fig. 5) or OR (ref. Fig. 6). The "min" and "max" function on the membership function is defined as follows:

$$\boldsymbol{mean} = \boldsymbol{min}\left\{ \mathbf{p}(\mathbf{x})\mu_{function1}(x), \mathbf{p}(\mathbf{x})\mu_{function2}(x) \right\} \tag{16}$$

$$\boldsymbol{union} = \max\left\{ \mathbf{p}(\mathbf{x})\mu_{function1}(x), \mathbf{p}(\mathbf{x})\mu_{function2}(x) \right\} \tag{17}$$

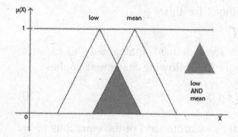 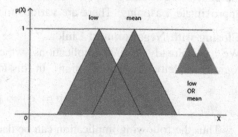

Fig. 5. Average formation of two linguistic variables "low" and "medium" over "min"-function [own study]

Fig. 6. Union formation of two linguistic variables [own study]

Of course, there are other ways to define the intersection operators. For example the Dubois-Prade operators:

$$\text{average operator:} \quad \frac{ab}{max\{a,b,\alpha\}} \, with \, \alpha \in (0,1) \tag{18}$$

$$\text{union operator:} \quad \frac{a+b-ab-min\{a,b,\alpha\}}{max\{1-a,1-b,\alpha\}} \, with \, \alpha \in (0,1) \tag{19}$$

Fuzzy-Systems

Fuzzy-Systems are not systems that use unsafe information. Rather, they are systems based on Fuzzy-Logic. Only the core of a Fuzzy-System, the so-called Fuzzy-Inference, works with vague information, which is why fuzzification and a defuzzification must be switched before, in order to convert precise values into fuzzy values and vice versa.

Fuzzification

A Fuzzy-System is initially given a safe value to work with. However, the system itself works with vague information, which must be generated in the first step of the Fuzzy-System. This step is called fuzzification: for a given linguistic variable ("strong", "fast"...), membership functions are defined for their linguistic terms. Then, for a sharp input value with the aid of the Fuzzy-Sets, the degrees of membership with respect to all linguistic terms are determined. Here, Fuzzy-Sets of linguistic terms are usually defined such that for a sharp input value the sum of the degrees of membership to all linguistic terms of a linguistic variable is one. This overlaps linguistic variables and ensures that no gaps occur in the description. Thus, for each fixed input value, it is possible to unambiguously calculate the degree to which it can be assigned to a linguistic variable. This distinction also means that several values exist for different linguistic variables.

Fuzzy-Inference

The Fuzzy-Inference is the core of a Fuzzy-System. Fuzzy-Rules are used to work with vague information. In order to apply a rule of the form already described to vague information, a rule is needed for the case that the premise is more or less true, and thus, the conclusion is more or less valid. This evaluation is also called Fuzzy-Closure or approximate reasoning. There are various methods for this:

Closing with Negation and Link

We have already learned implications, which were defined analogously to classical logic (complement, section, union). In this logic the following statement applies:

$$(\mathbf{p} \Rightarrow \mathbf{q}) \Leftrightarrow (\neg\mathbf{p} \vee \mathbf{q}) \tag{20}$$

Thus the following implication can be deduced with the aid of the equations of the negation

$(\neg A := 1 - \mu_A(x))$ and the union $(A \vee B := max\{\mu_A(x), \mu_B(x)\})$:

$$\mathbf{A \Rightarrow B := max\{1 - \mu_A(x)p(x), \mu_B(x)p(x)\}} \tag{21}$$

This implication is not completely sent, since if $\mu_A(x) = 0$ then the problem is that the conclusion has the value 1. This has disadvantages when linking to other rules. Therefore, there are more sophisticated variants.

Mamdani-Implication

The Mamdani implication is based on the idea that the truth content of the result should not be higher than the truth value of the premise:

$$_{\mathbf{A \Rightarrow B}}(\mathbf{x}, \mathbf{y}) := \mathbf{min\{\mu_A(x)p(x), \mu_B(x)p(x)\}} \tag{22}$$

With this the above-described problem is circumvented.

<u>More Implications</u>

There are much more possibilities for implications. Here the algebraic product is to be mentioned as a frequently used implication as well as the Zadeh-implication of the inventor of the Fuzzy-Logic:

$$Algebraic\, product: \mu_{\mathbf{A \Rightarrow B}}(\mathbf{x}, \mathbf{y}) = \mathbf{p(x) \cdot \mu_A(x) \cdot \mu_B(x)} \tag{23}$$

$$Zadeh-Implication: \mu_{\mathbf{A \Rightarrow B}}(\mathbf{x}, \mathbf{y}) = \mathbf{max\{min\{\mu_A(x)p(x), \mu_B(x)\}, 1 - p(x)\mu_A(x)\}} \tag{24}$$

Defuzzification

As a result of an inference operation, we obtain again a fuzzy set, with other words, we get a fuzzy information. A fuzzy system must convert this information back into sharp output values in order to be able to control the end devices. For this transformation, there are different methods. I will explain two of them.

Maximum-Method

To determine the exact output value, only the rule with the highest degree of completion is used for the maximum method. There are three different variants:

- <u>Selection of the mean value</u>
  The left and right vertices of the maximum are added and divided by two.
- <u>Selection of the left edge</u>
  Only the left corner point is used.
- <u>Selection of the right edge</u>
  The right corner is used accordingly.

In the MAX-PROD variant, all three variants provide the same result. However, this method has advantages and disadvantages, which must be estimated, because

- the calculation effort for a exact output value is low
- the determined value is independent of the actual degree of fulfillment of the rule with maximum degree of fulfillment
- If there are several rules with the same degree of fulfillment as the maximum, the mean value should be chosen, because different conclusions could be drawn.

Focus-Method

Above all, the last point can lead to complications in the maximum method why one likes to use the focus method. The resulting Fuzzy-Set is considered as a whole. Several rules can be considered, since the sharp starting point is the abscissa value of the focus of the area below the Fuzzy-Set. Unfortunately, the calculation effort is significantly higher, but the linking of several rules is not a problem. The output value may have little or no membership. However, this should be very unlikely with a suitable choice of rule base. In addition, the possible value range of the output variable is not fully exhausted. A rander extension can help here. In order to reduce the computation time, there are some variants, e.g. the focus method with SUM-MIN inference or the height method.

4 Conclusion

Humans spend most time in solving problems or coming to a decision, these regards especially to businessmen and manager. Decision making is attendant on precariousness. Unknown parameter like future changes, fuzzy described or even unknown requirements can increase the precariousness. But this cannot stop human in making decisions. Fuzzy-Logic represents the mathematical structure to get used to this vagueness. Diffuse speech is often used in Fuzzy-Logic like warm, hot, cold, fast, slow, etc. The users were helped with these words, because they can describe their experiences without handle complex mathematical algorithm in computer. This makes human machine interfaces simpler.

This summary can only describe a small part of the Fuzzy-Logic-Theory an only one example is presented. However, I hope that thinking behind the Theory and the potential of Fuzzy-Logic are recognizable. The most successful application of Fuzzy-Logic is the Fuzzy-Control, which is mostly used in industries to control complex constructions. The investments to solve a complex non-linear control-problem were reduced normally by using fuzzy-control. It sacrifices not the precision of classic mathematic models, but only the unused excessively precision. When using a naturally imprecise sensor, it is excessive to look for exact models. Now used Fuzzy-Methods in industry can be changed to conventional, mathematical and informatics methods. But the difference is the Fuzzy-Solution will be easier, more priceless, simpler to develop and simpler to implement. The Solution might not be perfect, but keep in mind, that the last 10% of precision costs 90% of the investment. This makes Fuzzy-Systems useful and economically justifiable. In other fields of application like the Fuzzy-Expert-Systems the solution without Fuzzy-Logic will not be possible. Until now the Fuzzy-Logic is used in few systems, but the successful using in other fields of application might be probably.

Instead of understanding the Fuzzy-Theory as a theory itself, the "fuzzification" can be seen as a method that can be used generally to transfer sharp theories to a fuzzy form. There is a generalization in automata theory that reflects the context as well: fuzzy cognitive maps (FCMs). Even other theories were fuzzificated, for example the fuzzy-differential-equation.

At least the Fuzzy-Logic is not a cure-all or magic bullet. But there are many possible applications. It is much to be hoped, that the Fuzzy-Logic (perhaps in combination with neural networks) leads to a more pleasant and more comprehensible interface between technical systems and humans.

References

1. Bodendorf, F.: Freimut. Daten- und Wissensmanagement, 2nd edn. Springer, Berlin (2006)
2. Böhme, G.: Fuzzy-Logik; Einführung in die algebraische und logische Grundlagen. Springer, Berlin (1993)
3. Frank, H.: Fuzzy Methoden in der Wirtschaftsmathematik. Eine Einführung. Vieweg, Decatur (2002)
4. Huckle, T., Schneider, S.: Numerische Methoden, 2nd edn. Springer, München (2006)
5. Laudon, K., et al.: Wirtschaftsinformatik: eine Einführung. Pearson Studium, München (2006)
6. Lippe, W.: Soft-Computing mit Neuronalen Netzen; Fuzzy-Logic und Evolutionäre Algorithmen. Springer, Berlin (2005)
7. Teschl, S., Teschl, G.: Mathematik für Informatiker, 3rd edn. Springer, Wien (2008)
8. Thomas, O.: Fuzzy Process Engineering: Integration von Unschärfe bei der modelbasierten Gestalltung. Gabler Edition Wissenschaft (2009)
9. Topolski, M.: The use of cluster analysis and the theory of mathematical records in the process of planning the production-warehouse flow. Res. Logist. Prod. (2015)
10. Topolski, M.: Planowanie optymalnej trasy przejazdu transportu samochodowego z wykorzystaniem miękkich metod obliczeniowych Autobusy. Technika, Eksploatacja, Systemy Transportowe (2015)
11. Topolski, M.: Komputerowy model rozmytego klasyfikatora Bayesa w ocenie rentowności produkcji, Logistyka (2014)
12. Traeger, D.H., Dirk, H.: Einführung in die Fuzzy-Logik. Vieweg+Teubner Verlag, Stuttgart (1994)
13. Zadeh, L.A.: Fuzzy sets. Inf. Control 8, 338–353 (1965)

Failure Effects Analysis by Multiple Random Variable

Karol Rástočný, Mária Franeková, and Jozef Balák[✉]

Faculty of Electrical Engineering, University of Žilina,
Univerzitná 1, 01026 Žilina, Slovak Republic
{karol.rastocny,maria.franekova,
jozef.balak}@fel.uniza.sk

Abstract. The safety-related systems are typically resisting against dangerous faults. Failure effects on the system can be determined directly by monitoring the original system installation, by simulation of the system operation using its model, or by computing or theoretical reasoning. The process of system ageing can be described with the help of the random failure time. If the system contains n elements, generally, the ageing process can be characterised as the n-dimensional random process with time-dependent random variables of the n-dimensional random vector. The probability density of the failure occurrence of the i-th system element is represented by the i-th random variable of the random vector. If the analysis of the safety integrity of the safety-related electronic system is used method FTA, that application of knowledge in the field of multiple random variable can greatly simplify the computation of dangerous failure (top event) rate.

Keywords: Multiple random variable · Dangerous failure · FTA

1 Introduction

Various methods have been successfully utilized to perform safety analysis of a control system. Usually more than one method is applicable and after thorough consideration and in dependence of particular case of analysis one appropriate method or combination of methods shall be chosen. State of the art methods, whether widely used or methods under development but meant for evaluation of safety, could be divided into two categories:

- Complex modelling methods that are able to cover functional and technical safety. Such model is applicable for risk assessment and analysis, functional requirements specification and verification of safety properties of a system [1].
- Combination of individual methods. Specific methods are used for risk analysis (risk graph, risk matrix, HAZOP,...) [2], another formal and semi-formal methods are used for modelling and specification of functional properties of a system (UML, state machines, Z - language, Petri nets,...) [3, 4] and some methods are advisable to model RAMS (reliability, availability, maintainability, safety) parameters of a system (FMEA, FTA, RBD, Markov chains, Petri nets,....) [5–10].

© Springer International Publishing AG 2017
J. Mikulski (Ed.): TST 2017, CCIS 715, pp. 424–435, 2017.
DOI: 10.1007/978-3-319-66251-0_34

Those methods are supported by a wide range of software tools (i.e. BQR CARE reliability software [11], Windchill Quality Solutions [12], ITEM software [13]), that could provide not only distinguishing features to create and edit model, but also implement tools for simulation followed by presentation of achieved results.

Basic disadvantage of most common methods (FMEA, RBD and FTA) is lack of possibility of covering up more than one or two factors affecting the safety of a system. Simultaneous influence of more than two factors can be for example modelled by means of Markov chains.

Modelling of the safety related properties of a system is done with the aim for probability of the dangerous state, expressed as a function of factors influencing the safety of a system, e.g.:

$$P_D(t) = g(\lambda, \mu, c, \ldots, t). \tag{1}$$

In specific situations, determination of dangerous failure λ_D instead of probability of the dangerous state is required (Fig. 1).

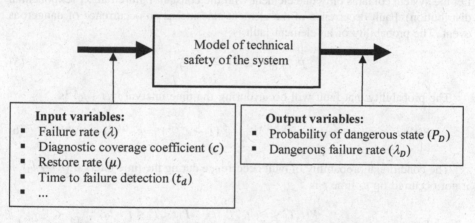

Fig. 1. General idea of the technical safety modelling [own study]

A non-standard approach can be used to the evaluation of simultaneous influence of various factors on the safety integrity of SRES, too. This approach is based on the using of a multiple random variable.

2 Probability Model of Random Failure Effects

Let the system contains n mutually independent elements and let the system ageing process is characterised as a n-dimensional random process with time-dependent random variables of the n-dimensional random vector. The probability density of the failure occurrence of the i-th system element is represented by the i-th item of the random vector. Consider the n-dimensional random vector $T = (T_1, T_2, .., T_n)$ with n random variables. Let the distribution function of the random vector T is

$$F(t_1, t_2, \ldots, t_n) = P(T_1 \leq t_1, T_2 \leq t_2, \ldots, T_n \leq t_n). \tag{2}$$

Let the random vector $\mathbf{T}$ has a distribution of the continuous type, then

$$f(t_1, t_2, \ldots, t_n) = \frac{\partial^n F(t_1, t_2, \ldots, t_n)}{\partial t_1 \ldots \partial t_n}, \tag{3}$$

where $f(t_1, t_2, \ldots, t_n)$ is a probability density of the random vector $\mathbf{T}$.

The probability of the vector $\mathbf{T}$ being in the area of the n-dimensional space, bounded by inequalities $a_1 < T_1 \leq b_1, \ldots, a_n < T_n \leq b_n$, is

$$P(a_1 < T_1 \leq b_1, \ldots, a_n < T_n \leq b_n) = \int\limits_{a_1}^{b_1} \ldots \int\limits_{a_n}^{b_n} f(t_1, t_2, \ldots, t_n) dt_1 \ldots dt_n. \tag{4}$$

2.1 Failure of One Element

Let the system contains only one element with the constant failure rate λ_1 (exponential distribution). Fault occurrence of the element also means an occurrence of dangerous event. The probability of an element fault is

$$P(T_1 \leq t) = \left(1 - e^{-\lambda_1 t}\right). \tag{5}$$

The probability that fault will occur during the time interval $\langle t, t + t_0 \rangle$ is

$$P(t < T_1 \leq t + t_0) = \left(1 - e^{-\lambda_1(t+t_0)}\right) - \left(1 - e^{-\lambda_1 t}\right) = e^{-\lambda_1 t} \cdot \left(1 - e^{-\lambda_1 t_0}\right). \tag{6}$$

The conditional probability of fault occurrence during the time interval $\langle t, t + t_0 \rangle$, if it not occurred up to time t is

$$P(t < T_1 \leq t + t_0 | T_1 > t) = \frac{P(t\langle T_1 \leq t + t_0, T_1 \rangle t)}{P(T_1 > t)} = \frac{P(t < T_1 \leq t + t_0)}{P(T_1 > t)} = \left(1 - e^{-\lambda_1 t_0}\right). \tag{7}$$

For the element with exponential distribution of the failure occurrence it holds good that if the element (system) functioned failure-free up to time t, then the probability of failure occurrence within the next time interval $\langle t, t + t_0 \rangle$ does not depend on time t but only on time interval $\langle t, t + t_0 \rangle$.

Information about the fact that up to time t no fault of the element occurred (the element was functional in time t) has an effect on the probability of failure occurrence within the time interval $\langle t, t + t_0 \rangle$, but there is no effect on failure rate of the element. Failure rate of dangerous (undesired) event in the system is

$$\lambda_D = \lambda_1. \tag{8}$$

2.2   Failure of Two Element

Let the system contains two mutually independent elements (element 1 and element 2) with the constant failure rates λ_1 and λ_2. Consider the 2-dimensional random vector $T = (T_1, T_2)$ with 2 random variables. Let the distribution function of the random vector T is

$$F(t_1, t_2) = P(T_1 \le t_1, T_2 \le t_2) = \left(1 - e^{-\lambda_1 t_1}\right) . \left(1 - e^{-\lambda_2 t_2}\right). \tag{9}$$

The marginal distribution functions of random variables corresponding to a probability of failures of individual elements:

$$F(t_1) = P(T_1 \le t_1) = F(t_1, \infty) = \lim_{t_2 \to \infty} P(T_1 \le t_1, T_2 \le t_2) = \left(1 - e^{-\lambda_1 t_1}\right), \tag{10}$$

$$F(t_2) = P(T_2 \le t_2) = F(\infty, t_2) = \lim_{t_1 \to \infty} P(T_1 \le t_1, T_2 \le t_2) = \left(1 - e^{-\lambda_2 t_2}\right). \tag{11}$$

From (4) results that

$$
\begin{aligned}
P(a_1 < T_1 \le b_1, a_2 < T_2 \le b_2) &= P(T_1 \le b_1, T_2 \le b_2) - P(T_1 \le b_1, T_2 \le a_2) \\
&\quad - P(T_1 \le a_1, T_2 \le b_2) + P(T_1 \le a_1, T_2 \le a_2) \\
&= F(a_1, a_2) + F(b_1, b_2) - F(b_1, a_2) - F(a_1, b_2),
\end{aligned} \tag{12}
$$

where a_1, a_2, b_1, b_2 are points in time. Their definition in 2-dimensional time space is shown in Fig. 2.

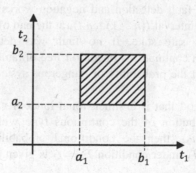

Fig. 2. Two-dimensional time space for calculation of probability $P(a_1 < T_1 \le b_1, a_2 < T_2 \le b_2)$ [own study]

Then the probability of the simultaneous occurrences fault of the element 1 in the time interval $(a_1, b_1\rangle$ and element 2 in the time interval $(a_2, b_2\rangle$ can be expressed:

$$P(a_1 < T_1 \le b_1, a_2 < T_2 \le b_2) = \left(1 - e^{-\lambda_1 b_1}\right).\left(1 - e^{-\lambda_2 b_2}\right) + \left(1 - e^{-\lambda_1 a_1}\right).\left(1 - e^{-\lambda_2 a_2}\right)$$
$$- \left(1 - e^{-\lambda_1 b_1}\right).\left(1 - e^{-\lambda_2 a_2}\right) - \left(1 - e^{-\lambda_1 a_1}\right).\left(1 - e^{-\lambda_2 b_2}\right). \tag{13}$$

System in view of fault occurrence can be in four states. These states are mutually exclusive (disjoint) and form a complete set of events.

State 1: Let $a_1 = a_2 = t$, $b_1 = b_2 = \infty$. Then

$$P(t < T_1 \le \infty, t < T_2 \le \infty) = P(T_1 > t, T_2 > t) = e^{-(\lambda_1 + \lambda_2)t}. \tag{14}$$

State 2: Let $b_1 = a_2 = t$, $a_1 = 0$, $b_2 = \infty$. Then

$$P(0 < T_1 \le t, t < T_2 \le \infty) = e^{-\lambda_2 t}.\left(1 - e^{-\lambda_1 t}\right). \tag{15}$$

State 3: Let $a_1 = b_2 = t$, $a_2 = 0$, $b_1 = \infty$. Then

$$P(t < T_1 \le \infty, 0 < T_2 \le t) = e^{-\lambda_1 t}.\left(1 - e^{-\lambda_2 t}\right). \tag{16}$$

State 4: Let $a_1 = a_2 = 0$, $b_1 = b_2 = t$. Then

$$P(0 < T_1 \le t, 0 < T_2 \le t) = \left(1 - e^{-\lambda_1 t}\right).\left(1 - e^{-\lambda_2 t}\right). \tag{17}$$

If the safety system has a mechanism of fault detection and negation, then this fact must also be respected in the failure effects analysis.

Let the mechanism of fault detection and negation works in such a way that if a fault was detected in time interval $\langle (n - 1).t, n.t \rangle$, at the end of this interval the system will get into predefined safe state. If no fault was detected in time interval $\langle (n - 1).t, n.t \rangle$, the system continues its operation free of faults, where $n = 1, 2, 3, \ldots$ From the safety viewpoint the probability of dangerous system state during this time interval must be known.

Generally, it holds good that if 2-dimensional random variables with the vector $T = (T_1, T_2)$ has a distribution of the continuous type with combined probability function $P(T_1 = t_1 | T_2 = t_2)$, then the conditional probability function $P(T_1 = t_1 | T_2 = t_2.)$ of the variable T_1 under condition $T_2 = t_2$ is given as:

$$P(T_1 = t_1 | T_2 = t_2) = \frac{P(T_1 = t_1, T_2 = t_2)}{P(T_2 = t_2)}. \tag{18}$$

Conditional probability of the simultaneous fault occurrences of the elements 1 and 2 in time interval $(t, t + t_0)$ under assumption that up to time t neither fault of element 1 nor fault of element 2 occurred, is

$$P(t < T_1 \leq t + t_0, t \langle T_2 \leq t + t_0 | T_1 > t, T_2 \rangle t) = \frac{P(t \langle T_1 \leq t + t_0, t < T_1 \leq t + t_0, T_1 \rangle t, T_2 > t)}{P(T_1 > t, T_2 > t)}$$

$$= \frac{P(t < T_1 \leq t + t_0, t < T_2 \leq t + t_0)}{P(T_1 > t, , T_2 > t)} \qquad (19)$$

$$= (1 - e^{-\lambda_1 t_0}) . (1 - e^{-\lambda_2 t_0}),$$

where the probability of the simultaneous fault occurrences of the elements 1 and 2 in time interval $(t, t + t_0)$ is

$$P(t < T_1 \leq t + t_0, t < T_2 \leq t + t_0) = e^{-(\lambda_1 + \lambda_2)t} . (1 - e^{-\lambda_1 t_0}) . (1 - e^{-\lambda_2 t_0}). \qquad (20)$$

Failure rate of dangerous event in the system is

$$\lambda_D = \frac{\lambda_1 e^{-\lambda_1 t_0} + \lambda_2 e^{-\lambda_2 t_0} - (\lambda_1 + \lambda_2) e^{-(\lambda_1 + \lambda_2)t_0}}{e^{-\lambda_1 t_0} + e^{-\lambda_2 t_0} - e^{-(\lambda_1 + \lambda_2)t_0}}. \qquad (21)$$

If $\lambda_1 . t_0 \ll 1$, $\lambda_2 . t_0 \ll 1$, then

$$\lambda_D \cong 2 . \lambda_1 . \lambda_2 . t_0. \qquad (22)$$

2.3 Failure of Three Element

Let the system contains two mutually independent elements (element 1, element 2 and element 3) with the constant failure rates $\lambda_1, \lambda_2, \lambda_3$. Consider the 3-dimensional random vector $T = (T_1, T_2, T_3)$ with 2 random variables. Let the distribution function of the random vector T is

$$F(t_1, t_2, t_3) = P(T_1 \leq t_1, T_2 \leq t_2, T_3 \leq t_3) = (1 - e^{-\lambda_1 t_1}) . (1 - e^{-\lambda_2 t_2}) . (1 - e^{-\lambda_3 t_3}). \qquad (23)$$

The marginal distribution functions of random variables corresponding to a probability of failures of individual elements:

$$F(t_1) = P(T_1 \leq t_1) = F(t_1, \infty, \infty)$$

$$= \lim_{t_2 \to \infty, t_3 \to \infty} P(T_1 \leq t_1, T_2 \leq t_2, T_3 \leq t_3) = (1 - e^{-\lambda_1 t_1}), \qquad (24)$$

$$F(t_2) = P(T_2 \leq t_2) = F(\infty, t_2, \infty)$$

$$= \lim_{t_1 \to \infty, t_3 \to \infty} P(T_1 \leq t_1, T_2 \leq t_2, T_3 \leq t_3) = (1 - e^{-\lambda_2 t_2}), \qquad (25)$$

$$F(t_3) = P(T_3 \leq t_3) = F(\infty, \infty, t_3)$$

$$= \lim_{t_1 \to \infty, t_2 \to \infty} P(T_1 \leq t_1, T_2 \leq t_2, T_3 \leq t_3) = (1 - e^{-\lambda_3 t_3}). \qquad (26)$$

From Eq. (4) results that

$$P(a_1 < T_1 \leq b_1, a_2 < T_2 \leq b_2, b_3 < T_3 \leq b_3) = F(a_1, a_2, a_3) - F(a_1, a_2, b_3) + F(a_1, b_2, a_3)$$
$$+ F(a_1, b_2, b_3) - F(b_1, a_2, a_3) + F(b_1, a_2, b_3)$$
$$+ F(b_1, b_2, a_3) - F(b_1, b_2, b_3).$$

$$(27)$$

Conditional probability of the simultaneous fault occurrences of the elements 1, 2 and 3 in time interval $(t, t+t_0\rangle$ under assumption that up to time t neither fault of element 1 nor fault of element 2 nor fault of element 3 occurred, is

$$P(t < T_1 \leq t+t_0, t\langle T_2 \leq t+t_0, t\langle T_3 \leq t+t_0 | T_1 > t, T_2\rangle t, T_3\rangle t)$$
$$= \frac{P(t\langle T_1 \leq t+t_0, t < T_1 \leq t+t_0, T_1\rangle t, T_2 > t, T_3 > t)}{P(T_1 > t, T_2 > t, T_3 > t)}$$
$$= \frac{P(t < T_1 \leq t+t_0, t < T_2 \leq t+t_0, t < T_3 \leq t+t_0)}{P(T_1 > t, T_2 > t, T_3 > t)}$$
$$= (1 - e^{-\lambda_1 t_0}).(1 - e^{-\lambda_2 t_0}).(1 - e^{-\lambda_3 t_0}),$$

$$(28)$$

where the probability of the simultaneous fault occurrences of the elements 1, 2 and 3 in time interval $(t, t+t_0\rangle$ is

$$P(t < T_1 \leq t+t_0, t < T_2 \leq t+t_0, t < T_3 \leq t+t_0)$$
$$= e^{-(\lambda_1 + \lambda_2 + \lambda_3)t}.(1 - e^{-\lambda_1 t_0}).(1 - e^{-\lambda_2 t_0}).(1 - e^{-\lambda_3 t_0}). \qquad (29)$$

Failure rate of dangerous (undesired) event in the system is

$$\lambda_D = \frac{\lambda_1 e^{-\lambda_1 t_0} + \lambda_2 e^{-\lambda_2 t_0} + \lambda_3 e^{-\lambda_3 t_0} - (\lambda_1 + \lambda_2)e^{-(\lambda_1 + \lambda_2)t_0} - (\lambda_1 + \lambda_3)e^{-(\lambda_1 + \lambda_3)t_0}}{e^{-\lambda_1 t_0} + e^{-\lambda_2 t_0} + e^{-\lambda_3 t_0} - e^{-(\lambda_1 + \lambda_2)t_0} - e^{-(\lambda_1 + \lambda_3)t_0} - e^{-(\lambda_2 + \lambda_3)t_0} + e^{-(\lambda_1 + \lambda_2 + \lambda_3)t_0}}$$
$$+ \frac{-(\lambda_2 + \lambda_3)e^{-(\lambda_2 + \lambda_3)t_0} - (\lambda_1 + \lambda_2 + \lambda_3)e^{-(\lambda_1 + \lambda_2 + \lambda_3)t_0}}{e^{-\lambda_1 t_0} + e^{-\lambda_2 t_0} + e^{-\lambda_3 t_0} - e^{-(\lambda_1 + \lambda_2)t_0} - e^{-(\lambda_1 + \lambda_3)t_0} - e^{-(\lambda_2 + \lambda_3)t_0} + e^{-(\lambda_1 + \lambda_2 + \lambda_3)t_0}}.$$

$$(30)$$

If $\lambda_1.t_0 \ll 1$, $\lambda_2.t_0 \ll 1$, $\lambda_3.t_0 \ll 1$, then

$$\lambda_D \cong 3.\lambda_1.\lambda_2.\lambda_3.t_0^2. \qquad (31)$$

3   Application of Model with Method FTA

Let the top event of the fault tree represents dangerous event occurrence and the logical function expressing relationship between the basic events and the top event is given as the equation (minimal cut sets)

$$\psi(u) = \sum_{j=1}^{m} R_j(u),$$

(32)

where $R_j(u)$ is the logical function of the j-th minimal cut that is formed by such state indicators of basic events of the vector u, that if all primary events corresponding to these indicators occur, then the top event occurs, too; m is a number of the minimal cuts.

Let failure of the element represents the primary event occurrence and let probabilities of particular basic event occurrences are defined. Provided that the primary events are statistically independent, the probability of top event occurrence as a consequence of primary events occurrence of the minimal cut R_j (serial connection of basic events) can be expressed as

$$P(R_j) = \prod_{i=1}^{n} P(u_i),$$

(33)

where $P(u_i)$ is the probability of i-th basic event occurrence in the minimal cut R_j and n is a number of the primary events in the minimal cut R_j. If minimal cut R_j does not contain the basic event is not in the, then $P(u_i) = 1$.

Since every minimal cut leads to top event occurrence, the probability of top event is given as

$$P_D = P\left(\bigcup_{j=1}^{m} R_j\right) = \sum_{j=1}^{m} P(R_j) - \sum_{j=1}^{m-1} \sum_{k=j+1}^{m} P(R_j \cap R_k)$$
$$+ \sum_{j=1}^{m-2} \sum_{k=j+1}^{m-1} \sum_{l=k+1}^{m} P(R_j \cap R_k \cap R_l) \ldots + (-1)^{m-1} P\left(\bigcap_{j=1}^{m} R_j\right).$$

(34)

Because the probability of top event (34) has the properties of a distribution function, so the top event rate

$$\lambda_D(t) = \frac{\frac{dP_D(t)}{dt}}{1 - P_D(t)}$$

(35)

and is valid, that

$$\lambda_D(t) \le \sum_{j=1}^{m} \lambda_j(t),$$

(36)

where $\lambda_j(t)$ is the failure rate of minimal cut R_j.

If the minimal cut contains one basic event, then the failure rate of this cut can be calculated by (8). If the minimal cut contains two basic events, then the failure rate of

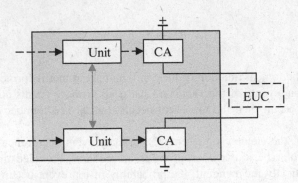

Fig. 3. Block diagram of a SRES [own study]

this cut can be calculated by (21), respectively (22). If the minimal cut contains three basic events, then the failure rate of this cut can be calculated by (30), respectively (31). If the minimal cut contains more than three basic events and their impact on the occurrence of top events is negligible.

For example, consider a safety-related electronic systems (SRES) with a dual-channel architecture whose simplified block diagram is shown in Fig. 3. The SRES consists of two identical control units (Unit A, Unit B), which controls two identical contactors (CA, CB) and implement safety function for equipment under control (EUC).

Impact of dangerous failure on the safety of the SRES can be described in a simple fault tree (Fig. 4).

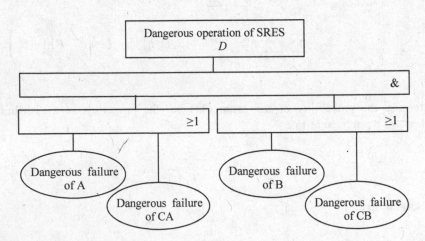

Fig. 4. Fault tree describing the failures effect on the safety integrity of SRES in Fig. 3 [own study]

The fault tree in Fig. 4 can be described by a logical function (by minimal cut sets)

$$D = A.B + A.CB + CA.B + CA.CB, \tag{37}$$

where D is the top event representing the dangerous failure of SRES, A (B) is the basic event representing a dangerous failure in the unit A (until B), CA (CB) is a basic event representing a potentially dangerous failure in the CA (CB) contactor.

The probability of dangerous failure of SRES (of the top event D) can be calculated from the equation

$$\begin{aligned} P_D(t) = {} & P_A(t).P_B(t).[1 - P_{SA}(t) - P_{SB}(t)] + P_{SA}(t).P_{SB}(t).[1 - P_A(t) - P_B(t)] \\ & + P_A(t).P_{SB}(t) + P_B(t).P_{SA}(t), \end{aligned} \tag{38}$$

where $P_X(t)$ is probability of basic event $X \in \{A, B, CA, CB\}$.

Failure rate of top event D can be calculated from (36) or and the dangerous failure rate of SRES (of the top event D) can be calculated using a simplified relation (36). Then

$$\lambda_D(t) \leq 2.\lambda_A.\lambda_B.t_0 + 2.\lambda_A.\lambda_{CB}.t_0 + 2.\lambda_{CA}.\lambda_B.t_0 + 2.\lambda_{AC}.\lambda_{CB}.t_0, \tag{39}$$

where $\lambda_A(\lambda_B)$ is a potentially dangerous failure rate the basic event A (B), $\lambda_{CA}(\lambda_{CB})$ is a potentially dangerous failure rate the basic event CA (CB) and t_0 it the detection-plus-negation time of failure.

A graph of the dangerous failure probability of SRES for the failure rate $\lambda_A = \lambda_B = 2, 5.10^{-6}$ h^{-1} and for the failure rate $\lambda_{CA} = \lambda_{CB} = 1.10^{-6}$ h^{-1} is shown in Fig. 5. Two graphs of the dangerous failure rate of SRES calculated by (35) and (39) are shown in Fig. 6. From Fig. 5 is clear, that if $\lambda.t \ll 1$, then (36) can be used to the calculation of the rate of the top event (the dangerous failure of SRES).

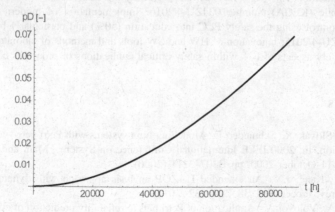

Fig. 5. Probability of SRES in Fig. 3 [own study]

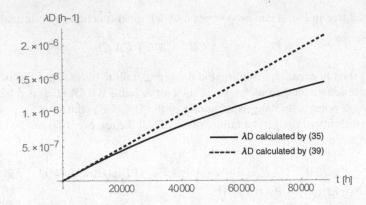

Fig. 6. Dangerous failure rate of SRES in Fig. 3 [own study]

4   Conclusion

The contribution deals with possibilities to apply the several methods within safety assessment of SRES with aim to simplify the process of safety integrity assessment. In the event of reduction it is necessary to consider whether difference between calculated value (probability of dangerous failure or difference dangerous failure rate) and real value is acceptable. If the analysis of the safety integrity of the safety-related electronic system is used method FTA, that application of knowledge in the field of multiple random variable can greatly simplify the computation of dangerous failure (top event) rate. Graphical results of the dangerous failure rate of SRES calculated by relations (35) and (39) are illustrated in Fig. 6. From result shown in Fig. 5 is clear, that if $\lambda.t \ll 1$, then (36) can be used to the calculation of the rate of the top event (the dangerous failure of SRES).

Acknowledgements. This paper has been supported by the Educational Grant Agency of the Slovak Republic (KEGA) Number 034ŽU-4/2016: Implementation of modern technologies focusing on control using the safety PLC into education (50%) and particularly by the project Number: 008ŽU-4/2015: Innovation of HW and SW tools and methods of laboratory education focused on safety aspects of ICT within safety critical applications of processes control (50%).

References

1. Einer, S., Slovák, R., Schnieder, E.: Modelling train systems with Petri nets—an operational specification. In: 2000 IEEE International Conference on Systems, Man, and Cybernetics, vol. 5, 8–11 October 2000, pp. 3207–3211 (2000)
2. Guo, L.J., Kang, J.X.: An extended HAZOP analysis approach with dynamic fault tree. J. Loss Prev. Process Ind. **38**, 224–232 (2015)
3. Klavner, A., Volovoj, V.: Application of Petri nets to reliability prediction of occupant safety systems with partial detection and repair. Reliab. Eng. Syst. Saf. **95**, 606–613 (2010)

4. Rástočný, K., Rástočný Jr., K.: UML—a part of an interlocking system development process. In: Mikulski, J. (ed.) TST 2012. CCIS, vol. 329, pp. 293–300. Springer, Heidelberg (2012). doi:10.1007/978-3-642-34050-5_33

5. Rástočný, K., Ždánsky, J.: Hazardous failure rate of the safety function. In: Mikulski, J. (ed.) TST 2015. CCIS, vol. 531, pp. 284–291. Springer, Cham (2015). doi:10.1007/978-3-319-24577-5_28

6. Rástočný, K., et al.: Quantitative assessment of safety integrity level of message transmission between safety-related equipment. J. Comput. Inform. **33**, 1001–1026 (2014)

7. Rástočný, K., Ilavský, J.: Effects of a periodic maintenance on the safety integrity level of a control system. In: Schnieder, E., Tarnai, G. (eds.) FORMS/FORMAT 2010—Formal Methods for Automation and Safety in Railway and Automotive Systems, pp. 77–85. Springer, Berlin (2011). doi:10.1007/978-3-642-14261-1_8

8. Skalný, P.: An application of graph theory in Markov chains. Reliability analysis. Adv. Electr. Electron. Eng. **12**(2), 154–159 (2014)

9. Shu, Y., Zhao, J.: A simplified Markov-based approach for safety integrity level verification. J. Loss Prev. Process Ind. **29**, 262–266 (2014)

10. Rástočný, K., et al.: Effects of diagnostic on the safety of a control system realised by safety PLC. In: 11th International Conference ELEKTRO 2016. Slovakia, 16–18 May, pp. 462–467 (2016)

11. http://www.bqr.com. Accessed 12 Feb 2017

12. http://www.ptc.com/product-lifecycle-management/windchill/product-risk-and-reliability. Accessed 12 Feb 2017

13. http://www.itemuk.com. Accessed 12 Feb 2017

The Concept of Tool to Support the Work of Air Traffic Controller in the Field of Aircraft Landing Scheduling in the TMA with Little Traffic

Daria Jagieło and Anna Kwasiborska[(✉)]

Faculty of Transport, Warsaw University of Technology, plac Politechniki 1,
00-661 Warsaw, Poland
{dja,akw}@wt.pw.edu.pl

Abstract. Recently, more and more are talked about the automation of air traffic control systems, which would allow relief work controllers. Methods or tools that allow for efficient and skillful management of air traffic are searched. The paper presents preliminary concept of the tool that would allow to schedule of aircraft landing at a given point and the scheduling is carried out taking into account the minimum separation between them. Separations are dependent on the types of aircraft. In addition, the program shows the routes by which the ship has to move in the area. The tool is dedicated to use in the TMA with little traffic, however, there are conducted further work, which would allow its use in sectors with far more traffic.

Keywords: Scheduling of aircraft · The tool supporting the work of air traffic controllers

1 Introduction

Intense air traffic growth increases the workload of air traffic controllers. It is therefore important to look for new methods and tools to organize traffic in a way that ensures the required level of security and the appropriate bandwidth of the aeronautical sector [2]. The most sensitive part of the airspace is the area around the airport, where two types of traffic are crossing, i.e., aerial and airplane operations. There are a large number of aircraft making complex maneuvers in very limited spaces. The dynamics of traffic in this area is large, so it is necessary to make quick decisions by those responsible for safe and effective traffic management [4].

Any inappropriate decision taken by air traffic controllers will result in dangerous situations. Traffic control allows for safe and efficient movement of ships in the sectors of aviation, and thus allows for increased airspace capacity. Therefore, we are constantly looking for tools to assist the work of those responsible for air traffic control, mainly in the control area of TMA.

The main problem for APP controllers is to decide which aircraft should land first, that is, in the case of one runway, this is the setting in the appropriate order of the

© Springer International Publishing AG 2017
J. Mikulski (Ed.): TST 2017, CCIS 715, pp. 436–446, 2017.
DOI: 10.1007/978-3-319-66251-0_35

landing planes. It is necessary to maintain minimum separation, both distance and time, and for several runways also prescribe a specific runway for a particular aircraft.

Currently it is said to include sequencing of aircraft arriving at an earlier stage. The enhanced Arrival Management (AMAN) arrivals and area navigation system allows for an increased air traffic planning horizon, approximately 180–200 NM, from the start point of descent, which is determined by the standard instrument flight procedures.

Currently used methods of maintaining proper separation between aircraft are built on the basis of the existing airways network as well as the control method initiated in procedural control, and therefore air traffic controllers are an indispensable part of the control process primarily due to their decision-making capacity on the basis of their decision- Collected a huge amount of information.

As already mentioned, tools are sought, systems that will support the work of the air traffic controller in decision-making. The first such system was the TCAS, which is an on-board collision prevention system that has been operating since the 1990s. Then an attempt was made to introduce systems with a short-term collision warning function. These are examples of alarm systems that are designed to prevent the dangerous consequences of misconceptions made by a person. Currently, systems are being sought to predict dangerous situations and propose solutions. Collision situations are to be detected on the basis of continuous monitoring of the traffic environment, data collection and processing.

In general, the above mentioned problem of landing aircraft in the correct order is reduced to the problem of assignment of tasks. The scheduling of tasks in air traffic itself has been presented in a purely theoretical sense. A review of scheduling methods and their relation to air traffic is presented in [2, 4].

The scheduling of landing aircraft may be related to a specific airport or where theoretically the multiple runways are the problem [8, 12]. Apart from identifying the appropriate sequence of landing aircraft, it is also important to devise such aircraft to maximize the desired effect while maintaining an adequate level of safety [9]. Furthermore, the appropriate creation of an optimal sequence of airplanes affects the capacity of airports and, consequently, the amount of delays [13], which is very important in the context of ever-increasing air traffic.

The article proposes a tool that, based on the traffic situation information that will take place within one hour, determines the sequence of planes approaching the landing in a given TMA. Sequence should maintain minimum separation between aircraft resulting from turbulence in the aerodynamic trajectory. In addition, aircraft are positioned in such a way as to avoid collision. Due to its uncomplicated construction, the tool is dedicated to low traffic TMA, which will allow for almost complete automation of air traffic control in the area. Of course, you can not completely exclude a person who should oversee the operation of the program and, in exceptional circumstances, make independent decisions.

Scheduling landing aircraft is not an easy task, it requires a great deal of information to be analyzed in a very short amount of time. An additional disadvantage is that the TMA area is very limited, and the dynamics in it are very large. We still have not yet created a powerful tool that could completely replace the air traffic controller, but we are constantly working on tools to support its work.

2 Analysis of Air Traffic in TMA

The analysis of air traffic was conducted at TMA Warsaw based on the time and place of arrival of the TMA in Warsaw and the route of the flight. The route was analyzed for its compliance with the applicable procedures, i.e. all derogations from the road rules were noted, i.e. assigned shortcuts, and hence the time and place of allocation and termination of the abbreviation. The most commonly chosen shortcut launching points were highlighted in color that are in effect in Fig. 2. Information was also collected on where the aircraft were already set in the landing queue.

Traffic analysis was done for the approach 33 in TMA Warsaw. Time of landing operations of aircraft and landings was measured. According to regulations, the route should be in accordance with the STAR procedures, i.e. the aircraft should move along the route set for each of the entry points.

As the volume of traffic in TMA Warsaw allows the use of abbreviations, thereby increasing the capacity of this sector, the following analysis focuses on the pointers where most of these abbreviations were allocated. Also important is where the shortcut ends. Analysis showed that for all directions short cuts ended at 10 or 12 NM, i.e., ERLEG (LITVO) and WA533.

Table 1. Percentage of operations scheduling, which were made at a given point for the inlet gate AGAVA [own study]

JOIN AT NM	Number of operations	Percentage of operations
4	2	1.60%
8	8	6.40%
9	3	2.40%
10	51	40.40%
11	3	2.40%
12	43	34.80%
14	1	0.80%
16	9	7.20%
17	1	0.80%
19	1	0.80%
20	3	2.40%
Sum	125	100.00%

One of the gates is AGAVA inlet and to the point measurements were made. Table 1 includes the percentage share of the scheduling operations that were performed at the point inlet for the AGAVA inlet.

Table 1 shows the first column "JOIN AT NM", where the aircraft were sequenced at the landing approach. The numbers in this column indicate the distance from the runway threshold in nautical miles. The number of operations, that is, the second column, indicates how many operations were performed, for which a point was chosen as the place of the ranking.

The percentage of operations, that is the last column, shows the percentage of all operations performed at TMA Warsaw in a given measurement period are those performed at a given point.

Table 2 presents the points where the shortcuts to reach the ERLEG point (LITVO) and point WA533 begin. Tables contain information about the start point of the shortcut, which indicates the point at which the shortcut was assigned. The next column is the number of operations that indicates the number of operations performed from the point indicated in the "Point of starting a shortcut" column. The end point of the shortcut indicates the indication of the point at which shortened routes were completed.

The sum of the operations from the beginning of the shortcut - indicates the number of all operations to which the shortcut was assigned at the point and ending in ERLEG (LITVO) or WA533.

Table 2. The most popular points create shortcuts intake routes for aircraft flown to the point AGAVA [own study]

Point of starting a shortcut	Number of operations	End point of the shortcut	Point of starting a shortcut	Number of operations	End point of the shortcut	Sum of operations from the beginning of the shortcut
XEMRO	10	LITVO (10 NM)	XEMRO	10	WA533 (12 NM)	20
ELKIR	5	LITVO (10 NM)	ELKIR	13	WA533 (12 NM)	18
AGAVA	18	LITVO (10 NM)	AGAVA	10	WA533 (12 NM)	28
OTMUL	9	LITVO (10 NM)	OTMUL	2	WA533 (12 NM)	11
WA564	2	LITVO (10 NM)	WA564	7	WA533 (12 NM)	9
BARNA	0	LITVO (10 NM)	BARNA	1	WA533 (12 NM)	1
AMSOS	0	LITVO (10 NM)	AMSOS	2	WA533 (12 NM)	2

3 Assumptions to the Program

The presented paper is the beginning of analyzing such a problem by the author, the article dealt with a simple motion situation for which an attempt was made to form a flow of traffic stream. The basic task of the created tool is to properly arrange the planes in the area concerned and to rank them at a given point, taking into account the weight class criterion, which in turn is related to maintaining proper separation between them.

In general, of aircraft, there are three basic types according to their maximum take-off mass, i.e. light (L), medium (M) and heavy (H). The exact classification is shown in Table 1.

The STAR procedure for RWY 33 in TMA Warsaw was used for the analysis (Fig. 1). A Standard Arrival Route (STAR) is a standard ATS route identified in an approach procedure by which aircraft should proceed from the en-route phase to an initial approach fix.

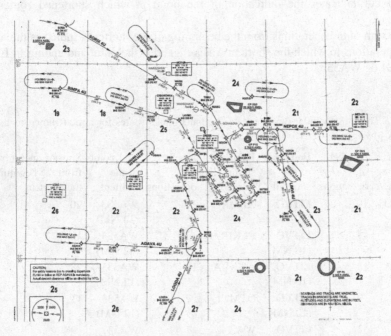

Fig. 1. Procedure STAR RWY 33 [own study]

The procedure shown in Fig. 2 is based on the theory of graphs. TMA is an acyclic network consisting of a set of nodes and arcs. The network also distinguishes routes that are nodes and arcs. There are four inlet gates, the green ones, and the one where the aircraft is aligned, the red one. White knots serve to designate alternative paths for moving aircraft.

For each of the inlet gates there are at least two pathways that the planes can move and have been created to avoid possible collision situations. When arranging landed planes they are assigned specific paths to reach the red node.

The curves indicate the distances between individual nodes. At this stage of development, it was assumed that all aircraft would move at constant speed. In addition, conditions were set that said whether the separation between aircraft was compromised, which was recorded as:

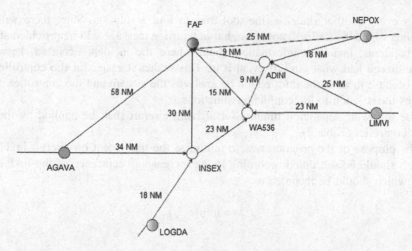

Fig. 2. Graph of procedure STAR [own study] (Color figure online)

$$t_{z_i}^{(r+1)_{m_l}} - t_{z_i}^{r_{m_l}} < p_g(m_l) \rightarrow \text{Reduced separation between aircraft, there is a conflict situation} \tag{1}$$

$$t_{z_i}^{(r+1)_{m_l}} - t_{z_i}^{r_{m_l}} \geq p_g(m_l) \rightarrow \text{Correct separation between aircraft, there is no conflict situation} \tag{2}$$

where:

$t_{z_i}^{r_{m_l}}$- moment of reporting r-th aircraft category m_l in i-th network node;
$p_g(m_l)$- g-th separation between aircraft depending on the type of aircraft m_l.

The green points are the TMA entry points, the white points are the places where the inbound routes can be assigned to the FAP/FAF point where the arrival planes are landed. The requested network is a directed and acyclic network.

Between the nodes a distance equal to dj was determined. We assume that the whole network of planes are moving at constant velocity. By imposing such constraint we can easily determine the time tp between nodes that will be equal [9]:

$$t_p = 3600 \times \left(\frac{d_{ij}}{v_s}\right)$$

It is assumed that the planes are moving according to a flight plan that is provided and introduced to the system some time in advance, e.g. before the controller starts operation. The task of the tool will be to arrange the aircraft at the FAP/FAF point with proper separation on each of the network nodes and assign the appropriate route for each flight. To indicate the appropriate route the tool must consider the conflict situations between the planes.

For every conflict situation, the tool tries to find a solution. Sure, there will be situations where the tool will not solve, but in return, a message will be provided about such incidents, including information about where the incident occurred, between which aircraft and what kind of event it is. This makes it easier for the controller to work because most of the information he analyzes the system and the controller itself remains the solution to the conflicting situations.

The following separation minima for arriving aircraft shall be applied by the air traffic controller (Table 2).

The purpose of the program was to minimize the time spent on aircraft in TMA. Aircraft should be scheduled according to the increasing occurrence of the FAP/FAF node, which should be recorded as:

$$t_{wy}^{r_{m_l}} = \min\left(t_{i_{FAF}}^{r_{m_l}}\right) \tag{3}$$

4 Realization of the Program

The program was implemented using Python, which is a high-level programming language of general purpose.

The operation of the program consists in performing the following sequence of steps:

1. Load data from a file.
2. Generate all possible paths for each trip.
3. Create a first timetable that does not yet take account of the separation between aircraft.
4. Finding a collision based on the distribution created in point 3.
5. Try to fix all collisions.
6. Generate flight schedules in which aircraft are ranked.
7. Generate timetables where aircraft are sorted, together with an indication of collision situations remaining solved.

The program operates on the basis of data entered into it. Based on this, a first decomposition is generated to search for collisions between aircraft. Seeking collisions is done by comparing the routes and times of transit on the routes, i.e. for each route the common points are searched for, and then checking whether the required separation has been maintained here.

If there are collisions in TMA then the program generates a message that contains all collision information, i.e. the collision number between which flights and aircraft types and where the node in the network collides. The example collision message looks like this:

```
KOLIZJA: 2
('KLM65F', 'M', [('AGAVA', 542), ('FAF', 1586)])
('UPS9', 'H', [('LIMVI', 1021), ('ADINI', 1471), ('FAF', 1633)])
[('FAF', 1586, 1633, 'M', 'H')]
```

The message pointed to two flights, the KLM65F call sign and the UPS call sign. Between them there is a collision situation at the FAF. The situation is to reduce the separation between two ships of different types. In this case, when the first type M (medium) and followed by the H (heavy) ship, the minimum separation between them should be 75 s, in which case there is 47 s between them, i.e. 28 s too little.

The next step in the program is to try to resolve the conflicts. The function defined in the source code chooses random paths for the two aircraft that are in a collision situation. For each of them a path is drawn from all possible paths beginning at the entry point and ending at the FAF (Final Approach Fix). The program ends when a new path selection fixes a collision. Otherwise the same steps are taken N times, where we can assume that N = 100, which seems to be a reasonable value for such a simple TMA scheme.

The justification will be to consider the situation where two different airplanes (one each gate) arrive at the TMA from two different entry gates and each of them has the ability to navigate three different routes, thus allowing them to fly in 9 different ways. Because if one of them can navigate on routes A, B, or C and the other on routes D, E or F, they can fly together in the following configurations: AD, AE, AF, BD, BE, BF, CD, CE, CF. If this attempt is made 100 times, it will probably hit each of the above pairs. If during this test you cannot find a pair that does not contain a collision, you must be sure that all the pairs have been repeatedly checked and none of them can be resolved. This is why it is important here to highlight the role of human beings in the workings of such systems, because in such a situation the control should take over the person and he or she should resolve the conflict, for example by changing the speed of one aircraft, which is one of the basic techniques of air traffic control.

5 Testing the Program

One of the most important steps in creating a program is testing, which allows you to check the functionality and correctness of the program. Two types of separation were tested, and three samples containing predicted traffic in a series of hours, each of which had a different traffic density, i.e., 15, 29 and 34 aircraft per hour. The sample included data such as order number, call sign, place and time of entry into TMA, and type of aircraft. Example of sample is shown in Table 3.

Table 3. Category/type of aircraft [own study]

The ICAO wake turbulence category	Maximum certificated take-off mass (kg)
Heavy (H)	MTOW $\geq$ 136,000
Medium (M)	7000 $\leq$ MTOW $\leq$ 136,000
Light (L)	MTOW $\leq$ 7000

Scheduling aircraft in the TMA was entering the data. During scheduling, a route and directions were given for the time at which the aircraft appeared at the individual nodes and at the FAF. Example scheduling is shown in Tables 4, 5 and 6.

Table 4. Minimum separation between aircraft [own study]

Following aircraft	Leading aircraft	Minimum distance	Minimum time
Light (L) Medium (M) Heavy (H)	Light (L)	4 NM	75 s
Medium (M) Heavy (H)	Medium (M)	4 NM	75 s
Heavy (H)	Heavy (H)	4 NM	75 s
Light (L)	Medium (M)	5 NM	90 s
Medium (M)	Heavy (H)	5 NM	90 s
Light (L)	Heavy (H)	6 NM	105 s

Table 5. Planned aircraft at TMA in 12:00–13:00 [own study]

Lp.	Call sign	TMA entry point	TMA entry point	AC type
1	LOT438M	AGAVA	159	M
2	SPRWL	LIMVI	250	L
3	SRN7700	NEPOX	320	L
4	KLM65F	AGAVA	542	M
5	LOT33DM	AGAVA	807	M
6	UPS9	LIMVI	1021	H
7	LOT3848M	LIMVI	1320	M
8	LOT3950	NEPOX	1590	M
9	QGA949HL	LOGDA	1824	M
10	AUA631WM	LOGDA	2179	M
11	SPMXHL	LIMVI	2520	L
12	LOT434M	LOGDA	2725	M
13	DLH2PNM	AGAVA	2955	M
14	LOT38080M	LOGDA	3261	M

Table 7 shows the results of the program for each sample.

By analyzing the last column of the table, it is noted that the program does not show difficulty scheduling aircraft when the traffic is low. With the increase in traffic, more unresolved collision situations continued to exist, which did not allow for the creation of a sequence of airplanes ensuring safe movement in the area in question. Therefore, it is necessary to supervise the traffic controller, who will monitor the situation and take steps to ensure the safe organization of traffic.

Table 6. Sample scheduling of aircraft in the FAF for a sample of 15 aircraft in one hour [own study]

Lp.	Callsign	AC type	TMA entry point	TMA entry time	Route	FAF time
0	ENT092PM	M	LOGDA	46	INSEX	910
1	SRN7700	L	NEPOX	320	NEPOX	1040
2	SPRWL	L	LIMVI	250	ADINI-WA536	1132
3	LOT438M	L	AGAVA	159	INSEX	1311
4	KLM65F	M	AGAVA	542	–	1586
5	UPS9	M	LIMVI	1021	WA536	1705
6	LOT33DM	H	AGAVA	807	–	1851
7	LOT3848M	M	LIMVI	1320	ADINI	1932
8	LOT3950	M	NEPOX	1590	–	2310
9	QGA949HL	M	LOGDA	1824	INSEX	2688
10	AUA631WM	M	LOGDA	2179	INSEX	3043
11	SPMXHL	L	LIMVI	2520	ADINI-WA536	3204
12	LOT434M	M	LOGDA	2725	INSEX	3589
13	DLH2PNM	M	AGAVA	2955	–	3999
14	LOT38080M	M	LOGDA	3261	INSEX	4125

Table 7. The result of the program for each sample [own study]

Separation	Sample	Number of collisions	Number solved collisions	Number of collisions remaining to be resolved
Distance	15	3	3	0
	26	9	7	2
	34	21	10	11
Time	15	4	3	1
	26	9	6	3
	34	29	0	29

6 Conclusion

The article presents a tool for sequencing aircraft operating in the selected area around the airport. Scheduling aircraft in air traffic is an important issue, because the proper sequence of them can help to increase the capacity of the runway and airspace sectors.

The created program can be a basic tool to support the work of an air traffic controller. The program fulfilled its purpose, that is, it allowed the sequencing of aircraft. In analyzing the results, it can be seen that a much better result was obtained for a traffic situation where the traffic was low. Difficulties in resolving conflict situations in traffic increase cases were only due to the restrictive constraints imposed by the program.

The controller's input is essential here. At present, it is not possible to completely replace a human being in decision making during the traffic control and management process. It is only possible to propose tools that would allow automated traffic control to be considered only as tools to assist data analysis and decision making.

In order to streamline the program and allow greater control of air traffic control in the selected area, further expansion is planned in the next steps to include the modules responsible for determining the variable speed of the aircraft and the automatic redirection of the aircraft in emergency situations in the waiting area. Short-term and minimum separation will also help to avoid collisions. Such actions can be taken with an adequate level of air traffic safety.

References

1. Florowski, A., Skorupski, J.: Ocena niezawodności wybranego algorytmu szeregowania lądujących statków powietrznych, w: Prace Naukowe Politechniki Warszawskiej. Transport, Oficyna Wydawnicza Politechniki Warszawskiej, nr 113, pp. 163–170 (2016)
2. Kwasiborska, A., Skorupski, J.: Metody szeregowania zadań jako narzędzie rozwiązywania problemu sekwencjonowania samolotów. Transport 101, 55–62 (2014). Prace Naukowe Politechniki Warszawskiej, Warszawa
3. Kwasiborska, A., Stelmach, A.: Pre-departure sequencing method in the terms of the dynamic growth of airports. J. KONES Inst. Aviat. 23(4), 253–260 (2016)
4. Kwasiborska, A., Markiewicz, K.: Metody listowego szeregowania samolotów lądujących jako narzędzie wspomagania kontrolera w podejmowaniu decyzji. Transport 104, 21–32 (2014). Prace Naukowe Politechniki Warszawskiej, Warszawa
5. Kwasiborska, A.: Multicriteria pre-departure sequencing of aircraft using the greedy scheduling algorithm. In: Jerzy, F. (ed.) CLC 2013: Carpathian Logistics Congress—Congress Proceedings, pp. 512–517. TANGER Ltd., Ostrava (2015)
6. ICAO, DOC 4444, Procedures for Air Navigation Services: Air Traffic Management, Fifteenth edition (2007)
7. ICAO, DOC 8168, Procedures for Air navigation Services: Aircraft Operations, ICAO, DOC 8168, Procedures for Air navigation Services: Aircraft Operations (2006)
8. Beasley, J., et al.: Scheduling aircraft landings—the static case. Transp. Sci. 34(2), 180–197 (2000)
9. Beasley, J., Havelock, P., Sonander, J.: Scheduling aircraft landings at London Heathrow using a population heuristic. J. Oper. Res. Soc. 52(5), 483–493 (2011)
10. Skorupski, J., Florowski, A.: Koncepcja implementacji systemu oceny procesu szeregowania samolotów lądujących, w: Zeszyty Naukowe. Transport Politechnika Slaska 87, 5–10 (2015)
11. Skorupski, J., Florowski, A.: Method for evaluating the landing aircraft sequence under disturbed conditions with the use of Petri nets. Aeronaut. J. 120(1227), 819–844 (2016)
12. Malarski, M., Piątek, M.: Kierunki rozwoju nowoczesnych systemów zarządzania ruchem lotniczym na świecie, Prace Naukowe Politechniki Warszawskiej. Transport (2009)
13. Company, T.B.: The generalized arrival planer: modeling and analysis for arrival planing. In: 28th International Congress of the Aeronautical Sciences, ICAS (2012)

Application of Reversible Logic in Synthesis of Traffic Control Systems

Roman Pniewski$^{(\boxtimes)}$, Piotr Bojarczak, and Mieczysław Kornaszewski

Faculty of Transport and Electrical Engineering,
University of Technology and Humanities in Radom,
Malczewskiego 29, 26-600 Radom, Poland
{r.pniewski, p.bojarczak, m.kornaszewski}@uthrad.pl

Abstract. Reversible gates allow for the creation of fault tolerant digital devices. Application of these gates enables for realization of safe control systems. At present, control systems based on relays are replaced by systems based on computer technology. This technology allows for the implementation of safe systems. Because this system is composed of many elements, it is very difficult to obtain high safety level for it. Application of reversible logic in synthesis of digital systems constitutes an alternative approach. The paper presents reversible gates along with examples of the usage of them in synthesis of digital systems. The main advantage of reversible logic is the possibility of synthesis of self testing and fault tolerant circuits. Application of circuits based on reversible logic allows developing safe control systems. The paper also presents the proposal for the application of reversible logic in simple control systems. Model and simulation of the proposed axle counting systems based on reversible logic have also been presented.

Keywords: Traffic control system · Reversible logic · Simulation

1 Introduction

Geographical distribution of railway traffic control systems forces the developers to put emphasis on more tough requirements for local control devices. They include controlling algorithms, information processing and reliability in data transmission. In 70th of 20-th century electronic circuits were used in railway traffic control systems. Railway traffic control systems based on keys and relays are substituted by electronic devices and in particular digital systems. An increase in integrity of integrated circuits allowed for the development of railway traffic control systems with more and more sophisticated functionality. An appearance of Programmable Logic Controllers (PLC) and industrial computers along with real time operating system allowed for the implementation of railway traffic control algorithms in software [1]. In contemporary digital railway traffic control systems algorithms for control, processing and data storage are mainly implemented in micro-computers systems as a software (program). Implemented algorithms are executed sequentially in accordance with the program stored in the computer memory. Recently, instead of solutions based on computer systems, hardware and hardware-software (System on Chip solution) can be applied. Modern railway traffic control system are more and more often based on custom integrated circuits. The main goal of railway traffic control systems is to ensure the safety. Thus, development of

© Springer International Publishing AG 2017
J. Mikulski (Ed.): TST 2017, CCIS 715, pp. 447–460, 2017.
DOI: 10.1007/978-3-319-66251-0_36

digital circuits for traffic control differs substantially from the traditional methods for synthesis of digital circuits. While developing traditional digital circuits, main emphasis is put on logic functions minimization. However, during development of railway traffic control systems the maim goal for the developer is to envisage behavior of the system for all possible cases. Irrespective of algorithms realized by digital circuits, the contemporary railway traffic control systems have to meet some safety norms. Modern railway traffic control systems have to meet following standards:

- PN-EN50126: railway applications – the specification and demonstration of reliability, availability, maintainability and safety;
- PN-EN50128: railway applications – communication, signaling and processing systems - software for railway control and protection systems;
- PN-EN50129: railway applications – communication, signaling and processing systems- safety related electronic systems for signaling.

These standards define most of requirements for solutions based on hardware, software and hardware-software configuration. The hardware and software using in railway traffic control systems should satisfy PN-EN50126, PN-EN50129 standards and PN-EN50128 standard, respectively [2].

Presented standards do not specify requirements for computer-aided design and integrated circuit fabrication using in railway traffic control systems. Alternatively, instead of traditional digital circuits, we can apply reversible logic which allows us to control the state of digital system.

2   Fault Tolerant Reversible Logic Circuits

Energy dissipation is a crucial issue which should be taken into account during design of VLSI circuits. Combinatorial logic circuits dissipate the heat of the volume kT ln 2 Joules per 1 bit of lost information, where k denotes Boltzman constant and T temperature of the circuit [3]. These losses happen when it is impossible to unambiguously determine input vector based on output vector. Reversible logic allows us to eliminate heat dissipation because there is no any information loss. Therefore, reversible logic will be more and more often applied in modern digital circuits. Synthesis of reversible logic circuits differs significantly from this used in traditional circuits. In reversible logic technology only one input of gate can be connected to the output of the subsequent gate. Additionally, it performs one-to-one mapping between vectors of inputs and outputs. However the final circuit has to be acyclic. Any reversible gate carries out permutation of input signals and executes reversible functions. The gate with k inputs (therefore also k outputs) is called reversible gate k * k. Every reversible circuit is only composed of reversible gates. All redundant outputs of such circuit (keeping only for reversibility) are called "garbage outputs" and inputs with assigned constant value are called "constant inputs". While designing reversible logic circuits the number of garbage outputs is minimized. Parity checking is a method which is often used to correct errors in digital circuits and communication systems. It results from inability of arithmetical functions to preserve parity. If during calculations the number of bits on the inputs is always even, then it is not necessary to check the parity on the path connecting these inputs with the output of the circuit. Implementation of reversible logic circuit based only on parity preserving gates ensures parity condition [6]. The paper presents parity preserving gates as well as their application in simple but important circuits.

2.1 Reversible Logic Gates

Many examples of reversible gates can be found in literature- from the basic developed a few decades ago to new versions allowing for parity preserving [4–6]. Next two subsections present both types of gates. There are following basic reversible gates: Feynman gate (FG), Peres gate (PG), Toffoli gate (TG) and Fredkin gate (FRG). Dimension of FG is equal to 2 * 2 and dimension for the rest of gates equal to 3 * 3. All of them have been thoroughly examined for last decades. However their simple structure and quantum cost of realization caused that many approaches to reversible logic gate development and tools for them appeared (Figs. 1, 2, 3 and 4).

A	B	$P = A$	$Q = A \oplus B$
0	0	0	0
0	1	0	1
1	0	1	1
1	1	1	0

Fig. 1. Reversible Feynman gate 2 * 2 along with its truth table [own study]

A	B	C	$P = A$	$Q = A \oplus B$	$R = AB \oplus C$
0	0	0	0	0	0
0	0	1	0	0	1
0	1	0	0	1	0
0	1	1	0	1	1
1	0	0	1	1	0
1	0	1	1	1	1
1	1	0	1	0	1
1	1	1	1	0	0

Fig. 2. Reversible Peres gate 3 * 3 along with its truth table [own study]

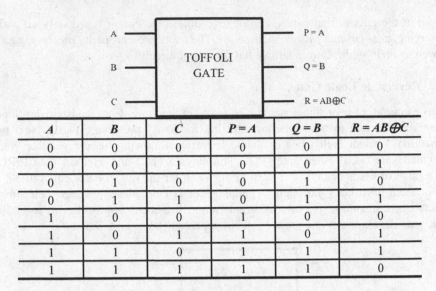

A	B	C	P = A	Q = B	R = AB⊕C
0	0	0	0	0	0
0	0	1	0	0	1
0	1	0	0	1	0
0	1	1	0	1	1
1	0	0	1	0	0
1	0	1	1	0	1
1	1	0	1	1	1
1	1	1	1	1	0

Fig. 3. Reversible Toffoli gate 3 * 3 along with its truth table [own study]

A	B	C	P = A	Q = A'B⊕AC	R = A'C⊕AB
0	0	0	0	0	0
0	0	1	0	0	1
0	1	0	0	1	0
0	1	1	0	1	1
1	0	0	1	0	0
1	0	1	1	1	0
1	1	0	1	0	1
1	1	1	1	1	1

Fig. 4. Reversible Fredkin gate 3 * 3 along with its truth table [own study]

2.2 Reversible Parity Preserving Gates

Fault tolerant circuits ensure correct operation of system when errors or damage of its elements occur. If the system is composed of fault tolerant elements then it is easily to detect and correct errors occurring in it. In many cases (for example in communication systems) detection and correction of errors are realized by parity checking. Fault tolerant reversible logic based on parity preserving gates was presented by Parhami [7]. In these gates, parity of inputs corresponds to parity of outputs. In other words, execution of exclusive–or operator (XOR) for all inputs (A⊕B⊕C for the gate 3 * 3) and execution of the same operator for all outputs (P⊕Q⊕R for the gate 3 * 3) gives the same result. A closer look at above presented truth tables shows that Fredkin gate 3 * 3 is the only basic gate which preserves parity. Next, examples of elements which also preserve parity are presented.

In 2006 Parhami presented double Feynman gate (F2G) [7]. It arose as a serial connection of two Feynman gates for 3 inputs (Fig. 5).

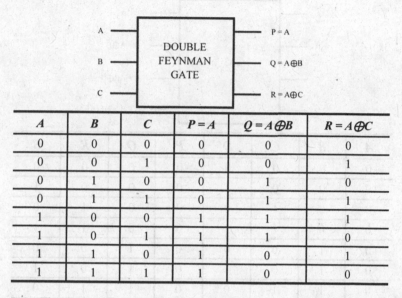

A	B	C	$P = A$	$Q = A \oplus B$	$R = A \oplus C$
0	0	0	0	0	0
0	0	1	0	0	1
0	1	0	0	1	0
0	1	1	0	1	1
1	0	0	1	1	1
1	0	1	1	1	0
1	1	0	1	0	1
1	1	1	1	0	0

Fig. 5. Double Feynman gate 3 * 3 along with its truth table [7]

In 2008 Haghparast presented a new gate so-called a novel fault tolerant reversible gate (NFT), designed for nanotechnology based systems [8] (Fig. 6).

A	B	C	$P = A \oplus B$	$Q = BC' \oplus AC'$	$R = BC \oplus AC'$
0	0	0	0	0	0
0	0	1	0	1	0
0	1	0	1	0	0
0	1	1	1	0	1
1	0	0	1	1	1
1	0	1	1	1	0
1	1	0	0	1	1
1	1	1	0	0	1

Fig. 6. A Novel Fault Tolerant gate 3 * 3 along with its truth table [8]

In 2009 Saiful Islam, Rehman, Hafiz and Begum developed a new fault tolerant adder, in which a new gate 4 * 4 so-called Islam gate (IG) [9]. Similar to basic gates, it performs one-to-one mapping. It means that one of input is passed to the output without the change. It is also universal, in other words, any logic function can be realized using this gate (Fig. 7).

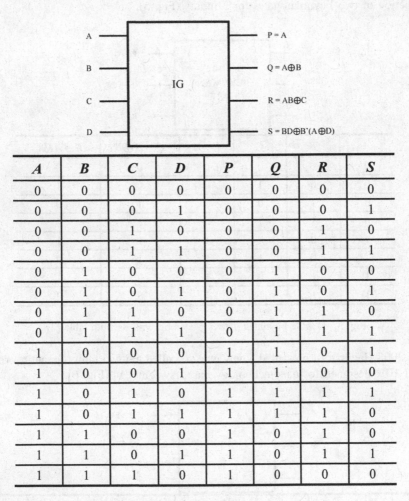

A	B	C	D	P	Q	R	S
0	0	0	0	0	0	0	0
0	0	0	1	0	0	0	1
0	0	1	0	0	0	1	0
0	0	1	1	0	0	1	1
0	1	0	0	0	1	0	0
0	1	0	1	0	1	0	1
0	1	1	0	0	1	1	0
0	1	1	1	0	1	1	1
1	0	0	0	1	1	0	1
1	0	0	1	1	1	0	0
1	0	1	0	1	1	1	1
1	0	1	1	1	1	1	0
1	1	0	0	1	0	1	0
1	1	0	1	1	0	1	1
1	1	1	0	1	0	0	0

Fig. 7. Islam gate along with its truth table [9]

2.3   Application Examples

Fault Tolerant Toffoli Gate

Toffoli gate 3 * 3 has been thoroughly studied in literature. However, it is not fault tolerant, thus it does not lend itself to fault tolerant circuits. Luckily, some modification of this gate make it fault tolerant. There exist three modifications presented in [8, 9] (Figs. 8, 9).

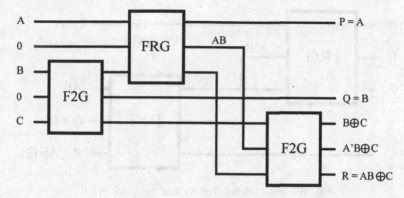

Fig. 8. Fault tolerant Toffoli gate presented in [7]

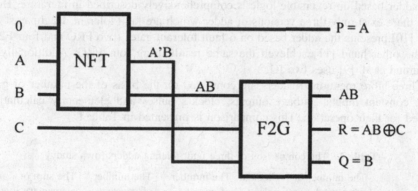

Fig. 9. Fault tolerant Toffoli gate presented in [8]

A closer look at presented solutions shows a significant difference in structures and the number of gates used. Solution presented in [8, 9] uses only two fault tolerant gates. The complexity of Toffoli gate for different realization is also presented. If we denote the basic logic operators as:

α = two inputs XOR operator;
β = two inputs AND operator;
δ = (NOT) operator;
T = the sum of all logic operators.

The sum of all logic operators for circuit realized in [7] equals $T = 6\alpha + 4\beta + 2\delta$, and for the circuit from [8] equals $T = 5\alpha + 3\beta + 2\delta$, and for the circuit from [9] equals $T = 4\alpha + 4\beta + 1\delta$. Therefore, the circuit from Fig. 10 has the simplest structure.

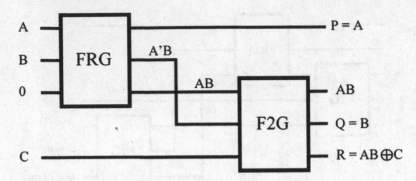

Fig. 10. Fault tolerant Toffoli gate presented in [9]

Fault Tolerant Adder

Full adder based on reversible logic is comprehensively described in literature. However, there exist only three versions of adder which are fault tolerant. Haghparast and Navi [10] presents the adder based on 6 fault tolerant gates (two FRG and four F2G). On the other hand [11] achieved the same result using four FRG. Additionally Al Mahamud et al. [9] uses two IG.

Three afore mentioned adders are compared on the basis of the number of gates used, constant inputs, garbage outputs, clock's pulses and elementary calculations needed for their operation. This comparison is presented in Table 1.

Table 1. The comparison of three fault tolerant adders [own study]

	The number of gates used	Clock's pulses	The number of garbage outputs	The number of constant inputs	The sum of elementary logic operation
ADDER [9]	2 IG	2	3	2	$8\alpha + 6\beta + 2\delta$
ADDER [11]	4 FRG	4	3	2	$8\alpha + 16\beta + 4\delta$
ADDER [10]	2 FRG + 2 F2G	6	6	5	$12\alpha + 8\beta + 2\delta$

We can choose one of three available versions of fault tolerant full adder depending on our special requirements. We can spot that the most effective structure from the operation point of view uses new developed IG gates presented in Fig. 11.

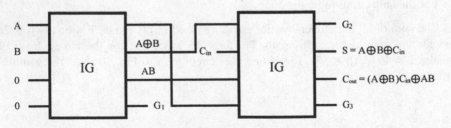

Fig. 11. Fault tolerant full adder presented in [9]

3 Modeling and Simulation of Reversible Logic Circuits

Software QUCS (Quite Universal Circuit Simulator) is a free electronic circuits simulator. Its first version runs only in Linux. At present, its new version runs in both Linux and Windows operating systems. The source code for this simulator is also available on website, what in turn allows for the modification of this software. QUCS can simulate both analog and digital circuits. The simulation of analog circuits is carried out using SPICE kernel, but the simulation of digital circuits is performed in multi-stage process. The simulator based on schematic generates a netlist and writes it to the netlist.txt file. The netlist is written using VHDL (VHSIC Hardware Description Language). Next, the conversion to C language is performed using FreHDL. To compile the obtained source code the compiler Mingw was used. Simulation process is controlled by commands from the batch file "qucsdigi.bat" [12]. Modification of this file allows for any control of simulation. This approach allows the user to include own programs to the Simulator. Thanks to it, it is possible to include to circuit diagram modules described in Verilog and VHDL. Symbol for appending file is automatically created on the basis of description of interface. Figure 12 shows print-screen for schematic editor of the simulator along with reversible gates being placed on the schematic sheet. Figure 13 presents example of simulation. QUCS contains a symbol editor which allows for edition of symbol assigned to the element. This editor has been used to create the library with reversible gates with the symbols.

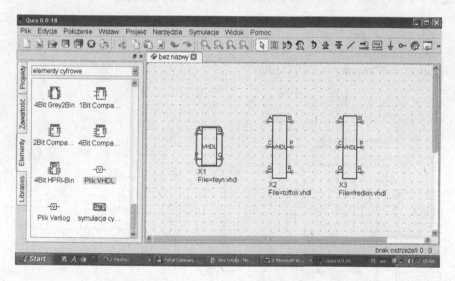

Fig. 12. Schematic editor for QUCS [own study]

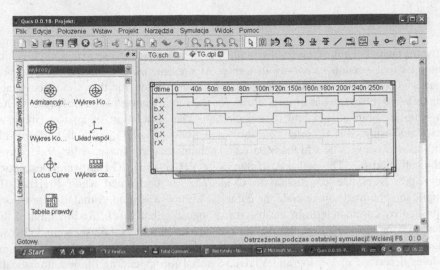

Fig. 13. Simulation of Toffoli gate [own study]

4 Laboratory Prototype for an Axle Counting System

Authors tried to implement an axle counting system in FPGA using reversible logic circuits. The system (Fig. 14) is composed of following modules:

- wheel sensor module (UCK)
- latch register (FF)
- control module (US)
- bidirectional counter (COUNTER)

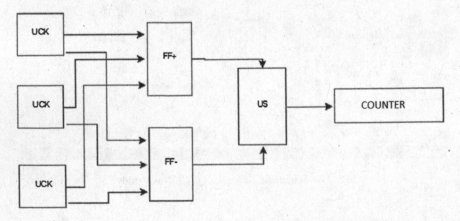

Fig. 14. Block diagram for the axle counting system [own study]

Modular structure of this system allows for its adaptation to customer's requirements (for example the change in the number of sensors).

Wheel sensor module is connected to the inductive sensors. They generates pulses when the wheel moves past sensors. Values 00, 01, 10, 11 coming from sensors correspond to different position of axle during train movement. Depending on the direction of the train movement, the pulse on either up or down input of the counter is given, what in turn causes either an increase or decrease in counter's contents (the number of axles). Flow chart for Finite-State Machine (FSM) which realizes an axle counting algorithm is presented in Fig. 15.

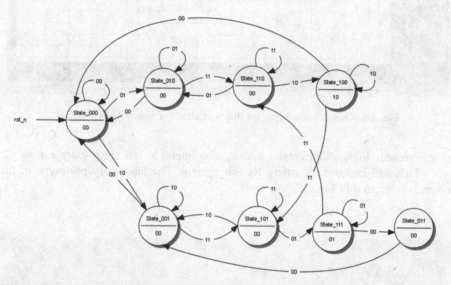

Fig. 15. Flow chart for FSM realizing an axle counting algorithm [own study]

Figure 16 presents output waveforms for the wheel sensor. This figure shows the behavior of system for the most complicated case, when the wheel stops in front of the sensor. Section A corresponds to the wheel's movement past the sensor. Section B presents system reaction when the wheel stops in front of the sensor. Signal change in C2 (caused by leaving the active area of the sensor by the wheel) does not generate extra pulses (only one pulse is fed to the input of the counter). Section C presents the behavior of the system after changing the direction of the movement. After moving the wheel past sensors C1 and C2, the pulse is generated and fed to the counter. Turning back (the change in sequence C1 C2) generates on second output an additional pulse which allows for the correction of counter's contents.

Axle counter (classic bidirectional counter) has been implemented using reversible logic [13–15].

Control module US and latch register have been realized using classic gates (this part of system is unsafe), in next version of the system authors will try to implement it

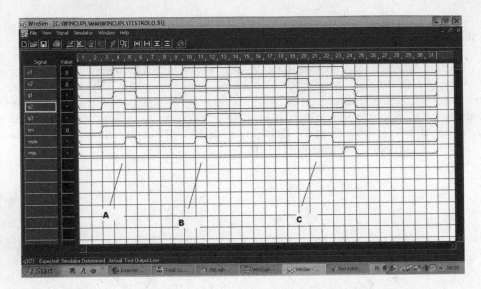

Fig. 16. Output waveforms for different states of sensors [own study]

using reversible logic. Additionally authors also intend to introduce extra control circuits which will increase the safety for this system. The laboratory prototype of this system is presented in Fig. 17.

Fig. 17. The laboratory prototype for the axle counting system [own study]

5 Conclusion

Presented in this paper an axle counting system allows for its implementation in FPGA. Authors have developed so far the following modules (based on reversible logic):

- wheel sensor module (UCK)
- latch register (FF)
- bidirectional counter (COUNTER)

Development of control module (US) based on reversible logic and adding comparators allow creating safe axle counting system, which can be easily implemented in FPGA structure.

Acknowledgments. This material is based upon work supported by Polish National Center for Research and Development under Grant No. PBS3/A6/29/2015.

References

1. Kawalec, P., Szydłowski, J., Mocki J.: Realizacja wybranych algorytmów działania urządzeń srk w programowalnych strukturach logicznych. In: International Scientific Conference Transport of the 21st Century, Warsaw (2001)
2. Pniewski, R., Kornaszewski, M., Chrzan, M.: Safety of electronic ATC systems in the aspect of technical and operational. In: Proceedings of the 16th International Scientific Conference Globalization and its Socio-Economic Consequences, Part IV, pp. 1729–1735. University of Zilina, The Faculty of Operation and Economics of Transport and Communications, Department of Economics, Rajecke Teplice, Slovak Republic, October 2016
3. Landauer, R.: Irreversibility and heat generation in the computational process. IBM J. Res. Dev. **5**(3), 183–191 (1961)
4. Taha, S.M.R.: Reversible Logic Synthesis Methodologies with Application to Quantum Computing. Springer, Cham (2016)
5. Peres, A.: Reversible logic and quantum computers. Phys. Rev. A **32**, 3266–3276 (1985)
6. Saravanan, M., Suresh, K.: Reversible logic circuit based twiddle factor generation. Int. J. Recent Innov. Trends Comput. Commun. **2**(9), 2895–2897 (2014)
7. Parhami, B.: Fault-tolerant reversible circuits. In: Proceedings of 40th Asilomar Conference on Signals, Systems and Computers, pp. 1722–1726, Pacific Grove, CA (2006)
8. Haghparast, M., Navi, K.: A novel fault tolerant reversible gate for nanotechnology based systems. Am. J. Appl. Sci. **5**(5), 519–523 (2008)
9. Al Mahamud, A., et al.: Synthesis of fault tolerant reversible logic circuits. In: Proceedings of IEEE International Conference on Testing and Diagnosis, pp. 1–4, Chengdu, China (2009)
10. Haghparast, M., Navi, K.: Design of novel fault tolerant reversible full adder for nanotechnology based systems. World Appl. Sci. J. **3**(1), 114–118 (2008)
11. Bruce, J.W., et al.: Efficient adder circuits based on conservative reversible logic gates. In: Proceedings of IEEE Computer Society Annual Symposium on VLSI, pp. 83–88, Pittsburg, PA (2002)
12. Pniewski, R.: Metoda oceny bezpieczeństwa cyfrowych systemów automatyki kolejowej. UTH, Radom (2013)

13. Selim Al Mamun Md., Karmaker, B.K.: Design of reversible counter. Int. J. Adv. Comput. Sci. Appl. **5**(1), 124–128 (2014)
14. Swaroop, V.G.S., et al.: Implementation of optimized reversible sequential and combinational circuits for VLSI applications. Int. J. Eng. Res. Appl. **4**(4), 382–388 (2014). (Version 1)
15. Babu, T.N., et al.: A low power adder using reversible logic gates. IJRET **01**(3), 244–247 (2012)

The High Level Risk Assessment of Security of Weight in Motion Systems

Wiktoria Loga[1] and Artur Ryguła[2(⊠)]

[1] University of Economics in Katowice, ul. 1 Maja 50, 40-287 Katowice, Poland
wiktoria.loga@edu.uekat.pl
[2] University of Bielsko-Biala, Willowa 2, 43-309 Bielsko-Biala, Poland
arygula@ath.bielsko.pl

Abstract. The development of modern transport technologies, such as Intelligent Transport Systems, Internet of Things and Big Data solutions, entailed also new threats. One of the consequences of digitization and automation of transport processes is the identification and continuous increase of the telematics risk. To effectively eliminate the threats, it is necessary to conduct detailed analyzes which will provide an information about critical processes in the system, the current security level and the stocks of the most vulnerable components. In this paper the authors present a high-level risk analysis of IT systems on the example of weighing in motion system. The authors identified critical elements of the system based on assessment of their default ICT security level. In addition, the paper contains recommendations for changes to in the current security, which can significantly reduce the risk of attack for the presented system.

Keywords: Telematics security · Weight-in-motion · High Level Risk Assessment

1 Introduction

Intelligent Transport Systems (ITS) are now a key component of road traffic. Their application is a solution, which is increasing the efficiency of transport systems and also enables the optimization of the traffic due to the ecological and social aspects. As highlighted in Directive [1] ITS are advanced applications which are the carrier medium used in order to provide innovative services related to transport. It should also be noted that these systems often are used in control and management process of traffic flows, thus determining the level of safety in transport networks. Considering the above and regarding to the threats that may exist in the information and telecommunication systems, it is advisable to search for tools to assess the risk levels of individual components of ITS systems. The authors of this study, according to the approach described in [2], proposed the HLRA (High Level of Security Risk Assessment) method to assess the level of risk in Weigh-in motion system (WIM) as an example of the ITS solution.

© Springer International Publishing AG 2017
J. Mikulski (Ed.): TST 2017, CCIS 715, pp. 461–470, 2017.
DOI: 10.1007/978-3-319-66251-0_37

2   Weigh-in-Motion System

Weigh-in motion system is the elementary tool for record and assessment of vehicles parameters passing through a measuring station. As shown in [3], their primary purpose is to protect the road infrastructure from against destruction and degradation associated with vehicles, which are oversized or overweight. WIM solutions combine a number of measuring, control and communication devices. An example of data exchange in WIM system along with a key processes and objects is shown in Fig. 1.

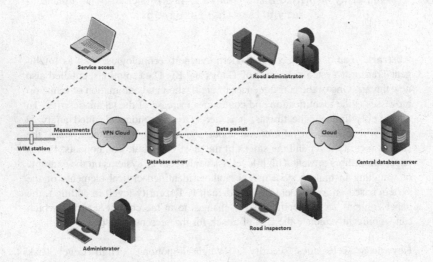

Fig. 1. An example of data exchange in WIM system [own study]

WIM station using the measuring devices performs weighing and identification of all the vehicles passing the station located at particular lane of road. Direct access to the system and the ability to perform calibration of the sensors is possible with the service's or administrator's level. Vehicle information are transmitted via a wireless link (cloud computing) and the VPN tunnel. Database server stores and archives compressed records and information about each vehicle. This database is used by road administration, road inspectors as well as the administrator and service. Particular collected data are transmitted to a central nationwide database server. Data transfer is done via internet network, using cloud solutions, but this time without a VPN tunnel. In order to protect the confidentiality of data, the encryption of license plate numbers is used.

3   The High Level Risk Assessment

3.1   Main Assumptions

Due to the implementation of IT solutions in the society and the business, information is one of the most valuable assets. The key role of this factor may be observed not only in organisations, but also in individual systems and requires the proper protection. Often, the security of the whole system depends on the security of information. When defining the security from the ICT point of view, it is important to acknowledge that the general term consists as follows [4]:

- physical security – verified based on precise risk, threat and vulnerability analysis,
- legal security – conformity with the law regulations concerning data storage, data processing and the information exchange process (the legal acts in this case are Personal Data Protection Act, Access to Public Information Act, Copyright and Related Rights Act),
- personal and organisational security – standards and procedures inside the organisation, managing the human recourses responsible for system handling,
- ICT security – protection of information and related actions in the system from possible threats using IT tools.

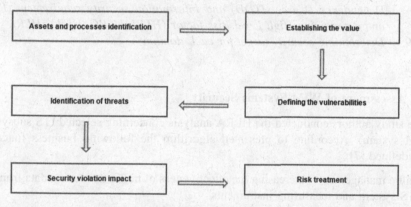

Fig. 2. Risk assessment process general scheme [own study based on [5]]

The base for providing and managing system security is conducting the detailed risk analysis. Analysis are powered by quantitative or qualitative approaches. Due to the drawbacks of quantitative method (lack of statistical data, lack of threat effects or intentional risk analysis), more often the qualitative approach is used. Standard [5] guidelines define qualitative method as basic risk management algorithm. The purpose of HLR analysis is presenting a consolidated and prioritized list of possible risks in a comprehensive, transparent and suited to system needs format. The assessment outcome enables to sort the risks into priority order with the highest risk level. Standard [5] points out the High Level Risk Assessment as a tool to identify the parts of system where the baseline security is enough and those that need special attention and additional protection. The general scheme of risk assessment process is presented in Fig. 2 [6].

High Level Risk Assessment requires system environment analysis along with business and societal functions and ICT recourses engaged in the system, what makes the risk research time-absorbing. In the HLRA analysis authors have used the following algorithm:

Step 1:   *Identification of system business functions.*

Step 2:   *Identification of corresponding business processes.*

Step 3:   *Processes division based on Criticality for Organisation level (C4O).*

Step 4:   *Identification of the assets engaged in the processes based on IT Dependency Degree and HR Dependency Degree (ITDD/MDD).*

Step 5:   *Identification of business processes needs that concern the protection of availability, integrity, authenticity, confidentiality and accountability as Protection Needs (PNI, PNC, PNA, PND, PNR).*

Step 6:   *Impact analysis based on loss of availability, integrity, authenticity, confidentiality and accountability as so called Business Impact (BII, BIC, BIA, BIR, BID).*

Step 7:   *Processes protection need assignation regarding availability, integrity, authenticity, confidentiality and accountability defined as Information security requirement) – (ISRI, ISRC, ISRA. ISRR, ISRD).*

Step 8:   *Domain orderliness including Criticality for Organisation level (C40), IT Dependency Degree (ITDD) and Information security requirement (ISRx) and determining High Level Risk factor (HLRI, HLRC, HLRA, HLR).*

Step 9:   *Determining security target for each domain*

3.2   Assessment of WIM Systems Security

In the study authors conducted the HLRA analysis concerning selected ITS subsystem (WIM system). According to presented algorithm the following business functions were defined [7]:

- traffic management – increasing the effectiveness of traffic flow using data from the ITS system and measuring instruments,
- transportation system safety and security – providing adequate security standards at cyber, as well as physical level,
- dynamic and reliable system handling – ability to provide rapid and effective reactions to the system operators,
- relevant system environment – creating system-friendly environment along with ensuring the sufficient amount of energy, computing power and data storage.

Based on presented business functions authors defined the processes inside the WIM system with the engaged entities and evaluated criticality level using four-level scale C4O. Each process was assigned with relevant value (Very high-4, High-3, Medium-2, Low-1), where the highest measure stands for critical process (in terms of proper system functioning) while the lowest measure stands for subsidiary process [2]. It should be emphasised that assigned values of C4O reached maximum level in majority of cases. This situation is linked to ITS systems specificity. WIM solution is the source of sensitive data (drivers personal data, vehicle number plate etc.) also the conducted weight preselection is basis of vehicle stoppage for further heaviness control and imposing a penalty payment. The criticality values are presented in Table 1.

Table 1. Criticality levels for WIM processes with engaged entities [own study]

ID	Process	Engaged entities	C4O
P1	Access to WIM application	Administrator, Inspectorate of Road Transport (ITD), General Directorate of Domestic Roads and Motorway (GDDKiA), Service	4.0
P2	Real-time data presentation	Administrator, ITD, GDDKiA	4.0
P3	Preselection and data registration	ITD, Administrator	4.0
P4	System and application validity check	Administrator, Service	4.0
P5	Managing the system	Administrator	4.0
P6	Measurement calibration	Service	4.0
P7	Data import and data export	Administrator, ITD, GDDKiA	3.0
P8	Data archiving	Database server	3.0
P9	Data pre-analysis	Administrator, ITD, GDDKiA	2.0
P10	Data transmission in the system	Administrator, ITD, GDDKiA, Service	4.0

The identification of critical processes enabled the selection of assets dependency degree. Analysis takes under consideration dependency on IT (ITDD) assets and human recourses (MDD). To match dependency value authors used three-level scale (High-3, Basic-2, Low-1) where highest value corresponds to full dependency while lowest value stands for processes the can be fulfilled using alternative assets [2]. System dependency level is presented in Table 2.

Table 2. IT (ITDD) and human resources (MDD) dependency degree [own study]

ID	Process	ITDD	MDD
P1	Access to WIM application	3.0	1.0
P2	Real-time data presentation	3.0	2.0
P3	Preselection and data registration	2.0	2.0
P4	System and application validity check	3.0	3.0
P5	Managing the system	3.0	3.0
P6	Measurement calibration	2.0	3.0
P7	Data import and data export	3.0	1.0
P8	Data archiving	3.0	1.0
P9	Data pre-analysis	3.0	1.0
P10	Data transmission in the system	3.0	1.0

High values of ITDD measure indicates strong cybernation of WIM system and in consequence higher ICT risk vulnerability. MDD measure reach maximum level in processes dependent on system operators or service operations

Presented analysis is conducted in basis of selected safety attributes, with are the foundation of information security management system. In the context of ITS the safety attributes are defined as follows [8]:

- Availability – ability of the ITS systems to provide the ITS services continuously. Continuity of action is crucial element. ITS solutions must provide 100% availability and data security, all failures in this area may cause serious complications.
- Integrity – Integrity information is generated and sent by the sender to the receiver without any unauthorized changes or entities during the transmission. The receiver must be sure that the information is authentic and did not changed during the transmission.
- Authenticity – information received by an ITS station is coming from an trusted and authentic source. Standard solution in this case is two-way authentication exchanging security certificates of the ITS system elements.
- The authorization requires safety management system and authentication process.
- Confidentiality – information should be available only for ITS stakeholders that are authorized to receive it. An example of sensitive data in the system is identification of specific vehicle or person.
- Accountability – every action taken by the ITS application functionality or message should be assigned to responsible asset. In case the event causes damage in the system identify answerable asset such as application, application user, system owner or operator. One example could be that a driver trusts a driver assisting ITS.

For each process authors selected measures concerning its protection needs in matters of integrity (PNI), confidentiality (PNC), availability (PND), accountability (PNR) and information authenticity (PNA) and the mean value for each process (PNM). To match protection needs values four-level scale was used (High-3, Basic-2, Low-1, None-0), where highest value corresponds to extensive protection needs while lowest stands lack of information protection needs [2]. The result of analysis is presented in Table 3.

Table 3. Information protection needs in WIM system [own study]

ID	Process	PNI	PND	PNC	PNR	PNA	PNM
P1	Access to WIM application	2.0	3.0	3.0	2.0	3.0	2.6
P2	Real-time data presentation	2.0	3.0	3.0	1.0	3.0	2.4
P3	Preselection and data registration	3.0	3.0	3.0	3.0	3.0	3.0
P4	System and application validity check	2.0	3.0	3.0	3.0	3.0	2.8
P5	Managing the system	3.0	3.0	3.0	3.0	3.0	3.0
P6	Measurement calibration	3.0	3.0	3.0	3.0	3.0	3.0
P7	Data import and data export	3.0	2.0	3.0	2.0	3.0	2.6
P8	Data archiving	3.0	2.0	3.0	2.0	3.0	2.6
P9	Data pre-analysis	3.0	2.0	2.0	1.0	3.0	2.2
P10	Data transmission in the system	3.0	3.0	3.0	3.0	3.0	3.0

3.3 Results

The estimation of discussed safety and security parameters enabled to perform the safety attributes analysis using the "What-if" method. For each process the possible threat scenarios were assigned. The result of analysis was to obtain business impact measures of integrity (BII), confidentiality (BIC), availability (BID), accountability (BIR) and information authenticity (BIA) and the mean value for each process (BIM). The measures are presented in Table 4.

In the study following consequence scenarios were considered [8]:

– law and contract violation,
– personal privacy violation,
– exposing to danger of loss of life or health,
– obligation or assignment realisation violation,
– system image and public relations deterioration,
– direct financial loss.

Table 4. Safety attributes analysis using "What-if" approach [own study]

ID	Process	BII	BID	BIC	BIA	BIR	BIM
P1	Access to WIM application	1.7	2.3	2.7	2.5	2.7	2.4
P2	Real-time data presentation	1.8	2.0	2.5	2.2	1.5	2.0
P3	Preselection and data registration	2.7	2.5	2.5	2.7	2.7	2.6
P4	System and application validity check	2.5	2.5	2.7	2.5	2.5	2.5
P5	Managing the system	2.0	2.7	2.3	2.3	2.7	2.4
P6	Measurement calibration	2.3	2.5	2.7	2.5	2.5	2.5
P7	Data import and data export	2.5	2.0	2.5	2.7	2.2	2.4
P8	Data archiving	2.7	2.2	2.7	2.5	1.8	2.4
P9	Data pre-analysis	1.7	1.7	2.0	2.2	1.8	1.9
P10	Data transmission in the system	2.7	2.7	2.7	2.7	2.7	2.7

The Information security requirement level was calculated in the matters of integrity (ISRI), confidentiality (ISRC), availability (ISRD), accountability (ISRR) and information authenticity (ISRIA), using the equation. Obtained values are presented in Table 5.

$$ISRx = PNx \cdot Bix \tag{1}$$

where:

ISRx – security requirement level of integrity (ISRI), confidentiality (ISRC), availability (ISRD), accountability (ISRR) and information authenticity (ISRIA).
PNx – protection need for selected safety attribute.
Bix – business impact of looping selected safety attribute.

Table 5. The level of protection requirements for system processes [own study]

ID	Process	ISRI	ISRD	ISRC	ISRR	ISRA	ISRM
P1	Access to WIM application	3.3	7.0	8.0	5.3	7.5	6.2
P2	Real-time data presentation	3.7	6.0	7.5	1.5	6.5	5.0
P3	Preselection and data registration	8.0	7.5	7.5	8.0	8.0	7.8
P4	System and application validity check	5.0	7.5	8.0	7.5	7.5	7.1
P5	Managing the system	6.0	8.0	7.0	8.0	7.0	7.2
P6	Measurement calibration	7.0	7.5	8.0	7.5	7.5	7.5
P7	Data import and data export	7.5	4.0	7.5	4.3	8.0	6.3
P8	Data archiving	8.0	4.3	8.0	3.7	7.5	6.3
P9	Data pre-analysis	5.0	3.3	4.0	1.8	6.5	4.1
P10	Data transmission in the system	8.0	8.0	8.0	8.0	8.0	8.0

The final measure of HLRA analysis is HLRx value, The measure is calculated separately for each attribute according to the equation:

$$HLRx = ISRx \cdot C4O \cdot ITDD \qquad (2)$$

where:

HLRx – process high risk level.
ISRx – security requirement level for selected safety attribute.
C4O – criticality of the process.
ITDD – IT dependency degree.

Table 6. HLR risk values for the WIM system [own study]

ID	Process	HLRI	HLRD	HLRC	HLRR	HLRA	HLRM
P1	Access to WIM application	53.3	112.0	128.0	85.3	120.0	99.7
P2	Real-time data presentation	73.3	120.0	150.0	30.0	130.0	100.7
P3	Preselection and data registration	128.0	120.0	120.0	128.0	128.0	124.8
P4	System and application validity check	120.0	180.0	192.0	180.0	180.0	170.4
P5	Managing the system	144.0	192.0	168.0	192.0	168.0	172.8
P6	Measurement calibration	140.0	150.0	160.0	150.0	150.0	150.0
P7	Data import and data export	90.0	48.0	90.0	52.0	96.0	75.2
P8	Data archiving	96.0	52.0	96.0	44.0	90.0	75.6
P9	Data pre-analysis	40.0	26.7	32.0	14.7	52.0	33.1
P10	Data transmission in the system	128.0	128.0	128.0	128.0	128.0	128.0

The ultimate results collation presented in Table 6, using the Eq. (3) was represented in percentage values.

$$HLR = \frac{HLRx}{Smax} \, [\%] \tag{3}$$

where:

HLRx – High level risk value.
Smax – Maximum of *HLRx* value (in discussed analysis *Smax = 216*).

The outcome of transformation (Table 7) according to the authors may be interpreted as follows:

– HLR ≥ 75% very high risk level;
– HLR ≥ 50% high risk level;
– HLR ≥ 25% moderate risk level;
– HLR < 25% low risk level.

Table 7. Summary of HLRA results [own study]

ID	Process	HLR	Risk level
P1	Access to WIM application	46.17%	Moderate
P2	Real-time data presentation	46.60%	Moderate
P3	Preselection and data registration	57.78%	High
P4	System and application validity check	78.89%	Very high
P5	Managing the system	80.00%	Very high
P6	Measurement calibration	69.44%	High
P7	Data import and data export	34.81%	Moderate
P8	Data archiving	35.00%	Moderate
P9	Data pre-analysis	15.31%	Low
P10	Data transmission in the system	59.26%	High

The results obtained in high leve risk (HLRx) calculation, along with the outcome of protection needs (ISRx) analysis are presented in Fig. 3.

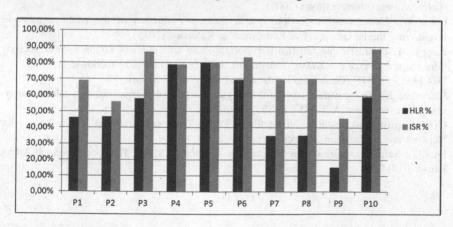

Fig. 3. Percentage distribution of protection needs and high level risk values in the WIM system [own study]

When discussing the general security of the WIM system, modules that require high protection are the processes, which have the direct impact on the correctness of measurements. Those processes are preselection, registration of vehicles and data transmission. In mentioned domains, it is necessary to assure the proper physical, legal, organisational and ICT security level. It is worth mentioning that discussed domains are external processes. The high rate of HRL parameter (cases where HLR was identified as "very high") was estimated for P4 (System and application validity check) and P5 (Managing the system). The results point out that P4 and P5 processes should be subjected for further analysis, where the specific vulnerabilities will be identified. Detailed analysis will allow to select adequate information security system on the cyber as well as the physical level. For all the domains the acceptable risk level should be preordained and continuously retained.

4   Conclusion

Example of the method HLRA applied to the telecommunication system (Weigh-in motion system) presented in work is a proposal for a risk assessment method in terms of ITS solutions. The proposed approach allows a systematic assessment of required level of protection and the risk level for a defined system processes. This methodology may also be related to the other ITS subsystems, where the risk of security attributes can potentially result in a direct threat to the participants of the transport system.

References

1. Directive 2010/40/EU of the European parliament and of the council of 7 July 2010 on the framework for the deployment of Intelligent Transport Systems in the field of road transport and for interfaces with other modes of transport
2. Białas, A., Karch, P.: Wysokopoziomowa analiza ryzyka - studium przypadku. In: Grzywak, A., Pikiewicz, P. (eds.) Nowe technologie w komputerowych systemach zarządzania - Internet w społeczeństwie informacyjnym (2005)
3. Loga, W.: Using weigh-in-motion system data in evaluation of the traffic conditions, Engineering thesis, University of Economics in Katowice (2015)
4. Pałęga, M., et al.: Bezpieczeństwo teleinformatyczne jako element kompleksowej ochrony informacji. In: Prace Naukowe Akademii im. Jana Długosza w Częstochowie (2013)
5. ISO 13335 GMITS (Guidelines for the Management of IT Security)
6. Liderman, K.: Oszacowania jakościowe ryzyka dla potrzeb bezpieczeństwa teleinformatycznego. In: Biuletyn IAiR WAT nr 19 (2003)
7. Cyber Security and Resilience of Intelligent Public Transport, European Union Agency for Network and Information Security, December 2015
8. Foss, T.: Safe and secure intelligent transport systems (ITS). In: Transport Research Arena, Paris (2014)

The Concept of Railway Traffic Control Systems Remote Diagnostic

Waldemar Nowakowski[✉], Tomasz Ciszewski,
and Zbigniew Łukasik

Faculty of Transport and Electrical Engineering,
Kazimierz Pulaski University of Technology and Humanities in Radom,
Malczewskiego 29, 26-600 Radom, Poland
{w.nowakowski, t.ciszewski, z.lukasik}@uthrad.pl

Abstract. The development of railway traffic control systems gives the new opportunities for remote diagnostics and events logging in the maintenance and diagnostics centres. This possibility is a result of use the information technology in modern systems. Despite the steady technological progress, there is the lack of common methods of acquiring diagnostic data, both in terms of the scope of collected data and the manner of their presentation. This results in the use of variety solutions by individual producers operating on the market. The authors of the article inspired by the Simple Network Management Protocol (SNMP) proposed the use of this network standard in the diagnostics of railway traffic control systems. Since the research are in the initial stage only Level Crossing Protection Systems are included in the study. For this study two Management Information Bases (MIB) were developed. In the first base SNMP traps, sent by the SNMP agent, are defined. The second one defines the scope of data that can be a part of the polling by management station. In the next step the authors developed the SNMP client software. This software allows you to change parameters of Level Crossing Protection Systems and makes them available at the request of the SNMP manager. In SNMP client software the possibility of sending information about the system faults in the form of SNMP traps are implemented. The conducted study have confirmed the usefulness of SNMP technology in the railway traffic control system diagnostics. There is also the need to extend the research to other devices and systems.

Keywords: Railway traffic control systems · LCPS · Remote diagnostics · SNMP · MIB

1 Introduction

Railway signalling and control systems have been using information technology more commonly because they allow to control railway traffic in train service stations from local control centres [1–4]. Within one local control centre there is equipment installed that is often made by different manufacturers. In order to keep control over these devices maintenance and diagnosis centres arise. A lack of a unified standard causes the fact that system manufacturers use various technical solutions which, subsequently, leads to difficulties, not only in the equipment's operation, but also in its maintenance [5–9].

© Springer International Publishing AG 2017
J. Mikulski (Ed.): TST 2017, CCIS 715, pp. 471–481, 2017.
DOI: 10.1007/978-3-319-66251-0_38

That is why, in this article, it has been proposed to use the Simple Network Management Protocol (SNMP) standard, known for computer network management in automatic system diagnostics in railway transport. That is why necessary research on Level Crossing Protection Systems (LCPS) has been conducted [10, 11].

2   Level Crossing Protection Systems (LCPS)

Because of the need to ensure safety, a particularly important place is a level crossing, which is an intersection where a railway line crosses a road at the same level. In order to protect the road users from the dangers related to the railway traffic, Level Crossing Protection Systems are created. Alerting and warning about hazards in these systems is accomplished with the use of road signals and acoustic signal generators (buzzers or bells). Additionally, level crossings can be equipped with barriers and warning shields (Fig. 1).

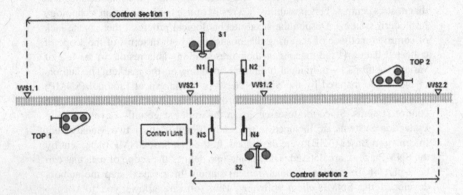

Fig. 1. An example structure of the LCPS (single-track railway) [own study]

A danger alert is turned on when a train is getting closer to an intersection. The train is usually detected when the train interacts with wheel sensors. Where it comes to the control section 1, it is the WS1.1 activation point, and for the control section 2 – WS2.2 activation point (Fig. 1). Information about the train is passed on to the control unit of the LCPS, which triggers the alert. The S1 and S2 road signals, acoustic signal generators and level crossing warning shields are turned on for a given direction. Another action is the closure of the barriers.

3   Simple Network Management Protocol (SNMP)

SNMP is a protocol serving as TCP/IP network management. Currently SNMP is a standard used by most manufacturers of network equipment [12–14]. Within this protocol there are two types of equipment: managers and agents. A computer plays a role of Network Management Station (NMS) when it runs a special program called SNMP manager. A device is being managed when it runs SNMP agent. Management takes place when the SNMP manager reads (or modifies) specific parameters, concerning the status of the device, which are stored by the SNMP agent. What is more, the agent can notify the manager about an unexpected event by sending a special message called SNMP trap, which includes information about the event (Fig. 2).

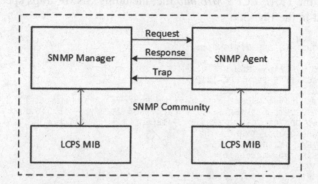

Fig. 2. Simplified architecture of the Simple Network Management Protocol [own study]

Information regarding the status of the device is being stored and shared by the agent, with the use of MIB (Management Information Base). In order to define MIB structure ASN.1 (Abstract Syntax Notation One) notation is required [15–17]. Every object in the base has its identifier (OBJECT IDENTIFIER), SYNTAX and ENCODING RULES. Objects are organized in a MIB base in hierarchies (tree structure). In order to gain access to an object representing a certain resource (manager data) one needs to present all names, separated by dots, from the root to the leaf. Names can be descriptive or numerical, e.g. *iso.org.dod.internet.private.enterprise* or *1.3.6.1.4.1*. ISO (International Organization for Standardization) is responsible for assigning names and numbers to nodes, this assures a unified nomenclature of objects. A benefit of the SNMP standard is a possibility to define new MIB bases. That is why the SNMP standard can be used in other areas, not only in computer network management. Own extensions can be added to the *private* subtree. It allows producers to create objects supporting specific parameters of their products, it also assures that these objects are detected by managing stations. This feature has been used by the authors of this paper who have proposed a concept to use the SNMP technology in diagnostics of equipment and railway traffic control systems.

4 MIB Specification of Level Crossing Protection System

Management station using SNMP has access only to the information defined in MIB bases. That is why, in order to adapt the SNMP standard to the diagnostics of the railway traffic control systems private MIB files have been created. The research included only the Level Crossing Protection Systems, which is why SNMP traps sent by the SNMP agent have been defined for them, also data that can be available to the management station has been determined. Where it comes to the SNMP traps, variables passed on in messages are the following parameters:

- *sysName*, *sysLocation*, which are system objects, defined in RFC 1213,
- *faultCode* INTEGER type, which is a code for any errors,
- *faultDesc* DisplayString type, which is a short description of an error.

Content of the *TRAP-LCPS-MIB.mib* file, including SNMP traps specification, has been presented in the Fig. 3.

```
TRAP-LCPS-MIB DEFINITIONS ::= BEGIN
IMPORTS
    DisplayString FROM RFC1213-MIB
    OBJECT-TYPE FROM RFC-1212
    enterprises FROM RFC1155-SMI;
-- (LCPS) Level Crossing Protection Systems
lcps OBJECT IDENTIFIER ::= { enterprises 2021 } -- 1.3.6.1.4.1.2021
lcpstrap OBJECT IDENTIFIER ::= { lcps 2 } -- 1.3.6.1.4.1.2021.2
faultCode OBJECT-TYPE
    SYNTAX  INTEGER (0..255)
    ACCESS  not-accessible
    STATUS  mandatory
    DESCRIPTION "Fault code
    1xx - 'I' category fault
    2xx - 'II' category fault"
    ::= { lcpstrap 3 }
faultDesc OBJECT-TYPE
    SYNTAX  DisplayString (SIZE (0..255))
    ACCESS  not-accessible
    STATUS  mandatory
    DESCRIPTION "Fault description"
    ::= { lcpstrap 4 }
lcps-trap TRAP-TYPE
    ENTERPRISE lcpstrap
    VARIABLES { sysName, sysLocation, faultCode, faultDesc }
    DESCRIPTION "The variables included in the LCPS trap"
    ::= 1
END
```

Fig. 3. Specification of the SNMP traps of the LCPS [own study]

In the second file, called *LCPS-MIB.mib*, variables describing the Level Crossing Protection Systems about which management station can inquire have been defined. Among the variables there are:

- *category* DisplayString type, category of level crossing (B or C),
- *nhbarriers* INTEGER type, number of half barriers (2 or 4 for B, 0 for C),
- *ntlines* INTEGER type, number of track lines (1 or 2),
- *rcp* INTEGER type, support the RCP (Remote Control Panel),

and the status board of the LCPS:

- *stWhen* DisplayString type, timestamp,
- *stState* INTEGER type, state of LCPS,
- *stSensor1*, *stSensor2* INTEGER type, state of rail wheel sensor,
- *stControl* INTEGER type, type of control,
- *stFaultI*, *stFaultII* INTEGER type, 'I/II' category fault,
- *stFaultONsensor*, *stFaultOFFsensor* INTEGER type, 'ON/OFF' rail wheel sensors fault.

A chosen fragment of a LCPS-MIB module has been presented in the Fig. 4.

```
LCPS-MIB DEFINITIONS ::= BEGIN
IMPORTS
    DisplayString FROM RFC1213-MIB
    OBJECT-TYPE FROM RFC-1212
    enterprises FROM RFC1155-SMI;
-- (LCPS) Level Crossing Protection Systems
lcps OBJECT IDENTIFIER ::= { enterprises 2021 } -- 1.3.6.1.4.1.2021
lcpsinfo OBJECT IDENTIFIER ::= { lcps 1 } -- 1.3.6.1.4.1.2021.1
-- LCPS parameters
    category OBJECT-TYPE
        SYNTAX  DisplayString (SIZE (0..1))
        ACCESS   read-only
        STATUS   mandatory
        DESCRIPTION "Category of level crossing (B or C)"
    ::= { lcpsinfo 1 }
{...}
-- State of level crossing
    stTable OBJECT-TYPE
        SYNTAX SEQUENCE OF STEntry
        ACCESS not-accessible
        STATUS mandatory
        DESCRIPTION "A table (concept) with information about the state
                     of LCPS"
        ::= { lcpsinfo 5  }
    stEntry OBJECT-TYPE
        SYNTAX STEntry
        ACCESS not-accessible
        STATUS mandatory
        DESCRIPTION "A single table entry (concept) with information
                     about the state of LCPS"
        INDEX { spIndex }
        ::= { stTable 1 }
    STEntry ::= SEQUENCE {
            stIndex INTEGER, -- Index
            stWhen DisplayString, -- When
            stState INTEGER, -- State of LCPS
            stSensor1 INTEGER, -- State of rail wheel sensor of 1 track
                                  line
            stSensor2 INTEGER, -- State of rail wheel sensor of 2 track
                                  line
            stControl INTEGER, -- Type of control
            stFaultI INTEGER, -- 'I' category fault
            stFaultII INTEGER, -- 'II' category fault
            stFaultONsensor INTEGER, -- 'ON' rail wheel sensors fault
            stFaultOFFsensor INTEGER -- 'OFF' rail wheel sensors fault
    }
(...)
    stIndex OBJECT-TYPE
        SYNTAX INTEGER (1..65535)
        ACCESS read-only
        STATUS mandatory
        DESCRIPTION "A unique value for each entry in the table with
information about the state of LCPS"
        ::= { stEntry 1 }
    stWhen OBJECT-TYPE
        SYNTAX DisplayString (SIZE (0..20))
        ACCESS read-only
        STATUS mandatory
        DESCRIPTION "The date and time of the event"
        ::= { stEntry 2 }
(...)
END
```

Fig. 4. Specification of SNMP objects of the LCPS [own study]

Correctness of private MIB files was verified with parsing, next it was compiled and added to the *MG-SOFT MIB Browser* environment which, in the research, plays a role of the SNMP manager (Fig. 5).

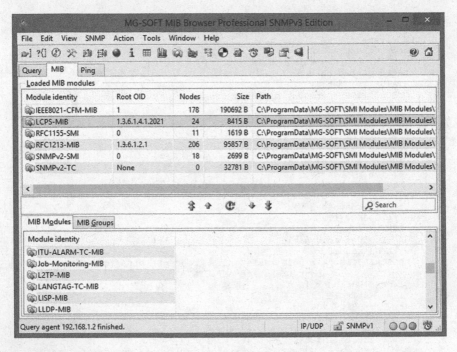

Fig. 5. SNMP manager after loading a private LCPS-MIB module [own study]

5 SNMP Agent – Simulator of Level Crossing Protection System

Private MIB bases served for developing the SNMP agent, which was at the same time a simulator of the Level Crossing Protection System. The agent, similarly to the SNMP manager, compiles *TRAP-LCPS-MIB* and *LCPS-MIB* bases. It also ensures interface for modification of all included in these bases objects. The agent's main screen has been presented on the Fig. 6.

Because of a simulation of real LCPS, the agent's software includes a possibility to create and modify system work history, which has been presented in the Fig. 7.

Both, the LCPS parameters and their work history can be read with the help of the SNMP manager which, in this case, was *MG-SOFT MIB Browser* (Figs. 8, 9 and 10).

The SNMP agent's software also included a possibility to send system error messages to the manager, in the form of SNMP traps (Fig. 11). Errors, in accordance with a definition in the MIB base have been divided into I category faults, which endanger the traffic safety directly, and II category faults, not endangering the safety.

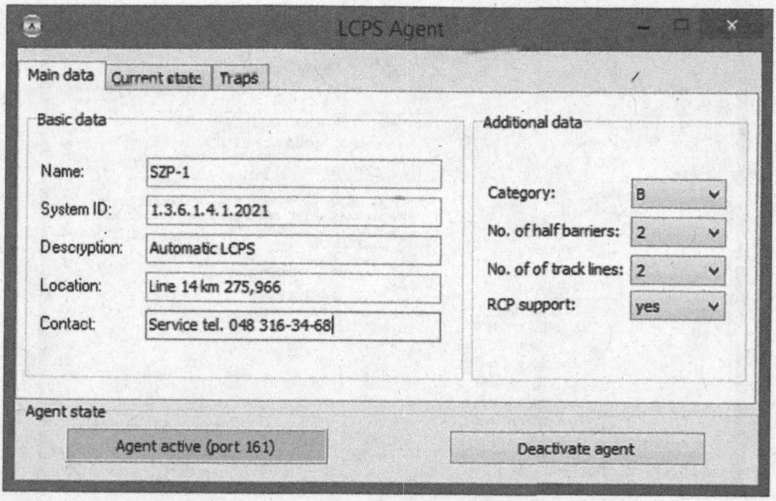

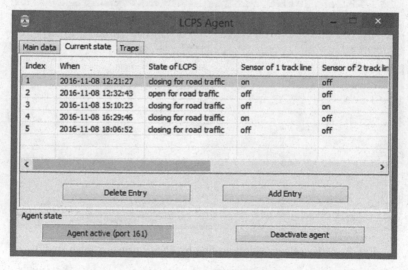

Fig. 6. SNMP agent's main screen [own study]

Fig. 7. SNMP agent's screen serving for simulation of LCPS [own study]

If a I category fault arises, the LCPS introduces a speed limit of the train, the speed which allows a safe stop of the train in case of an appearance of an obstacle on the level crossing. This category includes:

- WS1.1 or WS2.2 rail wheel sensor fault,
- S1 or S2 road signal fault,
- N1, N2, N3 or N4 barrier drive fault.

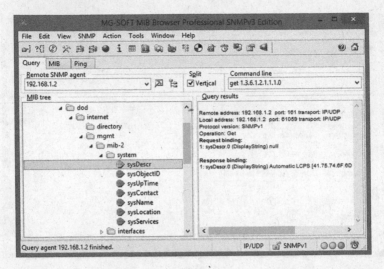

Fig. 8. MIB *sysDescr* object reading by SNMP manager [own study]

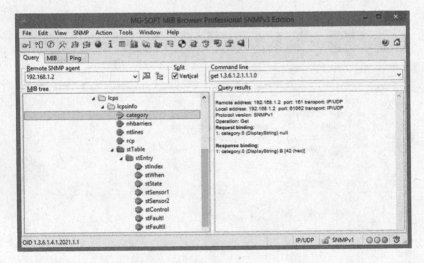

Fig. 9. MIB *category* object reading by SNMP manager [own study]

Faults, which do not endanger the traffic, thereby do not result in speed limits, are II category faults. These include:

- WS2.1 or WS1.2 rail wheel sensor fault,
- N1, N2, N3 or N4 barrier boom lighting fault,
- raising the closed N1, N2, N3 or N4 barrier boom,
- TOP white light fault,
- TOP orange light fault,
- power supply with battery backup,
- battery backup fault.

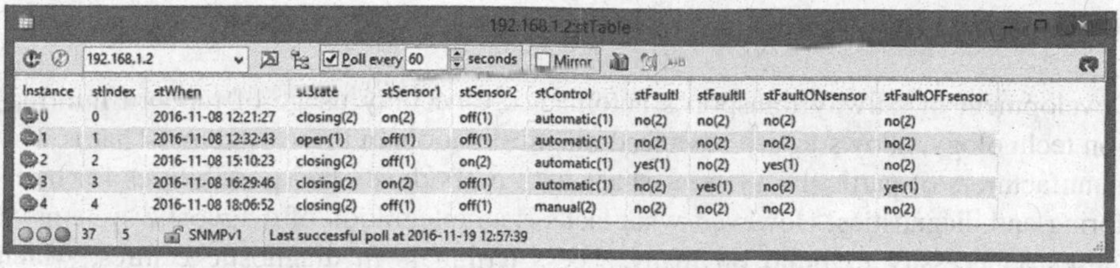

Fig. 10. Reading of LPCS work history [own study]

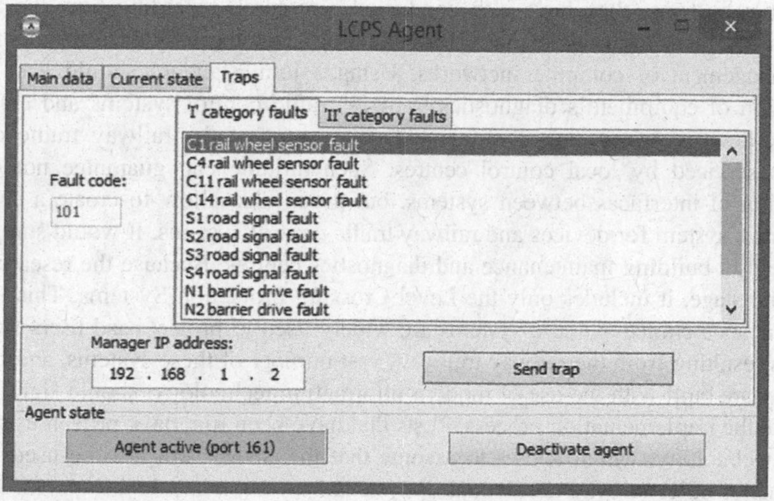

Fig. 11. SNMP agent's screen for sending warning messages [own study]

The SNMP manager receives and decodes SNMP traps. Figure 12 presents a console with example traps received by the *MG-SOFT MIB Browser* manager.

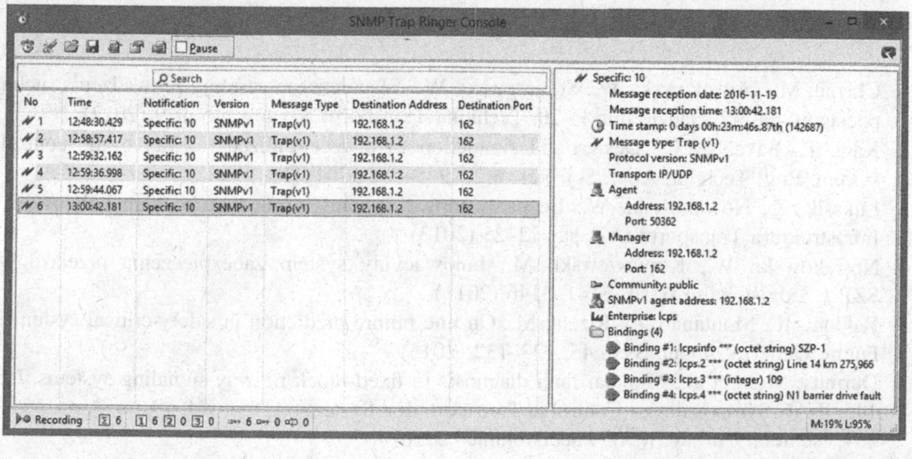

Fig. 12. Reception of traps with information about LPCS errors by SNMP manager [own study]

6 Conclusion

Development of railway transport's automatics, especially usage of modern information technology, allows to use advanced diagnostic tools in these systems. That is why manufacturers of particular systems share with users developed possibilities for monitoring and diagnostics. However, what lacks is an integration of diagnostic systems. It causes a necessity to build on many access terminals in diagnostic centres, which significantly restricts work of maintenance services. The authors of this article have noticed the need to unify methods of diagnostics and event registration in railway automatics systems, they have also conducted a research regarding how the SNMP standard can be used in this matter. This technology is commonly used for monitoring and management of computer networks. Using a joint standard should lead to the integration of equipment's diagnostics, railway traffic control systems and computer network devices, which are an integrated component of the railway traffic control systems serviced by local control centres. Such attitude can guarantee not only a unification of interfaces between systems, but it can also allow to create a common diagnostics system for devices and railway traffic control systems. It would surely be a great help in building maintenance and diagnostics centres. Because the research is in the initial stage, it includes only the Level Crossing Protection Systems. This has not been a chance choice as these systems are widely used to protect road users from the dangers resulting from the railway traffic. A vast number of these systems, and the fact that they are built with the use of modern information technology, should significantly facilitate the implementation process. Tests that have been run, have proven usefulness of this technology, which allows to assume that the SNMP can also be used for the diagnosis of other railway traffic control systems.

Acknowledgment. This material is based upon work supported by National Centre for Research and Development under Grant No. PBS3/A6/29/2015 entitled "The system for maintenance data acquisition and analysis of reliability and safety of traffic control systems".

References

1. Chrzan, M., Nowakowski, W., Sobiczewski, W.: Współczesne systemy prowadzenia ruchu pociągów na przykładzie linii E–20. Technika Transportu Szynowego nr 9 (2012)
2. Kara, T., Savas, M.C.: Design and simulation of a decentralized railway traffic control system. Eng. Technol. Appl. Sci. Res. **6**(2), 945–951 (2016)
3. Łukasik, Z., Nowakowski, W.: Bezprzewodowe systemy sterowania ruchem kolejowym. Infrastruktura Transportu nr 4, str. 22–25 (2013)
4. Nowakowski, W., Kornaszewski. M.: Innowacyjny system zabezpieczenia przejazdów SZP-1. Logistyka nr 6, str. 3141–3146 (2011)
5. Baldoni, R., Montanari, L., Rizzut, M.: On-line failure prediction in safety-critical systems. Future Gener. Comput. Syst. **45**, 123–132 (2015)
6. Durmus, M.S., et al.: Modular fault diagnosis in fixed-block railway signaling systems. In: 14th IFAC Symposium on Control in Transportation Systems (CTS), vol. 49, no. 3, pp. 459–464, Istanbul, Turkey, IFAC Papers online (2016)

7. Grimes, G., Adley, B.P.: Intelligent agents for network fault diagnosis and testing. In: 5th IFIP/IEEE International Symposium on Integrated Network Management, San Diego, Canada, Integrated Network Management V: Integrated Management in a Virtual World, pp. 232–244 (1997)

8. Mascardi, V., Briola, D., Martelli, M., Caccia, R., Milani, C.: Monitoring and diagnosing railway signalling with logic-based distributed agents. In: Corchado, E., Zunino, R., Gastaldo, P., Herrero, Á. (eds.) CISIS 2008. AINSC, vol. 53, pp. 108–115. Springer, Heidelberg (2009). doi:10.1007/978-3-540-88181-0_14

9. Łukasik, Z., Nowakowski, W., Ciszewski, T.: Definition of data exchange standard for railway applications, Prace Naukowe Politechniki Warszawskiej. Transport zeszyt 113, str. 319–326 (2016)

10. Garcia, M.: New methods for the condition monitoring of level crossings. Int. J. Syst. Sci. 46(5), 878–884 (2015)

11. Lewiński, A., Trzaska–Rycaj, K.: The safety related software for railway control with respect to automatic level crossing signaling system. In: Mikulski, J. (ed.) TST 2010. CCIS, vol. 104, pp. 202–209. Springer, Heidelberg (2010). doi:10.1007/978-3-642-16472-9_22

12. Duarte, E.P., dos Santos. A.L.: Network fault management based on SNMP agent groups. In: Proceedings of 21st IEEE International Conference on Distributed Computing Systems, pp. 51–56, Phoenix, USA (2001)

13. Hood, C.S., Ji, C.Y.: Proactive network-fault detection. IEEE Trans. Reliab. 46(3), 333–341 (1997)

14. Su, M.S., Thulasiraman, K., Das. A.: A scalable on-line multilevel distributed network fault detection/monitoring system based on the SNMP protocol. In: IEEE Global Telecommunications Conference (GLOBECOM 2002), pp. 1960–1964, Taipei, Taiwan (2002)

15. Łukasik, Z., Nowakowski. W.: Designing communication software for computer railway control systems with the use of ASN.1. Computer Systems Aided Science and Engineering Work in Transport, Mechanics and Electrical Engineering. Monograph No 122, pp. 391–398. Technical University of Radom (2008)

16. Łukasik, Z., Nowakowski, W.: ASN.1 notation for exchange of data in computer-based railway control systems. Transp. Probl. 4(2), 111–116 (2009). Silesian University of Technology, Gliwice

17. Łukasik, Z., Nowakowski, W.: Application of TTCN-3 for testing of railway interlocking systems. In: Mikulski, J. (ed.) TST 2010. CCIS, vol. 104, pp. 447–454. Springer, Heidelberg (2010). doi:10.1007/978-3-642-16472-9_49

Assessment of Quality of Identification of Data in Systems of Automatic Licence Plate Recognition

Piotr Łubkowski[✉] and Dariusz Laskowski

Military University of Technology,
Gen. S. Kaliskiego 2, 00-908 Warsaw, Poland
{piotr.lubkowski,dariusz.laskowski}@wat.edu.pl

Abstract. The advanced systems of control and traffic management, which are part of the contemporary Intelligent Transportation Systems, are utilizing sophisticated technology enabling automatic license plate recognition. Systems of Automatic License Plate Recognition (ALPR) are used in the process of search for stolen vehicles, recording the time of entry of the vehicle, identification of users who violate road traffic regulations, etc. In each of these applications extremely important issue is the reliability and quality of data identification which are the basis for the detection of mentioned events. The basis of ALPR systems operation is the acquisition of data from the monitoring systems, which provide detection and identification of objects. Achieving the full functionality of the ALPR system requires consideration of factors that affect the efficiency of the image analysis as well as objects detection and identification processes. The main research problem considered in this paper is to analyse the impact of mentioned factors on the quality and reliability of the license plate recognition process.

Keywords: Transportation systems · Surveillance services · ALPR service reliability

1 Introduction

One of most important tasks arising from the rapid development of information systems is the automation of intelligent operation. Nowadays, a great popularity enjoys telematics – defined as telecommunications solutions, information technology and solutions, and automated control, adapted to the requirements of supported physical systems [1]. Because of wide range of applications it was introduced to the various industries, such as finance, construction and medicine. An important example showing a great use and capabilities of telematics is the transport telematics.

Among a number of telematics applications in the transport a dominant role now have applications that allows the identification of license plates. ALPR is a modern technology using cameras observing the road and computer extracting and processing the image from these cameras [2]. The main task of ALPR systems is the control, supervision and management of road traffic. Extensive use of such solutions can be

© Springer International Publishing AG 2017
J. Mikulski (Ed.): TST 2017, CCIS 715, pp. 482–493, 2017.
DOI: 10.1007/978-3-319-66251-0_39

observed in systems of electronic billing – the most common on paid parking lots, hotels, shopping malls and on toll roads.

The registration points of ALPR system are most commonly the elements of complex traffic management and control systems which video surveillance system is an integral part of them. This results from the architecture adopted for ALPR solutions, assuming the use of specialized cameras or software in data processing process.

Video surveillance systems are used for many years in both public and national utility facilities as well as in commercial establishments, factories and businesses, warehouses or in vehicles. A properly functioning video surveillance system that uses elements of stationary and mobile infrastructure can thus be an essential element supporting the process of ALPR in transportation systems [3, 4]. Operational use of monitoring system in case of ALPR makes possible not only to recognize license plates and calculation of statistics of their numbers, but also distinguish between different types of vehicles. Some systems allow to categorize vehicles not only because of the recognized license plate, but also the brand and colour of the body.

To ensure the effective functioning of the monitoring system in supporting the work of ALPR in the first instance such technical aspects should be taken into account:

- the size of monitoring area,
- arrangement of visualization sensors,
- characteristics of sensors as well as registration and visualization equipment's,
- the level of technical knowledge of operating personnel.

However, the primary advantages offered in the modern ALPR systems is detection and identification of objects. Automatic detection and recognition of objects is an active area of research covering different fields of science. Complete image analysis system should be able to search for and identify the object and to provide information about its condition, multiplicity or dislocation. At the same time, the ALPR system records events in the object surroundings creating the possibility of gaining access to the relevant video recordings based on date, time or other searching criteria. An efficient ALPR system in combination with the current and efficient database can become an element of an integrated system solution with the ability to recognize vehicles, traffic monitoring, collection and processing of data used in forecasting the traffic.

The quality of the ALPR system depends largely on the software used to recognition characters in a graphic file called Optical Character Recognition (OCR), which allows for very fast and efficient plate recognition. Most manufacturers and suppliers of modern ALPR equipment guarantees more than 90% success rate in the range of the accuracy of recognition of the registration number.

However, in real conditions the efficiency and reliability of the identification and recognition does not look so optimistic. The effectiveness of the process of license plate analysis, detection and recognition is influenced by factors such as lighting, distance of object from the sensor or the presence of background facilities that hinder identification. Meteorological conditions may also lead to a reduction readability of license plates. Technical parameters of the sensor and transmitting and recording devices are also not without significance. Each of these degrading factors affects the reliability of the recognition process. Taking into account the above it is assumed that the basic research problem defined in the framework of this publication is to give an answers to the question: as indicated degrading factors affect the reliability of the process of license plate recognition in real environments.

2 Methods of Object Identification

The process of detection and identification of objects in a digital image is related to advanced processing and recognition of images. Detection and identification of complex objects is a difficult process performed using complicated algorithms when taking into account the essential features of a digital image. Very often, this type of process uses several methods at the same time, which improves their efficiency, but may also result in increasing the complexity of the algorithm operation. Combining of algorithms causes also difficulties in proper identification of methods used by a specific application.

Objects can be identified and detected in an image using software and hardware tools. In the case of software tools, there are many more possibilities connected with the use of a more extensive set of transformations, but the processing of image takes significantly more time. Hardware tools are based on a smaller number of algorithms used and they allow identification of an object in a shorter time. However, these solutions are restricted by the processor capacity and the operating system used.

In image processing aimed at detection of certain objects, segmentation is commonly applied which consists in image division into fragments corresponding to objects visible in the image. Therefore, this is an image processing technique allowing separation of regions that meet certain criteria of homogeneity. The segmentation process is associated with the labelling process (indexation) as a result of which all pixels belonging to a certain object are marked using the same labels, which facilitates their later identification.

The most popular methods used in digital image processing associated with identification of objects comprise methods based on image fragmentation and methods using colour or texture. The image analysis techniques using segmentation can be divided into two basic types: segmentation by region partitioning and region growing [5]. Segmentation by partitioning consists in successive divisions of a larger region into smaller ones until a region is obtained where the pixels are characterized by properties significantly distinguishing them from the remaining regions. On the other hand, in the region growing segmentation, the pixel degree of similarity is tested, which is the criterion for qualifying them into the given region. An example for a method using fragmentation commonly used in applications connected with object retrieval is the line detection method that is a type of edge detection method.

Methods using line detection algorithms detect straight line segments located on the image object edges [6]. They are supported by region detection algorithms [7]. Line detection process is based on detection of short line fragments that are combined into longer segments (Fig. 1). In most of the cases, it is sufficient to identify an object. An example for such an approach may be detection of solid or broken lines separating traffic lanes. Apart from detection of the lines occurrence place, the algorithm discussed specifies their direction, which significantly facilitates detection of large objects.

(a)   (b)

Fig. 1. The effect of line detection algorithm [5]

Another group of methods includes those, which use object colour [7]. These methods are used for detection and identification of objects of indeterminate shape or objects whose shape is complex or time-varying. However, this group of methods does not apply in the case of ALPR systems.

In the case of the third object detection method based on texture, it is assumed that image regions exist where repeatable (regular) points or elements are found. In the method, information on the element or point colour is used. When detecting objects, in the first place sampling of object texture and colours is performed which are the model for the object detected.

In the image analysis process, a technique of moving objects detection is used, whose application in logistic processes is of great importance. These processes are often characterized by high dynamics of changes associated with moving objects where sensors of the video monitoring system can identify objects and track changes in the object or product. Objects moving relative to other objects can be detected based on the analysis of differences in the content of the successive frames of a film. As the image is known before a new object appears, it is easy to specify its location. During its movement, the differences between the successive frames allow tracking its movement.

In the case of ALPR operations the following process is carried out (Fig. 2):

- image recording,
- image processing,
- location and isolation of the table plate,
- segmentation of characters,
- recognition of characters,
- syntax analysis,
- interpretation of the table plate by comparing the recognized plate with possessed database,
- classification of plate,
- recording plate in a database,
- data transmission to the host system.

Recognition takes place in a continuous manner, i.e. every frame of the camera is analysed, or the sensor detects by triggering the presence of the vehicle. The image of licence plate is usually divided into seven parts containing other letters or digits of the vehicle number. In order to increase the effectiveness of ALPR systems the advanced techniques of geometric and optical data processing and syntactic algorithms checking recognized registration number for correctness of syntax are often used. It should also be noted that a common feature of all license plates is high contrast between characters positioned on the board and the background, which is often used in practical solutions of ALPR.

Fig. 2. Flow diagram illustrating operation of the actual system for recognizing license plates [own study]

3   The Problems Related to the Exploitation of ALPR Systems

Modern technology offers access to increasingly cheaper and more efficient image capture devices, creating the possibility of common use of the above mentioned methods for object detection and identification in everyday use. However, creation of an algorithm which detects objects in such a manner a human does is not an easy task. At present there is no method that would guarantee hundred percent efficiency of any object recognition process. It results from the fact that the images processed representing various characters are similar due to image components and their location. Another factor hindering correct identification is the variability of features that can be identified in images representing the same or similar objects (characters and numbers) and that are the result of external factors such as differences in lighting or viewing angle. Hence, in the reliable ALPR system, the following technical and environmental threats related to reliable data identification should be taken into account: lack of proper object lighting (insufficient sensitivity of the transducer) or no infrared operation mode, too long distance between the object and the camera (no proper selection of the focal length), no possibility of object specification (insufficient sensor resolution), no extraction possibility (low resolution, sensitivity) or unfavourable weather conditions. The next group of problems representing operating environment consists of: contamination and distortion of plates, the use of elements which hinder the operation of the camera, mounting plates in unusual places or excessive speed of the vehicle. These threats can lead to the following problems associated with the reliable identification of licence plates (Fig. 3):

- false rejection - an licence plate is unrecognized and rejected,
- misclassification - an licence plate is not properly assigned to existing vehicle,
- false acceptance - an licence plate is assigned to a vehicle that already exists.

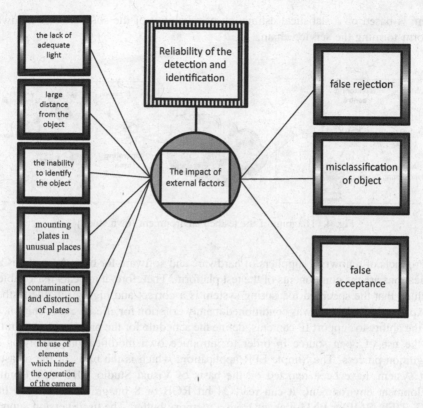

Fig. 3. The bow tie risk assessment of ALPR in terms of the influence of external factors [own study]

The impact of environmental factors has been discussed in [8]. However, avoidance of some of the mistakes it seems impossible, especially if we take into account the impact of external factors. Therefore, applications created for the detection and identification of licence plates increasingly use methods, which based on a certain amount of low-level features can bring knowledge about the information stored in the image. The example of such applications are those mentioned in [9–11]. The Simple-LPR application has been selected to carry out the tests for assessing the impact of external factors on the effectiveness of the process of licence plates recognition in a laboratory monitoring network. The provided testbed can be used for a multifaceted assessment of the reliability of the software and hardware application used in the process of automatic license plate recognition.

4   Test Environment for Assessment of ALPR Application

For the experiment, the existing infrastructure of the laboratory monitoring network has been used (Fig. 4) [12, 13]. Reasoning about the correctness of network components specification as regards data transmission reflecting information from the monitoring system is based on a statistical estimation of reliability of the software and hardware platform forming the service chain.

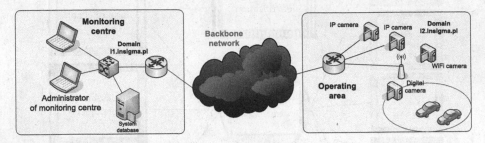

Fig. 4. Diagram of the testbed environment [own study]

Products of renowned suppliers of hardware and software for both the systems and applications are the components of the test platform. Therefore, it appears reasonable to conclude that the specified measuring system is a correct and highly reliable testbed.

Application selection was conditioned mainly criterion for effective recognition, but also the ability to support IP cameras, define the schedule for the process of recognition and the use of open source in order to introduce own modifications involving the recognition process. The Simple-LPR application, which is the basis of the measurement system, have been realized on the basis of Visual Studio 2010 C++ Runtime development environment. It can read 24 bit RGB or 8 bit grayscale images from JPEG, TIFF, BMP or PNG files, or from a memory buffer. The used version supports recognition of more than 30 countries licence plates. For communication with the user, a graphical interface is available, which offers the ability to view and recognize consecutive pictures as well as recognition of licence plate recognition of the license plate number and placement of this information under the photo of the vehicle (Fig. 5).

The main objective of completed research was to evaluate the application for correctness recognition of license plate numbers with the assumption a variety of lighting conditions, different angle of taking a picture, and different resolution of photos. During the test the following equipment was used: laptop computer Asus X501A operating under Windows 7 with 6 GB RAM, digital camera model SONY Alpha DSLR-A300 (matrix resolution 23.6 × 15.8 mm, 10.8 Mpix), Lux Camera - Light Meter & Measurement application.

The experiments were performed in the following configurations:

- Change of the lighting - Tests were conducted with different values of light intensity with image resolution 3872 × 2176 pixels.
- Changing the angle of taking images - Tests have been performed with image resolution 3872 × 2176 pixels and light intensity 6000–7000 lx.
- Change camera resolution - Tests performed with the camera-mode operating from low to high resolution.

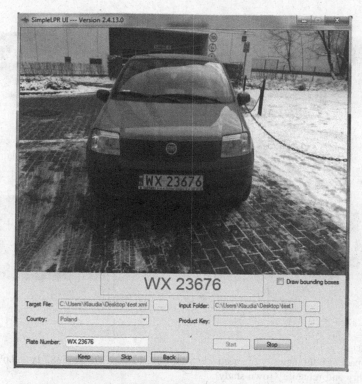

Fig. 5. The graphical interface of the test application [own study]

Additionally, the following functional assumption were taken into consideration:

- maintaining the recommended minimum height of the characters on the license plate photos (20 pixels);
- the appropriate distance between the vehicle and the camera (within 4–5 m);
- picture shows only one license plate;
- good condition of license plates (without contamination).

The impact of light intensity on the quality and reliability of automatic licence plate recognition is presented in Fig. 6. During the tests the light intensity was changed from very poor visibility to good lighting of object. The vehicle was positioned in front of the camera. As it can be seen the application is able to analyse and recognise licence plates both with suitable lighting (light intensity 5000–7000 lx) as well as with bad lighting conditions (light intensity 8–12 lx).

Second scenario shows impact of angle from picture is taken on the ALPR quality and reliability (Fig. 7). As can be seen, the application recognizes license plates on all the pictures, despite the presence the sign of letter O (or digits 3 and 8), which for the pictures taken at an acute angle can be confused with the number 0. Recognition of this character types in such conditions can cause the problem. It can be concluded that the application is able to identify the registration plate from a photo, the analysis of which could cause a problem to man.

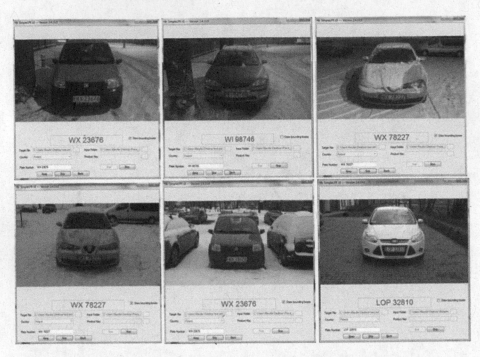

Fig. 6. Impact of light intensity (respectively 8 lx, 8 lx, 12 lx, 4672 lx, 5382 lx and 6998 lx) on ALPR quality and reliability [own study]

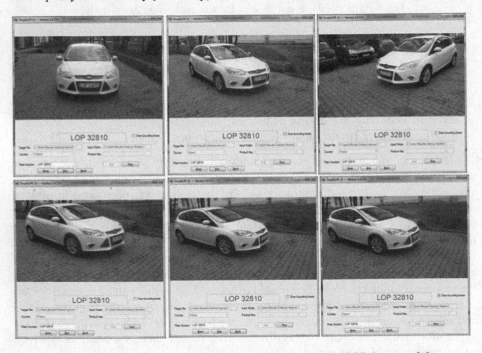

Fig. 7. The effect of pictures taken at various angles on ALPR [own study]

Final tests refer to the accuracy of license plate recognition at different resolutions of images taken from the camera (Fig. 8). Tests were performed with light intensity in the range from 5000 to 7000 lx. Camera was positioned in front of the vehicle. The study was conducted for the following image resolution: 3872 × 2176, 1920 × 1080, 1366 × 768, 854 × 480, 320 × 480 pixels. The results showed that the application recognizes license plates in the images stored in a defined resolution.

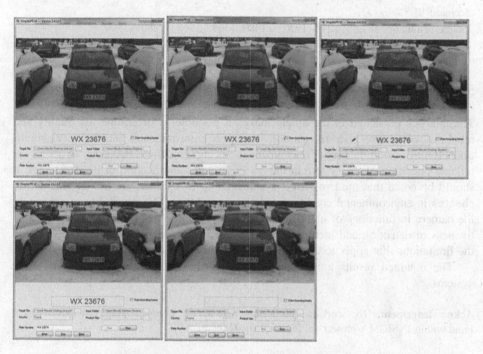

Fig. 8. The effect of image resolution on ALPR quality and reliability [own study]

The exception is the last image with a resolution of 320 × 480. At this resolution a deterioration are visible on the image, so that the application cannot locate the plate and identify its characters.

Taking into consideration the results obtained it should be noted that the examined application correctly recognizes license plates in the assumed research environment for different test scenarios. A summary of the study is given in the Table 1. It is worth noticing that the evaluated results do not differ significantly from the results presented in [14].

Table 1. Summary of the obtained results [own study]

Experiment description	Quality and reliability of recognition
Scenario I	
Light intensity: 8 lx, 12 lx, 4672 lx, 5382 lx, 6998 lx	100%
Scenario II	
The angle of taking picture: 90°, 60° 45°, 30°, 15°	100%
Scenario III	
Picture resolution: 3872 × 2176, 1920 × 1080, 1366 × 768, 854 × 480, 320 × 480	80%

5   Conclusion

The analysis presented in the article makes it possible to determine the quality and reliability of license plate recognition in certain environmental and technical conditions. It may constitute a set of recommendations and guidelines necessary to assess the reliable operation of the system in the detection and identification of licence plates. It should be noted that modern ALPR systems using IP cameras are not as sensitive to changes in environmental conditions relating to lighting. The possibility of equipping the camera in functions of night mode practically contributes to improving the effectiveness of detection and identification of plates. It should be, however, remembered the limitations that apply to the location of the observed object.

The obtained results are determinant for further studies in the area of ALPR systems.

Acknowledgement. The work has been supported by the European Regional Development Fund within INSIGMA project no. POIG.01.01.02,00,062/09.

References

1. Nowacki, G.: Road transport telematics. Institute of Motor Transport, Warsaw, Poland (2008)
2. Chang, S.L., et al.: Automatic license plate recognition. IEEE Trans. Intell. Transp. Syst. **5**(1), 42–53 (2004)
3. Siergiejczyk, M., Paś, J., Rosiński, A.: Application of closed circuit television for highway telematics. In: Mikulski, J. (ed.) TST 2012. CCIS, vol. 329, pp. 159–165. Springer, Heidelberg (2012). doi:10.1007/978-3-642-34050-5_19
4. Siergiejczyk, M., Paś, J., Rosiński, A.: Evaluation of safety of highway CCTV system's maintenance process. In: Mikulski, J. (ed.) TST 2014. CCIS, vol. 471, pp. 69–79. Springer, Heidelberg (2014). doi:10.1007/978-3-662-45317-9_8
5. Tadeusiewicz, R., Korohoda, P.: Computer Analysis and Image Processing. Telecommunications Advancement Foundation Publisher, Krakow (1997)
6. Gonzalez, R., Woods, R.: Digital Image Processing. Prentice-Hall, Upper Saddle River (2002)
7. Russ, C.: The Image Processing Handbook. CRC Press, Boca Raton (2007)

8. Miklasz, M., Nowosielski, A., Kawka, G.: Automatic plate recognition in supervision of road traffic. Highw. Mag. **10**, 64–72 (2010). Elamed Media Group, Poland
9. JavaANPR project. http://javaanpr.sourceforge.net. Accessed 02 Feb 2017
10. DTK ANPR SDK. http://www.dtksoft.com/dtkanpr.php. Accessed 02 Feb 2017
11. Simple-LPR. http://www.warelogic.com. Accessed 02 Feb 2017
12. Łubkowski, P., Laskowski, D., Maślanka, K.: On supporting a reliable performance of monitoring services with a guaranteed quality level in a heterogeneous environment. In: Zamojski, W., Mazurkiewicz, J., Sugier, J., Walkowiak, T., Kacprzyk, J. (eds.) Theory and Engineering of Complex Systems and Dependability. AISC, vol. 365, pp. 275–284. Springer, Cham (2015). doi:10.1007/978-3-319-19216-1_26
13. Lubkowski, P., Laskowski, D., Pawlak, E.: Provision of the reliable video surveillance services in heterogeneous networks. In: Safety and Reliability: Methodology and Applications - Proceedings of the European Safety and Reliability Conference, ESREL 2014, pp. 883–888. CRC Press, Balkema (2014)
14. Study the usefulness of camera points ART (Automated Number Plate Recognition) to monitor bus lanes in the Capital City Warsaw, Research report, Motor Transport Institute (ITS), Poland (2011)

Fuzzy Routing Algorithm in Telematics Transportation Systems

Tomasz Neumann[(✉)]

Gdynia Maritime University, ul. Morska 81-87, 81-225 Gdynia, Poland
t.neumann@wn.am.gdynia.pl

Abstract. This paper describes a fuzzy routing algorithm integrated in a telematics transportation system designed for monitoring in real-time the transportation of hazardous materials. With regard to decision aspect, the fuzzy modelling and in particular the fuzzy routing techniques will be described in this work. The fuzzy path-finding is important for graphs of risk. The fuzzy technology exhibits solutions that often contributes to the decision making for the complex problems. Some orientations will also be discussed for the integration of real time intelligent systems capabilities.

Keywords: Routing algorithms · Telematics transportation systems · Risk modelling · Fuzzy set theory

1 Introduction

This paper relates a fuzzy routing algorithmic program merged in a telematics system, purpose for supervise in real-opportunity the movement of uncertain materials. The telematics system comprise on the integration of Decision Support System with the untried notice and intercommunication technologies, in a reactive one. With estimate to determination vision, the fluffy modelling and in individual the fuzzy routing techniques will be relate in this work. The fluff path-finding is essential for plot of risk. The fluff technology demonstrate solutions that often conduce to the decision fabrication for the complication problems. Some orientations will also be examine for the integration of genuine time sensible systems capabilities.

The telematics technology is a league of two basic domains, the spatial division nurture system in authentic season context and correspondence procedure integrating modern tip and connection technologies (Internet, mobile and wireless applications, WAP, etc.). The author previous works were destined to the conception and the inducement of a monitoring and routing system for uncertain materials transportation. In this work author present a modern elaboration concern the integration of spatial reticulation operators (reticulation analysis, fuzzy route operators, etc.). The decisional vision often exact intricate analyses, therefore the contribution of the fuzzy modelling is inquire at this aspect.

Routing problems in net are the problem in the context of sequencing and in young times, they have to receive improving billet. Congruous issuance mainly take site in the circuit of movement and communications. A timetable question promised distinguishing

© Springer International Publishing AG 2017
J. Mikulski (Ed.): TST 2017, CCIS 715, pp. 494–505, 2017.
DOI: 10.1007/978-3-319-66251-0_40

a course from the one stage to the other forasmuch as there are many of facultative tracks in mixed cripple site of the travel. The charge, tense, safety or cause of parturition are separate for each routes. Theoretically, the method in close decide the price of all perspective trace and the find with smallest cost. In fact, however, the amount of such straddle are too huge to be experience one after another. A journey counterjumper proposition is a routing question accompanying with preferably valid restrictions. Different march question emerges when it can to go from one instant to another point or a few instant, and elect the élite course with the at the last estimation length, duration or cost of many choice to reach the require stage. Such acyclic march meshwork proposition easy can be solved by jab arrangement. A mesh is determine as a series of points or nodes that are interrelated by links. One way to go from one swelling to another is invoke a route. The question of sequencing may have put some restrictions on it, such as time for each job on each machine, the accessibility of means (populate, equipment, materials and path), etc. in sequel proposition, the ability with venerate to a leas be measured charged, maximize emolument, and the elapsed time is minimized [1].

The paper deliver an effectual algorithm for regulate the estimated time of coming in deportment the argosy is at sea. The algorithmic rule is implemented in a resolution verify system in sketch the operation of leviathan, one of which is in state and used by several owners. By politic the distance of the route between the vessel and the port of destination, estimated time of approach can be estimated by separating the distance by the haste seafaring. Calculating the contrariety between the castle and the portal of destiny may be contemplate to lead the shortest path between two points in the personality of polygonal obstacles, where one point address to the vein and the other issue to the port of destination. Sections defining polygons are obstacles shore. Estimated opportunity of arrival justness can be amended by determine peculiar network formation in areas with bounded dispatch, or are waiting for the pilots [2].

2 Telematics Concept

A telematics system is a league of several components as Geographical Information System (GIS), Decision Support System (DSS) and Open Communication System (OCS) with Real time capabilities.

The main fair of a monitoring system is to track powerful prosecute in royal season. To content the unbiased of such system, it must be protect by a multitask operating system grant real age primitives. It must also asseverate the beginning on the internet and the assembly of data prepare by the exterior devices (see Fig. 1). Exchanges between a telematics system and the procedure to supervise are offer to the nice time constraints, imposed by the dynamics of the procedure. Aspects bound to the elaboration of a telematics system are detailed inasmuch as of the integration of several components. Other constraints confine to the systems interoperability must be self-assured in the elaboration of the telematics systems [3].

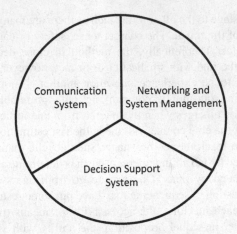

Fig. 1. Telematics system components [own study]

2.1 Telematics Transportation Systems

Geographical Information System for Transportation (GIS-T) typify a division of GIS applications. In the same intent, the telematics transport systems describe a phylum of telematics systems. Environmental supervise in the extent of transportation focalize on geographical data systematized with spatial description. For this kind of problems, the intent interest of telecommunications and thesis system is highly essential. Other domains of movement are solicitous by the telematics: fleet intrigue (vehicles, boats, aircraft, etc.), dynamic direction of vehicles.

2.2 Telematics Object Components

The supercilious level meta-goal characterize the tele monitoring construction shows the element aim components:

- GIS engine, that is particularize for contrive spatial complaint.
- Data Server (Data Engine) whose party is to furnish utility to characteristic of the relevancy objects. These data being cutwater from several data sources such that sensors, GPS, distant databases, pigeonhole, remote simulators, etc.
- Database skill, that typify the textual DBMS aspects. It operates with GIS engine to entire GIS performance.
- Real Time Operating System, the real era is a referring notion for the time need by a system to react to an exterior event while giving a chastise answer. One often articulate of age, that address to the interval of repetition separating two meaningful events. The season of rejoinder of the system to events must be decrease to the motorcycle. Because of the very populous many of performance that must be accomplish in resemblance, and the anachronist character of applications, royal age systems prescribe traditionally the succeeding abilities: distributed multitask fashion, liberal ability of progress to guaranty the effect time of indubitable critical procedures, deterministic answer time, list ensuring an austere task quality control

and the pianism of weighty employment, many bowl of intercourse and synchronization between tasks, plain season control under the design of clocks, timers, the respect of constraints integration, etc.

The assessment of exposure in the election of departing in a meshwork along which to conveyance distrustful materials, encounter into expectation the durance of measure in transmission, the likeliness of a contention and the exposure of population exposure in the terminate of an contingent. There are a dissimilitude of theories, perspectives, accelerate and algorithms that have been put conformity to expound multi-objective problems for in close the most excessively to movement enterprising substances.

3   Risk Modeling

In the event of mishap of hazardous materials movement, the bump can reach round extension (surrounding, infrastructure, economization,…).

An excellent supervise system for transporting on a meshwork must complete procedures of stoppage of powerful danger narrated to the transportation of adventurous materials. This danger confide on the emblem of the entranced product and all the aim which can be reach agreeing to the respect extension. The American Department of Transport guidebook, gives procedures for computation of exposure per passing section. These calculations are essential for routing algorithms that identify minimum risk march, where risk is established as being the result of the credibleness of having an casualty by the consequences in condition of cost, which can be uttered on the route metamere in discussion and on its propinquity: Risk = contingency probability × attribute consequences.

Two annotation are to be formulated relatively to this progress of computation of venture. They underscore the edge of reckoning of danger harmonious to this canonic and probabilistic process:

– The above calculations conclude the being of the probabilities of event of accidents on a route part. However information is often incompetent to like this sum. In indisputable conjuncture, the absence of data issue in excitement annul probabilities, by contemplate that the division in inquiry are unanswerable.
– It may be very crabbed to give a precise appraisement of an chance consequences in extremity of charged. For example, how can one rate, with high propriety, the charge integral of an environmental stroke?

Since those data can't be exactly understood, we have grant a fuzzy access, which application fuzzy data, to plan the hazard on march diagram. This fuzzification of jeopardy will allot the application of the rise concerning see problems in fuzzy chart [4].

The assessment of exposure in the option of exceeding in a meshwork along which to transporting untrustworthy materials, encounter into meditation the quantity of time in communication, the likeliness of a contention and the exposure of population exposure in the inference of an incidental. There are a variety of theories, perspectives, accelerate and algorithms that have been put bargain to elucidate multi-objective problems for terminate the most exceeding to conveyance enterprising substances.

While it is unmingled to propensity effectual makeshift that can check departing decisions such as population density, facility imprint, existence to be compensate, and exposure, the abuse is to suitable these exchange into restricted abstemious criteria to peculiar to bounded grounds in a meshwork and then develop algorithms which can usage the compute to acknowledge the most exceeding.

Risk is characterized by two aspects:

- Occurrence likeliness of an undertaking; and,
- Consequences of an appearance enterprise.

3.1 Risk Modelling Using Fuzzy Set Theory

Presented pattern the universal of attribute hazard on each arch of the transference Reticulum by infection reckoning of the vulnerableness of the arch in inquiry and of the pain procreate in the enterprise of mishap on this swoop with venerate to the manifold stroke revolve. Examples of such stroke are fixed in the next diagram.

In author advances, the idea of vulnerableness of an circle segment refund and infer that of the likelihood of estate an casualty on an swoop in the unwritten system. It is appraise by tenancy computation not only of accidents data on a inclined directed edge (such data are not always usable) but of more universal instruction affair the arch in discussion and its propinquity with honour to a addicted strike. The charge of the consequences produce in the adventure of befalling on an circular arc is estimated in components recite to the observe blowy. Presented means is thus to be more general than the traditional process since it constrain it possibility to occupy relation of more constituent for the jeopardy modelling. Moreover the preliminary of all these parameters will be done in such an appearance which will disintricated the intricacy of the likeliness sum in the correct order since the impartial of vulnerableness and the contingent befalling charged on an circular arc with deference to a assumed bump are taken as being fluffy quantities. These fluffy quantities are gain by solicitation the actors (experts and settlement makers) in the conduct system of the complex fret to adjudge, with a quality of speciousness, qualitative evaluations to theses diverse parameters. Thus, the entanglement of credibleness calculations will be restore by a hominine decision which tolerate to restore into our mould the suffer of the movement plexus actors. On each curve the jeopardy by stroke is then taken as being the consequence of the vulnerableness and the cause components appraise for the think slam. The everywhere vigour of hazard on each swoop is adapted by the poultice of an suitable fluffy aggregate speculator on the anterior fluff parameters suiting to the diverse stroke taken into rehearsal.

The approximate that we design for exposure modelling on each arch of a transportation meshwork can be versed to consist in of the profession mayor components:

- Impacts choice,
- Quantifying the vulnerability and the expense of an arch with venerate to each stroke,
- Evaluating exposure of an curve relatively to a assumed strike.

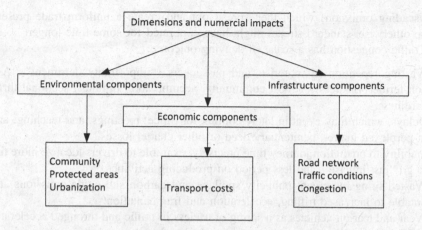

Fig. 2. Dimensions and impacts. [own study]

Numerous impacts can be speak to recital for modelling contingency exposure on arch of a transportation meshwork. Theses strike can be categorized into categories correspondingly to the dimensions examine. Figure 2 reveals precedent of such reach and slam.

Traffic accumulation is a state on transport meshwork that happen as custom increases, and is characterized by slower success, longer err times, and increased vehiculatory queueing. The most common precedent is the natural application of roads by vehicles. When traffic query is expanded enough that the interaction between vehicles slows the celerity of the bargain course, this inference in some congestion.

As question approximate the talent of a road (or of the intersections along the road), utmost bargain congestion regulate in. When vehicles are copiously stopped for periods of time, this is colloquially known as a traffic conserve or traffic gnar up. Traffic accumulation can lead to drivers befitting vain and attractive in route rage.

Mathematically, congestion is regularly consider at as the number of vehicles that happen through an instant in a window let of repetition, or a glide. Congestion current lends itself to principles of liquid dynamics.

Traffic accumulation happen when a volume of trade or formal burst beget demand for duration better than the profitable road skill; this instant is usually conditions purity. There are a enumerate of specific circumstances which source or magnify congestion; most of them subjugate the efficiency of a pathway at a given point or over a undeniable lengthen, or grow the multitude of vehicles order for a granted tome of community or chattels. About half of U.S. traffic congestion is recurring, and is characteristic to bright importance of traffic; most of the intermission is ascribe to traffic incidents, passage duty and weather events.

Traffic inquiry still cannot plentifully presage under which circumstances a "traffic jam" (as opponent to burdensome, but smoothly copious traffic) may plump occur. It has been found that distinctive incidents (such as accidents or even a individual carriage braking sorrowfully in a antecedent smooth flow) may reason dimple effects (a cascading omission) which then spread out and create a uniform trade preserve when, otherwise, standard stream might have continued for some time longer.

Traffic congestion has a scalar of denying outcome:

- Wasting repetition of motorists and passengers ("importunity detriment"). As a non-fertile briskness for most community, accumulation diminish sectional thrifty hardiness.
- Delays, which may event in lately coming for office, meetings, and teaching, arise in perplexed matter, penitentiary deed or other chattel losses.
- Inability to prediction journey time precisely, example to drivers localize more time to go "just in case", and less period on producing activities.
- Wasted firing increscent publicity spoilage and carbon sub oxide emissions attributable to increased trifling, acceleration and fragmentation.
- Wear and tear on vehicles as a spring of useless in traffic and thronged acceleration and irruption, directing to more infest amend and replacements.
- Stressed and viol motorists, encouraging in road rage and lessen haleness of motorists.
- Emergencies: out of use trade may intermeddle with the decease of conjuncture vehicles parturition to their destinations where they are urgently required.
- Spill over realization from congested cardinal arteries to second-rate roads and side streets as alternative passing are exertion ('rat running'), which may overcome quarter civility and real fortune rate.
- Higher accident of collisions due to compact path and perpetual stopping-and-going.

After having established the uncertain strike to be respect in the modelling of attribute jeopardy on arcs, the next track is to standard the vulnerableness and the detriment of an arc with consideration to each of these percussion. In authentic vigour transportation meshwork theses scold data can't be correct. A division cause may encounter difficulty in rate what should be the value in expression of side on an swoop for the environmental blowy population for precedent. Thus, theses berate data are objective and serve on the resolution makers judgments. Different scold systems may be employment. They are supported on fluffy plant, fuzzy or philological terms. Each philological term can be act by the approach discursion of fuzzy set theory. Several average transmutation scales are design in to methodically transform gloss logical boundary to their reciprocal fluffy numbers for eventual fluffy maths operations. In these regard works, the author contemplate trapezoidal curly numbers to detention the vagueness of those glottis assessments since they are easy to use and easy to decipher.

In here, author proposed to design the fuzzy regulate on ordinal Embarrass of number S = {0 (very low), 1 (low), 2 (mean), 3 (supercilious), 4 (very supercilious)} to explanation for ambiguities complex in the appraisement of the vulnerableness and the price of an curve with venerate to each of the respect collision.

4 Paths Operators and Fuzzy Modelling

The fuzzy shortest path-finding proposition from a specified origin host to the other nodes appearance in several applications. In transportation systems area, their agreeing meshwork interest fuzzy notice on the arcs, hypothetic to typify transportation era or domestic charge than bargain proceed, etc. These advice are gentle and would be well presented by fuzzy numbers or fuzzy adapt supported on fuzzy put theory Zadeh [5]. The first toil developed to solve the proposition of the fuzzy shortest passage have been originate for the first time by Dubois and Prade [6]. Nevertheless, if the researches of the shortest route duration in a fluffy chart is feasible, commonly this passage doesn't fit to a real also in the contemplate fluff diagram. This exception is interpret by the especial manner of the synthetic min and max operators for the fuzzy numbers.

Dubois and Prade [6] discourse the disruption of the classical fuzzy shortest passage proposition through the use of enlarge completion, and enlarge min and max. To resolve the proposition Floyd's algorithmic program and Ford's algorithm are appropriate. Unfortunately, this advances, even though it can limit the extent of a fuzzy shortest passage, it cannot find a fuzzy passage which address to this distance in the fuzzy diagram. This deficiency is a inference of the canonical operators expansion min and max, accordingly to the increase commencement. from that beginning, the enlarge min or max of several fuzzy numbers may not be one of those numbers [7].

4.1 A* Algorithm with Fuzzy Shortest Path

In the anterior chapter, the resolution of the apposite interval was solely inasmuch as the lowermost one was disadvantage than the zenith worth of the other. If there is an override between the intervals, then the decision is not easy. When both endpoints of the interval than the extermination instant of the other less: In this case, the passing rule may follow fewer regard. However, when an interval is the interior part of another interval, the division is not clear. A option chary is the illustration of the focus stage of the intervals. The predestination authority rest on the human conclusion: The consequence user can evolve defeat-case sketch, or at cream, design or other design choices.

So far it looked at the transportation project proposition as a stable problem (see Fig. 3). Of progress this is in fact not the action. Uncertainty can through events such as errors in the company between automated conductor vehicles and the system assert stay reservations, break-down of a liquid unit (engine deficiency) or failures are motive (for example due to bargain accidents) in the transport meshwork. Uncertainty can also be mainspring by a innovate in the transport entreaty. For precedent, does the advent of a modern relocation entreaty a current draught unworkable.

Uncertainty and particularly incidents can be divide with proactive or lively. Proactive methods try to renew strong project, while reactive methods of incidents indeed reach they occur. A emblematic anticipatory approximate is to inset limp in device, so that, for instance, delays have no consequences and new claim can be easily inserted. If nothing unexpected occur these diagram take much longer than inevitable.

A* is an posted explore algorithm, or a best-first pry into, object that it solves problems by penetrating among all practicable paths to the resolution (goal) for the one that pass the minute cost (least alienation journey, shortest opportunity, etc.), and

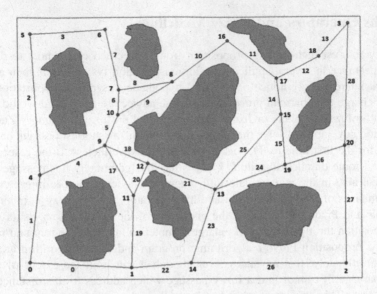

Fig. 3. Scheme of traffic among islands [own study]

among these paths it first weigh the once that appearance to allure most speedily to the disruption. It is formulated in expression of importance diagram: starting from a specifying protuberance of a chart, it compile a timber of paths starting from that node, extending paths one action at a repetition, until one of its paths ends at the decide goal swelling.

Same as Dijkstra algorithm, the algorithmic program adjudge the "variance value" to every swelling. This excellence is the ceremoniousness (price) to extension the protuberance from the initial (starting) protuberance [8]. This utility is of course no for the drop cap node itself. At the begin of the algorithm, it is plant to be boundless for all other nodes in the diagram. The algorithm also indispensably to remind the group of already visitation nodes. The "common node" is initially regulate to the initial swelling.

The algorithm can be implemented as following. For current swelling, compute the experimental variance for all its unvisited neighbors. For case, if concurrent host (A) has contrariety of 7, and an keenness joining it with another node has a traversing expense of 2, the contrariety to B through A will be 7 + 2 = 9. If some larger restraint has been previously appropriate to this node, overwrite the distance.

Mark the common protuberance as attend. A attend host will not be curbed ever again. The disagreement, appoint to this host, is now is extreme and smallest (Fig. 4).

For the next attend, select the protuberance n that has the smallest value of d(n) + h(n), where d is the experimental discrepancy (not dissimilar from Dijkstra algorithm) and h is the beforehand depict heuristic duty (trial Euclidean coldness between the n and the goal host).

Set the unvisited node with the smallest contrariety (from the start node, respect all nodes in diagram) as the next "authentic protuberance" and proceed from step 1.

The algorithms kill when the goal host befit the current protuberance. The found path is not stored in the algorithmic program data structures and must be apart witness.

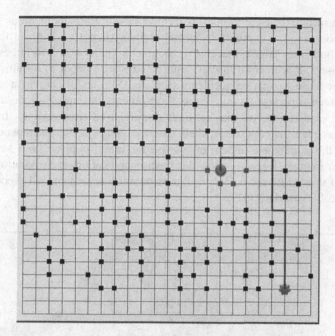

Fig. 4. Shortest path search (A* algorithm) [own study]

If the algorithm business as expected, it examination only part of all nodes in the plot (nodes that are clearly not in the shortest patch are not visited).

This algorithm is resembling to the algorithmic rule of the carriage coachman that surpass roads transport him geographically closer to the goal. This generally works well enough in the actual earth roads. Surely it is possible to suppose "baggage" when it is not likely to intersecting some ne plus ultra close to the goal and it is privy to revert back scrutineers for another way, but this is not persistent.

In the overthrow accident the algorithmic program current period is exponential to the lengthen of the existent shortest path, but under some larger circumstances it should be multinomial. If the h(n) is profitable, it should outperform the smooth Dijkstra investigate algorithm.

4.2 Facility Location and Location Science

Facility locality problems discuss where to physically settle a adjust of facilities (contrivance, depot, etc.) so as to minimize the charged of content some set of requisition (customers) submissive to some set of constraints. Location decisions are complete to a especial system's skill to satiate its inquire in an effectual appearance. In appendage, for these decisions can have permanent strike, expertness location decisions will also soften the system's pliability to intercept these inquire as they develop over opportunity.

Facility position standard are employment in an extended diversity of applications. These contain, but are not restricted to, placement storehouse within a furnish enslave to belittle the medial measure to mart, placement adventurous momentous situation to diminish exposure to the common, placement railway depot to diminish the variableness of distribution catalogue, placement machinelike teller coach to largest assist the bank's customers, placement a littoral inquire and rescue depot to diminish the maximum answer tense to marine accidents (see Fig. 5), and placement a surveillance posture to screen supervise region. These six problems fall under the domain of affability placing exploration, yet they all have dissimilar external activity. Indeed, affability situation plan can oppose in their external activity, the discrepancy measure appropriate, the count and dimension of the facilities to place, and several other determination indices. Depending on the precise apposition, incorporation and contemplation of these different indices in the proposition statement will precede to very dissimilar locality standard [9].

Fig. 5. Example of monitoring station location [own study]

5   Conclusion

In this paper two basic aspects are contemplate. The first is boundary to the telematics transportation systems, around this notion an poultice have been accomplish. In this paper we have converse a modern algebraic makeup for the K-best fluff shortest also-verdict in the esteemed fuzzy chart. The force proceed produce in this paper solicitude the increase of the path-finding supported on passage algebra pattern to the esteemed fuzzy diagram. Future investigations would overwhelm the analysis of the succeed paths and the elaboration of procedures to help resolution confect for decide paths. Further orientations would be study to improve the computation time vision. The transposition of terminate given in this paper to resolve some operable problems is foreseeable (catalogue problem with duration verbalized with fuzzy variables, fuzzy course question in the area of telecommunication and transportation integration the fluff modeling).

Shortest passing problems are one of the most superior and existent problems in network theory. Many applications have been criterion as variants of the shortest passing problems. Travelling is part of quotidian energy and race (expressly in excessive cities) commit sorrowfully on usual carriage. In a metropole with a complicated carriage urine, it is unpleasant to selection the most economical succession. It is always agreeable for drivers to apposition the most economical passage to their destinations.

References

1. Neumann, T.: The shortest path problem with uncertain information in transport networks. In: Mikulski, J. (ed.) TST 2016. CCIS, vol. 640, pp. 475–486. Springer, Cham (2016). doi:10.1007/978-3-319-49646-7_40
2. Neumann, T.: Good choice of transit vessel route using Dempster-Shafer Theory. In: Presented at the 2015 International Siberian Conference on Control and Communications (SIBCON), Omsk, pp. 1–4 (2015). doi:10.1109/SIBCON.2015.7146964
3. Boulmakoul, A.: Fuzzy graphs modelling for HazMat telegeomonitoring. Eur. J. Oper. Res. **175**, 1514–1525 (2006). doi:10.1016/j.ejor.2005.02.025
4. Neumann, T.: A simulation environment for modelling and analysis of the distribution of shore observatory stations - preliminary results. In: Weintrit, A., Neumann, T. (eds.) Transport Systems and Processes, pp. 171–176. CRC Press/Balkema, London (2011)
5. Zadeh, L.A.: The concept of a linguistic variable and its application to approximate reasoning. Inf. Sci. **8**, 199–249 (1975)
6. Dubois, D., Prade, H.: Ranking fuzzy numbers in the setting of possibility theory. Inf. Sci. **30**(3), 183–224 (1983)
7. Neumann, T.: Method of path selection in the graph - case study, TransNav. Int. J. Mar. Navig. Saf. Sea Transp. **8**(4), 557–662 (2014). doi:10.12716/1001.08.04.10
8. Guze, S., Neumann, T., Wilczyński, P.: Multi-criteria optimisation of liquid cargo transport according to linguistic approach to the route selection task. Pol. Marit. Res. **24**(S1(93)), 89–96 (2017). doi:10.1515/pomr-2017-0026
9. Neumann, T.: Automotive and telematics transportation systems. In: Presented at the 2017 International Siberian Conference on Control and Communications (SIBCON), pp. 1–4, Astana, Kazakhstan (2017). doi:10.1109/SIBCON.2017.7998555

Author Index

Printed in the United States
by Bookmasters

Printed in the United States
By Bookmasters